Mohammed Elmusrati

Modelling Stochastic Uncertainties

Also of Interest

Measurement Uncertainties
Error Propagation, Probabilistic Modelling, Statistical Methods
Michael Krystek, 2024
ISBN 978-3-11-145343-9, e-ISBN (PDF) 978-3-11-145371-2

Quantities and Units
The International System of Units
Michael Krystek, 2023
ISBN 978-3-11-134405-8, e-ISBN (PDF) 978-3-11-134411-9

Metrological Infrastructure
Edited by Beat Jeckelmann, Robert Edelmaier, 2023
ISBN 978-3-11-071568-2, e-ISBN (PDF) 978-3-11-071583-5
in *De Gruyter Series in Measurement Sciences*
Edited by Klaus-Dieter Sommer, Thomas Fröhlich
ISSN 2510-2974, e-ISSN 2510-2982

Classical Mechanics
Hiqmet Kamberaj, 2021
ISBN 978-3-11-075581-7, e-ISBN (PDF) 978-3-11-075582-4

Geodesy
Wolfgang Torge, Jürgen Müller, Roland Pail, 2023
ISBN 978-3-11-072329-8, e-ISBN 978-3-11-072330-4

Mohammed Elmusrati

Modelling Stochastic Uncertainties

From Monte Carlo Simulations to Game Theory

DE GRUYTER

Author
Prof. Mohammed Elmusrati
University of Vaasa
School of Technology and Innovations
Wolffintie 32
FIN-65101 Vaasa
Finland
mohammed.elmusrati@uwasa.fi

ISBN 978-3-11-158470-6
e-ISBN (PDF) 978-3-11-158505-5
e-ISBN (EPUB) 978-3-11-158545-1

Library of Congress Control Number: 2024943488

Bibliographic information published by the Deutsche Nationalbibliothek
The Deutsche Nationalbibliothek lists this publication in the Deutsche Nationalbibliografie; detailed bibliographic data are available on the Internet at http://dnb.dnb.de.

Cover image: gremlin / E+ / Getty Images
Typesetting: VTeX UAB, Lithuania

www.degruyter.com

Preface

During my master's degree program circa 1993, I undertook a course on measurement theory, which examined the subject through the lenses of uncertainty analysis and probability theory. I developed a deep appreciation for the subject, and it fundamentally changed my worldview, as I began to see a degree of uncertainty in almost everything, whether due to a lack of information or the limits of precision. My fascination with this field grew further when I encountered the captivating beauty and unparalleled elegance of the relationship between uncertainty modeling equations and their simulation on computers, and how they could yield remarkably similar results that supported each other.

For instance, I would hypothesize that there was a queue of cars at a crowded gas station, with each car taking a finite, random amount of time (relative to the others) before leaving the station. There were N fuel dispensers, and cars arrived at the station randomly, following a certain distribution. The question was, what is the average time a person would spend waiting at the station, or what is the average number of cars in the queue after, say, two hours? This problem can easily be solved using queueing theory, but what's fascinating is that when you use computer simulation, without relying on any of the modeling methods used in the mathematical solution, you still get almost the same result.

The success of simulation, given assumptions about the randomness and independence of car arrivals at the station, or the absence of abnormal events such as a dispenser malfunction, gives us the ability to experiment with more complex scenarios—such as using real observational data to predict the average queue length at any given time of the day. There are countless examples that we can analyze and simulate, such as the spread of epidemics in a community, energy consumption rates, traffic accidents, and even at the technical level, such as simulating signal reception in communications engineering or studying reliability of aircraft and nuclear power plants, as well as modeling social relationships or simulating international relations and predicting political futures.

Bringing the theater of events, with all its uncertainties and complex interrelationships, into a computer program that accurately simulates what happens in the real world is both an important and exciting topic. The more complex the system under study, the more difficult or even impossible it becomes to find precise mathematical solutions. Therefore, we may resort to making numerous assumptions to simplify the problem so that it can be solved mathematically. The further these assumptions are from reality, the less accurate and reliable the results will be. However, with vast amounts of real data, it's possible to build accurate stochastic models that represent the variability in the data and use these models to produce accurate simulations of reality. This leads us to what is known as the digital twin.

For a long time, the mathematical analysis of uncertainty and simulation was not only my preference, but also a hobby where I often challenged myself. I would imag-

https://doi.org/10.1515/9783111585055-201

ine some scenario of random events with hidden correlations or dependencies between their elements, then attempt to analyze the problem mathematically and come up with solutions. Afterwards, I would simulate the problem using Monte Carlo and compare the results. I felt immense satisfaction when the results were very close, and if they were far off, I would usually find that the problem was inherent in the mathematical model I had followed—such as mistakenly assuming the independence of a certain event.

When I became a lecturer, these topics dominated the courses I taught, such as digital communications. I encouraged my students in their graduation projects to think outside the box and to simulate their logical ideas even in the presence of uncertainty, identifying strengths and weaknesses. Mastery of uncertainty analysis and simulation tools is a powerful skill that students at all academic levels should be well-versed in. This book can serve as a textbook for a semester-long course on the subject or as a supplementary text for specialized courses like stochastic equations, measurement theory, digital communications, financial studies, and more.

This book took me a very long time to complete due to my lack of free time, as I would return to it periodically over many long months. I wanted to add more chapters to the book, such as some machine learning algorithms and their role in uncertainty recovery, as well as fuzzy modeling. However, I realized that this would delay the book for several more years, so I decided to settle with what I had and perhaps include these additions in the second edition.

I still remember from my early university days that textbooks contained only a few solved examples and many unsolved problems at the end of each chapter. This used to cause me a lot of frustration, because I couldn't verify whether my solutions were correct or not. At that time, it was very difficult to obtain solution manuals, which were only available to the professor. Learning by practice or through examples is a powerful method of training students to handle complex problems. For this reason, I have chosen not to include any unsolved problems in this book. Instead, the student will find more than 150 fully solved examples and dozens of simulation programs.

Part of this book was completed while I was a visiting professor at Aalto University in Helsinki, and I would like to thank Professor Riku Jäntti for his continued hospitality in his department. I would also like to thank everyone who supported me in completing this book, making it a distinctive scientific contribution to this branch of applied sciences.

I must acknowledge that much of this book was written during weekends and summer vacations, often at the expense of time with my family. Throughout this journey, my wife has been unwavering in her support, even when it meant sacrificing time and attention at home. It is with immense joy and gratitude that I dedicate this humble book to my beloved wife, Nagat, my wonderful daughters, Aia and Zakia, and my dear son, Salem.

Finally, it is my sincere hope that readers find this book to be a useful and enlightening resource on their journey to understanding and modeling stochastic uncertainties.

Contents

1 Uncertainty and the philosophy of life

You're probably familiar with Heisenberg's Uncertainty Principle in quantum physics, which highlights the inherent uncertainty involved in measuring the properties of microscopic particles. While it's commonly believed that Newton's laws precisely describe the motion of larger objects in a deterministic manner, this notion isn't entirely accurate. Newton's laws are deterministic only under specific ideal conditions and strict assumptions. Despite their capability to provide sufficiently accurate predictions of object motion, they can never achieve 100 % accuracy. It's important to note that the nature of uncertainty in quantum physics is different from that in Newton's laws, as we'll explore later. In a different context, consider the field of health information, where conflicting advice and results are commonplace. These discrepancies extend to matters such as dietary choices, the link between habits and diseases, and even previously held beliefs about the dangers of LDL cholesterol. The exaggerated impact of LDL cholesterol in older studies has been rectified by recent research, although this doesn't imply that high levels of LDL are beneficial. In the past, coffee was demonized for its supposed harm, even compared to alcohol in terms of negative effects. However, recent studies present a more favorable view of moderate coffee consumption, with numerous health benefits. Similarly, the perception of animal fats as highly dangerous has been challenged by recent research. These examples represent only a fraction of the conflicting results encountered, not only in the area of health but across various fields. Such inconsistencies are often attributable to the substantial uncertainties associated with the conventional tools, including measurement devices, modeling, and statistical analyses, which are employed for data study and analysis. So, what is uncertainty? It can be defined as the inability to achieve a 100 % perfect estimate or prediction, a concept we will delve into further within the pages of this book. Broadly speaking, uncertainty permeates numerous facets of our lives, extending beyond quantum realms and the healthcare system. It casts its shadow over economics, all branches of engineering, physics, chemistry, medicine, the social sciences, biology, and a myriad of other fields.

Let's delve into the myriad ways in which uncertainties shape our lives through a few common questions. Should you pursue subject A or B for your studies? What will the labor market for energy technology look like next year? Is it wiser to accept a new job offer or remain in your current position? In which project should you invest your hard-earned money? When and why will you reach the end of your life? Will you be involved in a car accident in the coming week? Is it a sound decision to purchase this product? Should you commit to marriage with this person? What truly lies beneath their concealed personality? What are their thoughts about you? Can we accurately diagnose illness based on available parameters, and should one opt for surgery or medication? Is it worth taking the risk on this particular project? How will the supply and demand for a specific product evolve next month? What intricate model governs the popularity of mobile games? What can we anticipate as the response of multi-agent service providers to a given project? What mysteries lie in the universe's history before 500 million years

https://doi.org/10.1515/9783111585055-001

ago? What secrets are hidden within black holes? How did life originate on Earth, and what is its purpose? Why does our world exhibit such a remarkable balance despite the complexities of human behavior? How will cancer cells react to specific chemotherapy treatments? Can we accurately model the behavior of a virus within a bacterial colony?

Countless similar questions and examples arise from the inherent uncertainties in our lives and the universe. There is always a level of uncertainty associated with any hidden event (either partially or completely hidden). Furthermore, there are uncertainties associated with any incomplete or distorted information. Almost every parametric or non-parametric model has an uncertain part. Uncertainties are present almost everywhere, in our personal life, in society, in business, in biology, in politics, in technology, in physics, in medicine, in religion, in history, etc. Risks and opportunities have no meaning without uncertainties. In fact, even our whole lives have no meaning without uncertainties. Uncertainty is, in fact, part of the logic of life. For example, if you know with certainty that you will have a car accident on a certain time/day, you will decide not to drive on that day in order to avoid the accident. But this would violate the fact that you must have an accident due to the lack of uncertainty. It is a paradox! And hence, the whole logic of life could be distorted. Furthermore, life would be rather boring. For example, would you enjoy playing a game where you try your luck by tossing a coin with identical heads on both sides? If the unseen future is disclosed in front of you and without any chance to change any of your movements or your strategies, what will life be like for you? Boring? Frightening? Sad? Depressed? It might be anything, but for sure there would be no inspiration, no ambitions, no motivation, no interest in living anymore (at least from our current perspective)! Everything would be known beforehand, including all the misfortunes and sorrows.

The level of uncertainty is dynamic, influenced by factors such as time, place, individual perspective, experience, models, and more. Consider this: I am privy to the balance in my bank account, but you remain unaware of it (unless you work for my bank)! Nevertheless, your interest in this information may be quite limited. Generally, it is not possible to completely eliminate uncertainty, but it is possible to reduce it. Reducing uncertainty means improving the predictability of a certain event. The most challenging decision a decision-maker can encounter is when confronted with a 50 % chance of success in a risky event. In the world of Bernoulli events, the 50–50 case represents the pinnacle of uncertainty. At this juncture, one might as well flip a coin to determine the outcome of the decision. Any deviation from this 50 % threshold, whether more or less, signifies a reduction in uncertainty, making the event more predictable. Consequently, altering the odds in either direction paves the way for a more systematic decision-making approach, as opposed to relying on pure randomness. To mitigate uncertainty, various approaches can be employed. These encompass gaining a more comprehensive understanding of the event, constructing accurate analytical or statistical models, delving into the historical outcomes of similar events, and more. In essence, the highest level of uncertainty is encountered when we lack any pertinent information (or perhaps lack the ability to use it effectively) to build a meaningful model for the uncertain event. For example, imagine a patient visiting a hospital presenting specific symptoms. The physician, using their expertise, initially considers various potential causes for these symptoms. Taking a broader view of the patient's overall health, including factors like age, medical history, weight, and lifestyle habits (such as smoking), the practitioner adjusts the ranking of these potential causes accordingly. Moreover, to address these uncertainties, they may request additional measurements or data. This could involve procedures like blood screening or a CT scan to provide a more precise assessment of the underlying condition. Figure 1.1 illustrates this process. The top figure depicts numerous potential causes based solely on the symptoms presented. However, as we gather more information and conduct a thorough analysis, the number of potential causes can be significantly reduced, thus mitigating uncertainty to a considerable extent. Ultimately, the goal is to arrive at a single, highly probable cause. The same example could be generalized for diagnosing technical failures or possible market directions, and so on.

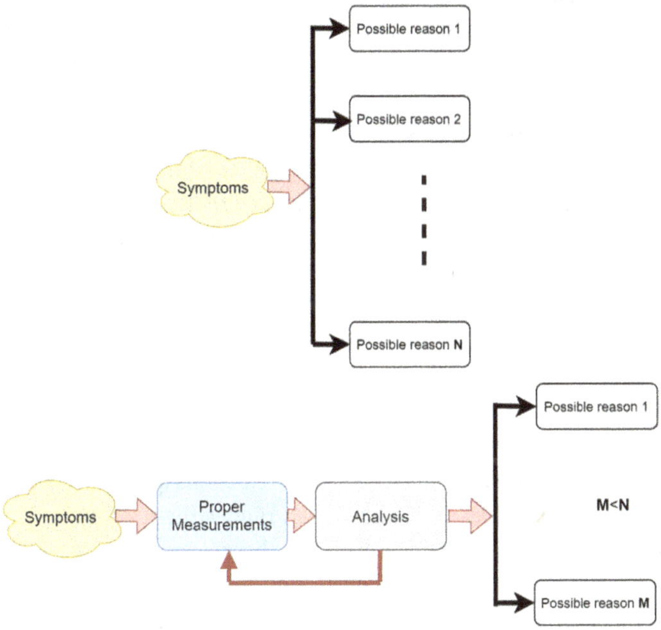

Figure 1.1: Diagnostic Example.

One might ponder, why do uncertainties exist? In essence, what gives rise to these inherent uncertainties? To address this concept, let's briefly outline a few of the underlying factors contributing to uncertainty, using elementary examples for illustration.

- Future (Time Barrier): The future outcomes of events with multiple random and uncertain possibilities are inherently uncertain. Can we ever achieve a high degree of accuracy in predicting such random outcomes? The answer is a complex interplay of "Yes" and "No." To illustrate this, let's consider the classic example of flipping a coin with two sides: heads and tails.

In the absence of any prior information about the coin flip, the best prediction is an equal 50 % chance of landing as heads or tails, representing maximum uncertainty. However, in theory, this uncertainty could be entirely resolved if the event were transformed into a deterministic one. Such a transformation becomes possible when we possess comprehensive knowledge of all the relevant parameters, often referred to as experiment attributes, at the moment of the coin toss (assuming we can control the toss timing). This comprehensive knowledge would entail understanding factors such as the initial orientation of the coin, the precise force and angle applied during the toss, the material, weight, size, and dimensions of the coin, ambient conditions such as air pressure and wind, the rigidity of the landing surface, and, perhaps most crucially, a flawless mathematical model of the tossing process denoted as $f(x_1, x_2, \ldots, x_n)$. These attributes collectively constitute the experiment parameters, and the model provides a binary result—either "Head" or "Tail." With exact values for all experiment attributes and a perfect model, the once-random coin toss becomes deterministic, enabling us to predict the outcome in advance. However, in practice, this level of precision is extremely challenging to attain, even in the case of a seemingly simple coin toss. More complex scenarios involving numerous overt and covert parameters, as well as intricate, non-linear, and dynamically changing mathematical models, make deterministic prediction virtually impossible. There exists another class of processes for which predictability remains elusive, regardless of the extent of available information. An illustrative example is thermal noise or any form of uncorrelated randomness. In these cases, randomness is intrinsic, not stemming from a lack of knowledge. Quantum mechanics presents another prime example of such inherent uncorrelated randomness. In summary, the ability to predict the future random outcomes of events depends on the availability of comprehensive, and completely accurate information as well as the nature of the underlying processes. While deterministic predictions are theoretically feasible in ideal conditions, practical limitations often render them unattainable, especially in complex scenarios. In certain processes, inherent randomness defies predictability, regardless of our level of knowledge.

- Lack of complete knowledge: In this case, the outcomes of a certain event have already happened in the past. However, there is no complete information about the outcomes. In some cases, the reason for uncertainty is not the barrier of time but rather the lack of information due to other barriers, such as the barrier of location, the barrier of visibility, the barrier of precision, and many other barriers. Suppose you have an opaque box that cannot be seen through, and all we know is that it contains a number of small balls of different colors. If we can't see inside the box, and we have no prior information, we are faced with a state of uncertainty about the number and colors of the balls. To this category, we may also add the lack of an exact process model, the lack of knowledge about all the possible influencing attributes

as well as their mutual correlations and dependencies, naturally limited precision, etc.
- Chaos effects. In some dynamic systems, predicting their response is a formidable challenge because these systems are incredibly sensitive to even the tiniest variations in their parameters and initial conditions. Small adjustments in these factors can trigger substantial alterations in the system's behavior, a phenomenon known as the "butterfly effect." However, we often lack precise information about the system's initial conditions, and we cannot attain absolute precision in measuring system parameters due to the limitations of our instruments and the presence of noise. As a result, chaos represents a type of uncertainty that remains fundamentally unresolved. In fact, it can be considered as an additional layer of uncertainty alongside the previously discussed lack of complete knowledge.

Uncertainty, in one form or another, is an inherent aspect of every facet of life, whether it's in the realm of technical or non-technical matters. This book serves as a guide to introduce key concepts of uncertainty and presents a unified technical approach to address them, with the aim of appealing to a wide range of readers. As such, the majority of the technical content within it requires only a fundamental understanding of university-level mathematics. Moreover, the subject matter is presented without delving into intricate philosophical discussions.

There is no doubt that the topic of uncertainties is very broad and can be presented in many different ways. However, in order to enhance our predictive ability (i. e., reduce the level of uncertainty), we always need to exploit any information we have about the system. In other words, we should study and understand the system and both its internal and external relations. However, we usually start by constructing a model of the system under study. Modeling is the human process used to study and understand reality. There are several approaches to modeling, depending on the nature of the system. For example, models can be parametric, non-parametric, abstract, descriptive, deterministic, stochastic, statistical, or involve computer simulations, among other methods. Nevertheless, it's crucial to recognize that no model, regardless of its sophistication, can completely encapsulate the complexity of the real systems. In other words, there is no such thing as a 100 % perfect model.

Nevertheless, delving into the study and comprehension of any natural, technical, or non-technical system is virtually impossible without the use of modeling. Natural systems encompass physical, biological, chemical, and meteorological processes. Technical systems include electrical engineering, civil engineering, computer science, and more. Finally, non-technical systems include economics, sociology, psychology, marketing, politics, and numerous others. For instance, consider the electrical properties of semiconductors, a natural phenomenon. We can construct a model for these properties, which can then be harnessed to explore the effects of introducing other materials, a process known as doping, to create new substances with desired behaviors (an engineering application). This, in turn, paves the way for developing transistors, the fundamental yet pivotal components powering advanced computing devices and revolutionizing information, computing, and communication technologies. The notion of a *perfect model* entails the ability to precisely predict the behavior of a real natural system with 100 % accuracy at any given moment. In practice, achieving this level of precision is extremely challenging. It is very challenging to count how many internal and external parameters might influence the system's behavior. Additionally, any instrument used to measure system behavior, whether it's measuring pressure, temperature, electrical current, or other factors, introduces a certain precision as well as level of distortion to the measured values. This means that the measured values are approximations, and the exact values remain unknown even after measurement. While the accuracy can be very high, say 99.9999 %, the perfect exact value remains unknown. Therefore, we have to accept the fact that achieving perfect and complete knowledge of any natural behavior is impossible. Consider the straightforward example of the relationship between current and voltage across a resistor, known as Ohm's law, which states that $v = R \times i$. Here, v is the voltage across the resistor terminals in Volts, R is the resistance value in Ohms, and i is the electric current in Amperes. This relationship appears simple and functions well in direct current (dc) and low-frequency applications. However, it falls short of capturing the actual relationship between voltage and current under various circumstances. The basic Ohm's law model doesn't account for ambient temperature, which is known to affect resistance, nor does it consider the impacts of humidity and pressure, both of which influence resistance. As current flows through the resistor, it generates heat, altering the value of the current. In alternating current (ac) situations, a skin effect comes into play, causing the current to concentrate on the outer cross-section of the resistor due to the reverse magnetic field at the conductor's center. The electric and magnetic fields spread across the resistor body, impacted by capacitance and inductance effects, and even mutual coupling with other nearby electrical elements. Furthermore, the exact value of the resistance isn't known with certainty. Even when measured, the value obtained is limited by the accuracy of the measuring instrument, and factors such as aging and corrosion further impact the resistor. Moreover, the resistor itself acts as a random voltage source due to the unpredictable movement of electrons within its molecules, a phenomenon known as thermal noise. It is impossible to predict this noise voltage because it is uncorrelated. Hence, we may conclude that Ohm's law is just a good model that approximates the voltage-current relationship in resistors at low frequencies. However, even with the advanced technology that we have today, expressing the model with 100 % perfect accuracy at any frequency remains an unattainable goal (fortunately, this level of perfect precision isn't necessary in electronic design). Therefore, it is better to express the model as $v = R \times i + \Delta v$. Here Δv is the model uncertainty. This uncertainty includes both incomplete knowledge and the inherently random noise. Every formulated model of any system has its uncertainty level. This is where probability theory becomes invaluable, offering a systematic means to describe randomness and uncertainty. As a result, probability and stochastic processes play a crucial role in numerous fields.

As previously mentioned, our primary approach to understanding reality revolves around modeling. Typically, models are tailored to represent specific aspects of the system under consideration. Effective models aim to strike a balance between two often conflicting objectives: 1) simplicity and 2) accuracy. This necessitates a compromise to ensure a practical and useful representation.

Models serve a variety of purposes, including gaining insight into reality, studying systems, analyzing performance, estimating unobserved parameters, making predictions and forecasts, and simulating systems under different conditions. In the majority of cases, the successful modeling of systems depends on the application of mathematics and statistical analysis, as depicted in the Figure 1.2.

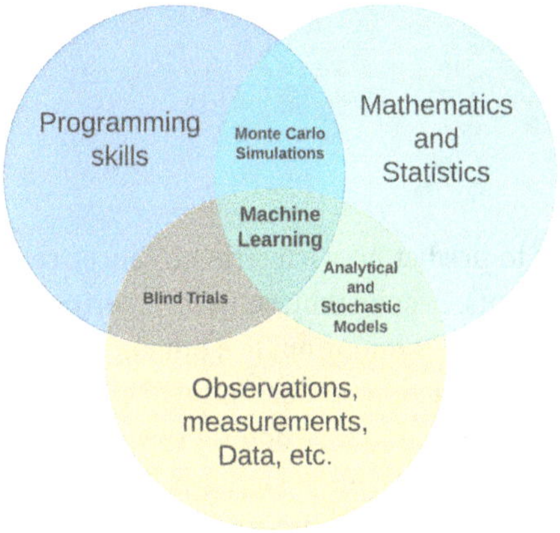

Figure 1.2: Some Modeling Tools and Methods.

We firmly hold the view that mathematics serves as a primary language for constructing models. Models, in themselves, are human creations that offer a means to comprehend a wide array of subjects and phenomena.

The principal goal of this book is to present a comprehensive approach to addressing uncertainties.

Figure 1.3 illustrates a potential approach for visualizing different kinds of uncertainties and the tools used to manage them.

To begin, we can categorize uncertainties into two classes: rational and irrational. By "rational," we refer to uncertainties surrounding the behavior or responses of intelligent, adaptive systems equipped with sensory feedback. A prime example of this is observed in living organisms. Human beings, representing the pinnacle of the animal kingdom, possess full consciousness, awareness, intentions, and decision-making abilities, among other attributes. Conversely, other organisms may exhibit simpler responses and reactions driven by their innate survival instincts. Nevertheless, even modeling the behavior of such primitive organisms can prove remarkably complex in many

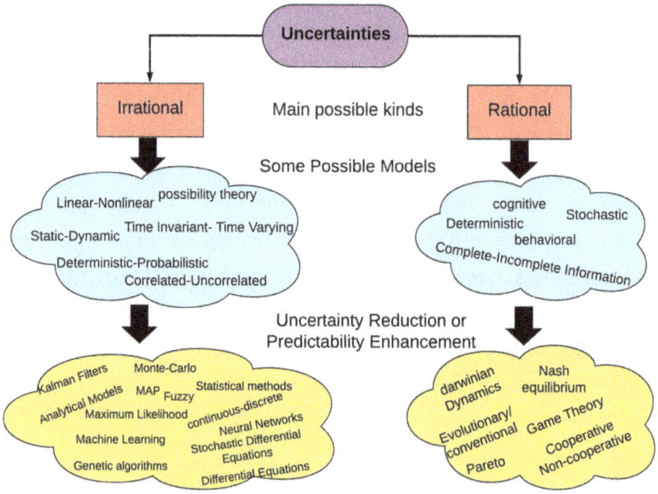

Figure 1.3: Uncertainty Manifestations.

scenarios. Consider, for instance, our attempts to predict how bacteria or cancer cells respond and adapt to counteract medical treatments. Despite thousands of experiments and a wealth of data, substantial uncertainties persist. Zooming out to a human context, forecasting specific social behaviors, predicting the reactions of decision-makers within companies to various business and market conditions, and envisioning strategies employed in multi-agent competitions with limited resources pose similarly challenging questions. Rational agents encompass not only biological entities but also man-made intelligent systems, such as robotics and autonomous vehicles.

Let's consider another example of wireless communication: *cognitive radios*. Below the 5 GHz frequency range, the radio spectrum is heavily congested, with most bands allocated to specific *primary* users. However, it's been observed that in certain locations or at particular times, many spectrum slots are underutilized by these primary users. In response, cognitive radios have emerged as intelligent devices designed to harness these opportunities, transmitting data in unoccupied *white* or lightly used *gray* spectrum slots. However, there's a crucial constraint: cognitive radios must operate without causing disruption or harmful interference to the communication of primary users. This necessitates spectrum sensing to monitor the status of primary traffic. Additionally, when multiple cognitive radio devices share the same available white spectrum, there's a need to consider how they interact with each other. These devices are equipped with an intelligent core comprising sensors and algorithms to facilitate responses to other radios, especially when competing against different agents.

To model rational behavior in this context, it's essential to define objectives, strategies, capabilities, opponents, the environment, and other pertinent factors. Game theory emerges as the conventional tool for analytical investigation, modeling, and management of such conflicts. Within the field of game theory, numerous models exist, including cooperative and non-cooperative games, as well as deterministic and stochastic games, among others.

The second category in Figure 1.3 pertains to uncertainties associated with irrational systems. Here, irrational implies that the study or modeling is based solely on observed behavior or data, without delving into the underlying rationale. For example, when predicting stock market prices, despite the multitude of factors influencing market fluctuations—many of which result from rational actions—our approach may center on interpreting observations as a random time series of data with certain statistical characteristics.

In practice, most types of uncertainty can be effectively addressed within the irrational class. This entails working with observed data or measured time series and striving to identify the most suitable model that can replicate this data. Generally, irrational systems can be categorized based on several attributes, including linearity vs. nonlinearity, static vs. dynamic behavior, and determinism vs. stochasticity.

To tackle uncertain systems, various techniques are at our disposal, including analytical methods employing equations (e. g., stochastic nonlinear differential equations), simulations (which bypass the need for constructing explicit analytical models), statistical approaches, machine learning, and more. In this book, we will explore a plethora of techniques for modeling and simulating uncertainty. Among these techniques, Monte Carlo simulation is of particular importance, serving as a vital method for studying, analyzing, validating, and predicting uncertainties within complex systems. This is especially valuable in cases where creating a comprehensive analytical model is exceedingly intricate or impractical.

With Monte Carlo simulation, there's no requirement to construct a detailed analytical model of the system. Instead, what's essential is the establishment of well-justified assumptions. These assumptions should be based on either empirical measurements or a thorough understanding of the system's behavior. For instance, let's consider a scenario where you need to determine the optimal number of service providers for a public service system, such as the number of staff employees at a tax office. Based on your observations, the arrival of individuals at service counters follows a random process, with an average rate of N persons per hour (a value that can be observed or extracted from historical data). Additionally, the time it takes for each person to complete their task is also a random process, with an average duration of T_s minutes (this can be measured as well).

The waiting area of the public service office can accommodate a maximum of M persons at any given time. In the system's design, the objective is to ensure that the total time a customer spends waiting and being served (T_{total}) remains below a specified limit, $T_{\max}$, with a certain level of confidence, say, p. Having more service providers than strictly necessary can enhance the system's efficiency by minimizing waiting times and queue lengths. However, it's not a cost-effective solution, as it would increase resource utilization and cost. Therefore, the aim is to keep the number of service providers at the minimum necessary to achieve the desired performance criteria, i. e., $T_{\text{total}} \leq T_{\max}$

and total queue length $\leq M$. Both targets should be met with a predetermined level of probability, for example, $p = 0.9$. Increasing the value of p implies reducing the risk of failing to meet the performance targets, but this would also necessitate employing more service providers.

Solving this problem analytically is relatively straightforward using well-established queuing models. However, when the analytical approach is not feasible or when the necessary assumptions for the arrival rate and service time probability distributions are hard to define, Monte Carlo simulation can be a valuable alternative. Analytical solutions often rely on certain assumptions such as the use of Poisson and exponential distributions for arrival rates and service completion times, respectively. Nevertheless, making incorrect assumptions about the distributions can introduce bias and errors into the model. A notable advantage of Monte Carlo simulation is its flexibility regarding the choice of probability distributions. You can use any distribution, or even actual recorded data, to study the system's performance. Additionally, Monte Carlo simulation can serve to validate the results of analytical uncertainty analysis. Proficiency in applying Monte Carlo simulations is indispensable in various research domains. This book delves into these concepts through a series of illustrative examples, offering valuable insights into their practical use. Although any programming language could be used, we mainly use Matlab-style programming in this book. Nevertheless, to give another option for students who cannot afford a Matlab license, we have used mainly the Matlab open-source alternative known as *Octave*. However, all the codes written for Octave in this book can be run using Matlab as well (sometimes with minimal modifications). Besides Octave, another free simulation environment called *Scilab* has also been used to solve only a few examples. The style of Scilab is slightly different, but it has a few additional features compared with Octave, like the *Xcos*, which is a graphical programming environment for modeling, simulating, and analyzing dynamical systems. Xcos uses similar concepts to those of Simulink in Matlab.

The uncertainty management techniques outlined in this book are versatile and can be employed with diverse datasets, irrespective of the specific context in which the data is interpreted. For instance, consider a scenario where we have a set of attributes, such as patient health data, and corresponding observations related to different types of diseases. In a different context, these attributes could pertain to weather data, including date and time, with observations linked to electricity demand. In yet another application, attributes might include personal details like age, salary, marital status, and gender, while the observations capture the customer's detailed monthly purchases.

These are just a few examples of different applications, but the beauty of these tools lies in their applicability regardless of the data's underlying meaning. Naturally, understanding and interpreting the results generated by the model remains essential, but the tools themselves can be effectively harnessed in a multitude of applications that transcend the specific semantics of the data.

Book outlines

This book is divided into six chapters, and we'll provide a brief overview of each one. In Chapter 2, we delve into the fundamental mathematical tools required for managing uncertainties, namely probability theory and stochastic processes. Our primary objective in this chapter is to maintain simplicity. As a result, we've steered clear of unnecessarily intricate derivations. Additionally, we've seamlessly integrated the concepts of Monte Carlo simulation, allowing for a natural progression in understanding. To facilitate comprehension, we've included numerous solved examples and provided comprehensive code samples for tackling specific problems, with detailed explanations of the code's functionality.

The majority of real-world systems exhibit dynamic behavior and can be effectively represented using either linear or nonlinear differential equations. Within these dynamic equations, uncertainties can be incorporated as integral components. Chapter 3 serves as an introduction to this topic, elucidating key concepts associated with modeling dynamic systems.

In Chapter 4, we provide a concise yet pivotal introduction to estimation theory. Estimation theory stands as a primary framework for addressing and managing uncertainty. In this chapter, we focus on imparting the fundamental principles and key foundations that underpin this theory.

The Kalman filter is known as the premier estimation algorithm for linear dynamic systems affected by additive Gaussian white noise. It stands as one of the most successful and widely employed algorithms for managing uncertainty. Chapter 5 is dedicated to an in-depth exploration of Kalman filters, encompassing their derivations from a statistical point of view and some simulations, acknowledging their significant utility and relevance.

Finally, in order to introduce the uncertainties surrounding the responses of intelligent organisms and rational entities, Chapter 6 provides a comprehensive exploration of crucial fundamentals within game theory. Additionally, this chapter offers a range of solved examples and codes to illustrate these concepts in practice.

In this book, our aim has been to avoid needlessly complex mathematical derivations and to emphasize practical applications, offering guidance on the effective utilization of tools and algorithms for managing uncertainty. Throughout, we've provided around 150 solved examples and numerous simulation codes to illustrate key concepts.

It's important to note that this book doesn't attempt to provide an exhaustive compendium of all available tools for handling uncertainty. Rather, it focuses on five core chapters: "Probability theory and stochastic processes", "Dynamic modeling", "Estimation theory", "Kalman filters", and "Game theory". Each of these topics represents a full-fledged field with extensive literature and dedicated texts.

However, this book strives to present these topics cohesively within a unified framework for managing uncertainties. After engaging with this material, readers will be well versed in these advanced techniques and equipped to tackle practical challenges. The book is particularly valuable for technical projects that necessitate simulations or emulations of real systems, making it a useful resource for undergraduate and master's level students embarking on graduation projects and theses.

This book is designed to appeal to a wide range of readers, serving as a valuable resource for both self-study and those seeking a reference in diverse fields. These fields encompass estimation theory, control theory, computer simulation, wireless communication, power systems, biotechnology, finance, and beyond.

2 Probability theory and stochastic processes

While this chapter doesn't serve as a comprehensive reference for probability theory, it does offer a significant introduction to fundamental concepts necessary for quantifying and evaluating uncertainties. Even if you believe you have a good grasp of probability theory and stochastic processes, I encourage you to work through the provided examples and explore the accompanying simulation codes for a deeper understanding.

While numerous excellent textbooks delve into this field, I personally favor Papoulis' book [1].

After completing this chapter and working through the provided examples, you will:

- Understand the key concepts of probability theory.
- Become familiar with random variables and various probabilistic models.
- Acquire knowledge about stochastic processes and applications.
- Become proficient in the use of Monte Carlo simulations, a valuable tool for tackling complex problems in a variety of research areas.
- Understand how to quantify and evaluate uncertainties

2.1 Probability theory

Consider an experiment with a minimum of two potential outcomes. Since we lack complete information about the experiment, there is uncertainty regarding the precise outcome before it occurs. In such scenarios, probability theory proves invaluable in quantifying our level of certainty about the experiment. Probability theory is a powerful mathematical tool that provides quantitative measures for random events and stochastic processes, making it essential for addressing uncertainty. A common approach to introducing probability theory centers around the concept of the *sample space*. The sample space is the comprehensive set encompassing all potential outcomes of a given experiment, with each subset of this sample space referred to as an *event*. To illustrate this concept, let's commence with a straightforward experiment of flipping three coins and examining the possible outcomes. This experiment will have 8 possible outcomes. However, there will be uncertainty about the exact outcome. The experiment can be performed by tossing one coin three independent times, or by tossing three coins at once. As we mentioned in the previous chapter, the reason for the uncertainty is the lack of complete information about the tossing process. We show a systematic method to quantify the uncertainty level of the experiment outcomes. The sample space of the three coins toss experiment can easily be represented by the following set

$\mathbb{S} = \{TTT, TTH, THT, THH, HTT, HTH, HHT, HHH\}$, where T refers to the *tails* and H to *heads*. This sample space contains all possible outcomes of the experiment. Are there any other potential outcomes? From this sample space, we may define many events (*how many?*). For example, we may consider the event *the first coin landed tails*. This event is a

https://doi.org/10.1515/9783111585055-002

subset containing the following outcomes $E_1 = \{TTT, TTH, THT, THH\}$. Another possible event is *the last two coins landed heads*. This event is represented by the subset $E_2 = \{THH, HHH\}$. Probability theory could be defined over this concept of sample space and events.

Let's introduce the axioms of probability theory. An axiom is a statement that is not subject to proof or disproof; its truth is considered self-evident. If we define the sample space as $\mathbb{S}$ and a certain event E_i, the probability theory is based on the following three axioms:

$$P(E_i) \leq 1, \tag{2.1}$$

$$P(\mathbb{S}) = 1. \tag{2.2}$$

For any disjoint events E_i and E_j, we have

$$P(E_i \cup E_j) = P(E_i) + P(E_j). \tag{2.3}$$

Next are some important theorems that can be easily proved based on the above axioms:

$$P(\mathbb{S} - E) = 1 - P(E) \tag{2.4}$$

$$P(\Phi) = 0 \tag{2.5}$$

$$P(E_1 \cap E_2) \leq P(E_1 \cup E_2) \leq P(E_1) + P(E_2) \tag{2.6}$$

$$P(E_1 \cup E_2) = P(E_1) + P(E_2) - P(E_1 \cap E_2) \tag{2.7}$$

$$E_1 \subseteq E_2 \Rightarrow P(E_2 - E_1) = P(E_2) - P(E_1). \tag{2.8}$$

Figure 2.1 shows the Venn diagram description of some set theory operations. The results of an uncertain experiment are referred to as *equally likely* when they have iden-

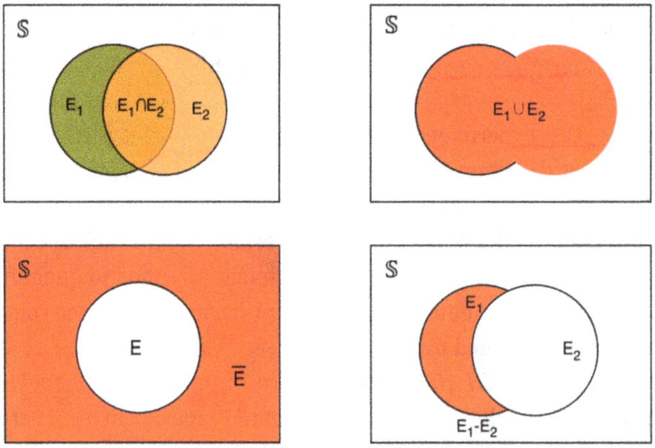

Figure 2.1: Venn diagram.

tical probabilities of occurring. For instance, a coin is considered equally likely if it has an equal chance of landing on either *heads* or *tails*. In the case of the previously mentioned experiment involving three coins, the sample space encompasses eight distinct events. All eight of these outcomes are *mutually exclusive*, signifying that the occurrence of one event or outcome precludes all others. Mutually exclusive events cannot occur simultaneously. For instance, if the result of the experiment is represented as $\{THT\}$, it becomes *impossible* for any other outcome to coexist simultaneously. Drawing from the axioms presented in (2.2) and (2.3), it can be asserted that:

$$P(\mathbb{S}) = P(\{TTT\}) + P(\{TTH\}) + \cdots + P(\{HHH\}) = 1. \tag{2.9}$$

If this experiment were fair, with equally likely outcomes, the probability of each event would naturally be $\frac{1}{8}$. This may appear straightforward and intuitive, but what happens if the tossing process is not fair (i. e., a biased experiment)? In such cases, it's not a requirement for all outcomes to have identical probabilities. Nevertheless, one principle remains universally true: the total probability across all distinct events in the sample space must equal One.

Another approach to estimating the probabilities is to observe the events of the experiment, as follows:

$$P(E) = \lim_{N\to\infty} \frac{N_E}{N}. \tag{2.10}$$

Where N represents the total number of repetitions of the same experiment, and N_E is a counter that is incremented each time the event E is observed. The convergence of this limit must be defined based on specific stochastic criteria. For instance, it can be proven that this limit converges with a probability of 1, although a detailed discussion of this will follow. The equation (2.10) serves as the foundation of Monte Carlo simulations.

We have briefly examined two approaches to assessing the probability of uncertain events. The first approach relies on logic, incorporating set theory and logical operations. The second approach is grounded in observations and measurements. Which of these approaches is more accurate? When the problem is well-defined in all aspects, as in the coin flip example, the logical approach is more accurate. However, when a problem is not well-defined or is exceedingly complex to model, the observational approach becomes the more suitable choice. It is not uncommon to employ both methods for the same problem. For instance, we use logic to construct an initial model of the problem and use observations or measurements to refine the model or estimate its parameters.

Example 1. In the example of tossing three fair coins, what is the probability of the middle coin landing as *heads*?

We may define the event of "*middle coin is H*" as the following set

$$E = \{THT, HHT, THH, HHH\}.$$

Since this event consists of four disjoint outcomes, then $P(E) = 4 \times \frac{1}{8} = \frac{1}{2}$.

Example 2. Let's begin with our first Monte Carlo simulation code. Write a versatile code to simulate the toss of three coins, allowing for a general probability of each coin landing as heads, denoted by p_i, where i represents the coin number, and $i \in (1, 2, 3)$. Note that if the probability of the ith coin landing as heads is p_i, then the probability of it landing as tails is $1 - p_i$. After simulating the experiment for a substantial number of trials, find

- The probability that all outcomes are heads
- The probability that the first coin lands heads
- The probability that the first two coins land heads and the third coin lands tails.

You may use $p_1 = 0.3$, $p_2 = 0.5$, and $p_3 = 0.8$. The simulation problem is coded first with Scilab and then with Octave (or Matlab). We will provide a step-by-step explanation for both codes.

```
// Scilab Code
clear
N=10000; // Number of repeating the experiment
p1=0.3; // the probability of the first coin to land heads
p2=0.5; // the probability of the second coin to land heads
p3=0.8; // the probability of the third coin to land heads
x=rand(N,3); // x is a matrix with N by 3 column.
// rand function generates uniform random numbers from 0 to 1
// Scaling the probability matrix in y
y=[p1*ones(N,1) p2*ones(N,1) p3*ones(N,1)];
z=sign(y-x); // z matrix with N by three columns
// and the items are +1 for heads and -1 for tails
// matrix z contains the repeated experiment results.
// It is possible now to calculate the probabilities.
// What is the probability P1 that we obtain all heads (HHH)?
P1=length(find(sum(z,2)==3))/N
// What is the probability P2 that the first coin is heads?
P2=length(find(z(:,1)==1))/N
// What is the probability P3 that we obtain (HHT)
P3=length(find(z(:,1)+z(:,2)==2 & z(:,3)==-1))/N
```

Understanding this simple code is an essential step toward learning how to write Monte Carlo simulation codes for more complex, uncertain systems. It is worth mentioning here that this is just one way to simulate this experiment with uncertain outcomes. It is possible to write the code in several other ways, which could be even more efficient. However, in this context, we prioritize simplicity and intuitiveness. In essence, our programming style reflects two key factors: our thought process and our proficiency with the programming language we are using, including its capabilities and features. Consequently, there can be various coding styles that all achieve the same task. There are different criteria for comparing and preferring one coding style over another, such as execution speed, ease of comprehension and modification, and resource consumption (e. g., memory). The subsequent steps describe the explanation of the simulation code in detail:

- The initial line of code begins with a double slash //. In Scilab, this notation signifies that everything after the double slash will not be executed. Therefore, it is employed to add comments to the code.
- Line 2: *clear* is the first command in the code. It is used to clear the variables' values that might be stored in memory.
- Line 3: $N = 10000$, is the number of repetitions of the simulated experiment. In general, increasing the number of experiment repetitions should lead to more accurate results. Nevertheless, in Monte Carlo computer simulations, there are typically some values of N beyond which we may not observe significant improvements in accuracy. Experiment with different values of N and compare the outcomes. You can also compare these results to the exact values that will be calculated later.
- Lines 4–6: $0 \leq p_i \leq 1$ for $i = 1,2,3$ represents the probability that the ith coin is heads.
- Line 7: Regarding the command $x = \text{rand}(N,3)$, Scilab generates a matrix with N rows and three columns containing uniformly distributed random numbers ranging from 0 to 1. Uniform random numbers imply that the probability of each value occurring between 0 and 1 is equal. In the Scilab language, if you wish to generate random numbers with a different distribution, you need to specify it within the *rand* command. For instance, to generate numbers with a normal distribution instead of a uniform one, you can use the command $x = \text{rand}(N,3,'\text{normal}')$. The choice of distribution function depends on the specific requirements of the simulation problem.
- Line 10: The matrix y has the same dimensions as x, meaning it also has N rows and 3 columns. The command $\text{ones}(N,1)$ generates a column vector consisting of N elements, each set to the value 1. Consequently, each column i in the matrix y contains a constant number corresponding to the probability of the ith coin landing as "heads."
- Line 11: The *sign* command yields a result of +1 if its argument is greater than or equal to 0, and −1 if its argument is less than 0. Consequently, the matrix $z = \text{sign}(y - x)$ is an array with dimensions $N \times 3$, where all its elements are either −1 or +1. To clarify this step, let's consider a uniform random number $0 \leq \alpha \leq 1$. Given its uniform distribution, the probability that $P(\alpha \leq \beta) = \beta$ for $0 \leq \beta \leq 1$. For instance,

$P(a \leq 0.3) = 0.3$. Now, what happens when you apply the function sign$(a - 0.3)$? Correct! It results in ±1! If $a < 0.3$, the argument of the sign function is negative, so the result is −1; otherwise, it's +1. Thus, the probability of obtaining sign$(a - 0.3) = -1$ is $P(\text{sign}(a - 0.3) = -1) = 0.3$! In the context of a coin-tossing experiment, we can assign +1 to the event "heads" and −1 to "tails". How can we simulate an experiment where the coin lands "heads" with a probability of 0.85? It's as simple as executing one of these two equivalent expressions: sign(rand(1) − 0.15) or sign(0.85 − rand(1)). To repeat the same experiment, for instance 1,000 times and store the results in a vector S, you can achieve it using a single command $S = \text{sign}(0.85 - \text{rand}(1000, 1))$. Now, the operation performed in the code $z = \text{sign}(y - x)$ should be more evident.

- Line 16: Observe that each row in the matrix z contains 3 columns with elements of ±1. Where +1 represents the simulated event *heads* and −1 represents the simulated event *tails*. If we are looking for the experiment outcomes where all the coins landed *heads*, we can do that in at least two ways. One by checking each element, i. e., performing the logic AND between all elements in each row. Another easier way is to look at the rows that have a sum of 3. What if we are looking for the rows that have 2 *heads* and 1 *tails*? We can do that simply by allocating the rows with the summation of 2. The interesting question now is how we allocate the elements in the matrix z. In any programming language, you may do that in several different ways. In this code we have used the command find(.). This command is available in *Octave* and *Scilab*. It returns the indices of the true argument. For example, for a vector $x = [-1, 0, 5, -6, 10]$ the result of the command find(x == 0) is 2, because the index of 0 is the second. The result of the command find($x > 2$) will be 3 and 5, because the argument is true for 5 and 6. They have indices' numbers 3 and 5, respectively. The probability that all outcomes are *heads* is the percentage where the sum of each row in the matrix z is 3. As shown in (2.10), one should count how many times the outcome is 3 divided by the total number of trails. The number of the required event can be computed using the command *length*, which gives the length of the vector. Finally, by dividing the number of times that the required outcome has occurred by the total number of repeating the experiment N, we obtain an estimation for the probability.
- Based on the same concepts, try to study how we compute P_2 in the code above.
- To compute the probability that the outcome will be *HHT*, we used an *AND* logic which is represented in Scilab by the command "&". By the way, we may use other alternatives for computing P3 such as P3=length(find(z(:,1)==1 & z(:,2)==1 & z(:,3)==−1))/N. Try it in the code above!

If you want to use Matlab or *Octave*, the difference in the code is marginal. In Octave, the simulation problem can be achieved in the following code:

```
% Octave code
clear all
```

```
N=10000; % Number of repeated experiment
p1=0.3; % the probability of the first coin to land heads
p2=0.5; % the probability of the second coin to land heads
p3=0.8; % the probability of the third coin to land heads
x=rand(N,3); % x is matrix with N by 3 column.
% rand generates uniform random numbers from 0 to 1
% Scaling the probability matrix in y
y=[p1*ones(N,1) p2*ones(N,1) p3*ones(N,1)];
z=sign(y-x); % z matrix with N by three columns
% and the items are +1 for heads and -1 for tails
% matrix z contains the repeated experiment results.
% We can now calculate the probabilities.
% What is the probability P1 that we obtain all heads (HHH)?
P1=length(find(sum(z,2)==3))/N
% What is the probability P2 that the first coin is heads?
P2=length(find(z(:,1)==1))/N
% What is the probability P3 that we obtain (HHT)
P3=length(find(z(:,1)+z(:,2)==2 & z(:,3)==-1))/N
```

Upon executing the code, the obtained results are as follows: $P1 = 0.1159$, $P2 = 0.3005$, and $P3 = 0.0298$. It's important to note that you might obtain slightly different results due to the inherent variability in random trials. However, over an infinite number of trials (at least theoretically), these values will converge with a probability of 1 to certain specific values. Now, let's calculate the exact results for the example addressed by the codes.

- The probability to have *HHH*: since all coin toss outcomes are independent, the probability will be $P(C_1 = H, C_2 = H, C_3 = H) = P(C_1 = H)P(C_2 = H)P(C_3 = H) = 0.3\times = 0.5 \times 0.8 = 0.12$. Where C_i is the outcome of coin i. It can be either H or T.
- The probability that $P(C_1 = H) = 0.3$ is already given.
- Finally, the probability that we obtain *HHT* is given by $P(C_1 = H, C_2 = H, C_3 = T) = P(C_1 = H)P(C_2 = H)P(C_3 = T) = 0.3\times = 0.5 \times (1 - 0.8) = 0.03$.

Additional analytical examples will be presented later to further illustrate this topic. It's evident that simulation results closely approximate the exact analytical outcomes. While this example is straightforward, it's essential to recognize that many real-world systems with stochastic behavior are significantly more intricate, making it challenging to derive closed-form mathematical solutions. As a result, simulation stands out as a valuable and efficient approach for tackling highly complex problems. Throughout this book, we have modeled numerous examples to demonstrate this.

Example 3. If you toss two fair and independent dice, what is the probability that the sum is 9? Check your answer by simulation.

The sample space of this experiment consists of 36 events such as $\mathbb{S} = \{(1,1),(1,2),\dots,(1,6),(2,1),\dots,(2,6),\dots,(6,6)\}$. The event that the sum is 9 could be represented as $E = \{(3,6),(4,5),(5,4),(6,3)\}$. Since all outcomes are disjoint and both dice are fair, then $P(E) = \frac{4}{36} = 0.11$. Utilizing the same principles elucidated in the previous three-coin example, we can proceed to draft the simulation code in Scilab and Octave, as demonstrated below. It is imperative to meticulously traverse the code, step by step, to ensure a thorough comprehension of its functionality.

```
clear
N=10000; // Number of repeated experiment
//generate N by 2 matrix with random integers between 1-6
x= grand(N, 2, "uin", 1, 6);
y=sum(x,2); //add the outcome of each die
P=length(find(y==9))/N //the probability
```

The Octave version is again very similar, and it can be expressed as:

```
% Octave Code
clear all
N=10000; % Number of repeated experiment
%generate N by 2 matrix with random integers between 1-6
x= unidrnd(6,N,2);
y=sum(x,2); %add the outcome of each die
P=length(find(y==9))/N %the probability
```

The Scilab and Octave codes are quite similar, with a notable distinction in their random integer generation functions. In Scilab, random integers with a uniform distribution are created using the *grand* function with the type set as *uin*. The dimensions are denoted as $N \times 2$, where N signifies the number of experiment repetitions, and it should be sufficiently large to ensure accuracy. We'll delve into the process of selecting an appropriate value for N later. Additionally, we set the minimum and maximum integer values in the function to 1 and 6, respectively. For Octave, we can employ the *unidrnd(x, a, b)* function, which generates random integers ranging from 1 to x with dimensions specified as $a \times b$. Executing either of these code segments will yield results very close to the analytical outcome, which, in this case, is $P = 0.1085$.

The sample space can include continuous intervals, such as location and time. For instance, imagine throwing an object onto a square table measuring 2×2 meters. We'll assume the presence of barriers along the table's edge to prevent the object from falling off. If we designate one corner of the table as the reference point, denoted as $(0,0)$, then the sample space for this experiment can be represented as $\mathbb{S} = \{[0,2],[0,2]\}$.

The potential outcomes of this experiment, denoted as $E = \{x,y\}$, where x and y lie within the interval $[0,2]$, exhibit an uncountable range of values. They can assume any value between 0 and 2.

Another illustrative example pertains to human age; some studies suggest that the maximum human lifespan could approach around 120 years. Consequently, the sample space for this scenario can be formulated as $\mathbb{S} = (0,120]$.

2.1.1 Conditional probability and Bayes theorem

Consider any two events A and B, if we knew that event A has occurred, it raises the question of how this newfound information influences the probability of event B occurring. If A has occurred, the probability that B has occurred depends on the intersection part between A and B, as illustrated in Figure 2.2. Mathematically, it is given by:

$$P(B|A) = \frac{P(AB)}{P(A)}. \tag{2.11}$$

Where $P(B|A)$ is called *the probability of B given that A has occurred*, or *the conditional probability of B given A has occurred.*

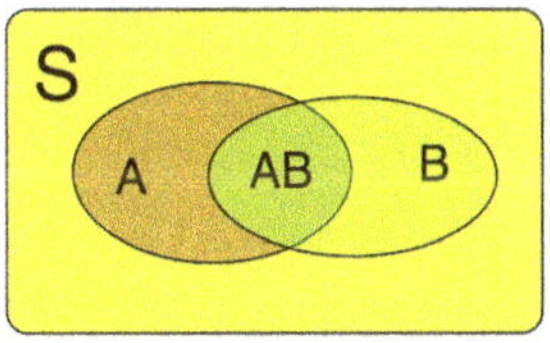

Figure 2.2: Venn diagram for conditional probability.

Example 4. Consider the sample space $\mathbb{S} = \{0,1,4,6,9,13\}$, and two events as: A = even numbers in $\mathbb{S}$ and $B = \{4,6,13\}$. Assume the outcomes are equally likely. Find the probability of A and the probability of A if B has occurred.

It is clear that $A = \{0,4,6\}$, and since the sample space has 6 possible outcomes with equal probability of occurring, then $P(A) = \frac{3}{6} = \frac{1}{2}$. If we knew that B has occurred, we determine the intersection between A and B such as $AB = \{4,6\}$. Therefore, $P(A|B) = \frac{P(AB)}{P(B)} = \frac{2\div 6}{3\div 6} = \frac{2}{3}$. It is clear that the available knowledge about a certain experiment may improve our certainty level about its outcomes.

There are several interesting insights stemming from conditional probability, including:

- If events A and B are disjoint (i. e., mutually exclusive), the intersection between them is empty, i. e., $A \cap B = \Phi$. From probability properties $P(\Phi) = 0$, therefore, $P(A|B) = P(B|A) = 0$. If you flip a single coin, what is the probability that you have *tails*, given that the result was *heads*? It is clear that it should be 0, because the outcomes of the coin are mutually exclusive.
- If A is a subset of B, the occurrence of B must include the occurrence of A as well. However, it's important to note that the converse may not always be true. Mathematically, if $A \subset B$, then, $P(B|A) = 1$. In the previous Example 4, let's define the event $C = \{4\}$. It is clear that the event $C \subset A$. Therefore, if C has occurred, this implies the occurrence of A, i. e., $P(A|C) = 1$. But the converse is not necessarily true. In this example, $P(C|A) = \frac{1}{3}$.
- Let's assume that events A and B are independent. In this context, independence implies that the occurrence or nonoccurrence of event A has no impact on the probability of event B. In this case, we have $P(B|A) = P(B)$. Substituting this result in the conditional probability equation, we obtain: $P(B|A) = \frac{P(A \cap B)}{P(A)} = P(B)$. Therefore, $P(A \cap B) = P(A) \times P(B)$, which is correct if and only if for independent events. For possibly dependent events, the relation is given by: $P(A \cap B) = P(B|A) \times P(A) = P(A|B) \times P(B)$.

What are your thoughts on mutually exclusive events? Do you consider them dependent or independent? It's important to note that they are, in fact, dependent, as the occurrence of one event precludes the possibility of the others.

Since $P(A \cap B) = P(B|A) \times P(A) = P(A|B) \times P(B)$, therefore, we may express the previous conditional probability form as:

$$P(B|A) = \frac{P(A|B) \times P(B)}{P(A)}. \tag{2.12}$$

This is the fundamental expression of Bayes' theorem, a highly valuable tool for addressing complex issues in stochastic analysis and estimation theory. We will elaborate on this further in subsequent sections and chapters.

Assume an event A which could be expressed by its intersections with n disjoint events B_i $i = 1, \ldots n$ that partition the sample space $\mathbb{S} = \cup_{i=1}^{n} B_i$ as shown in Figure 2.3 for 6 partitioning events. The probability of the event A can be expressed as $P(A) = P(A \cap B_1) + P(A \cap B_2) + \cdots + P(A \cap B_6)$. From the conditional probability formulation, we may express the probability as

$$P(A) = \sum_{i=1}^{n} P(A|B_i)P(B_i). \tag{2.13}$$

Based on equation (2.13), we can rephrase the Bayes' formula in (2.12) as:

$$P(B_k|A) = \frac{P(A|B_k)P(B_k)}{\sum_{i=1}^{n} P(A|B_i)P(B_i)}. \tag{2.14}$$

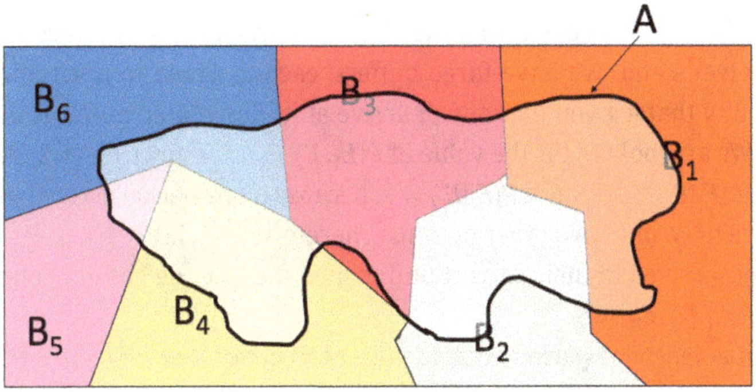

Figure 2.3: Venn diagram shows partitioned sample space.

Example 5. Two fair dice have been rolled. Consider the event A when the sum of the landed faces is less than 4, and the event B when the sum is exactly 5. Are A and B independent?

Solution. The events A and B are given as

$$A = \{(1,1), (1,2), (2,1)\}$$
$$B = \{(1,4), (2,3), (3,4), (4,1)\}.$$

Since the outcomes are equally likely (fair dice), $P(A) = \frac{3}{36}$ and $P(B) = \frac{4}{36}$. It is clear also that $A \cap B = \Phi$, and hence, $P(A \cap B) = 0 \neq P(A)P(B)$. Therefore, A and B are dependent. It's actually possible to determine this without calculating the probabilities. How, you ask? Well, it's evident that events A and B are mutually exclusive.

Example 6. Consider a scenario where three sensors transmit digital signals in the form of packets to a receiver. These sensors employ time-division multiplexing (TDM), where the first sensor sends its packet, followed by the second, then the third, and so on. The receiver is equipped with individual buffers for each sensor. Now, let's imagine a situation where the TDM central organizer experiences a failure, causing each sensor to initiate the transmission of its packets at random and independently of one another. It's important to note that the received packets do not overlap because the receiver possesses the capability to distinguish and separate them.

A What is the probability that the receiver will receive the packets in the correct sequence?

B What is the probability that the receiver will receive at least one packet in its correct order?

C What is the probability that none of the packets will arrive in the correct order?

Resolve the issue through analytical methods and validate your findings by conducting Monte Carlo simulations using both Scilab and Octave programming languages.

A Define E_i where $i = 1, 2, 3$, is the event that the ith packet arrives in its correct sequence. At the receiver's end, we have three buffers, each assigned to a specific sensor. The probability that a given packet will arrive at its designated buffer is as follows: $P(E_i) = \frac{1}{3}$. We are looking for the value of $P(E_1, E_2, E_3)$. For any $i, j \in \{1, 2, 3\}$, we have $P(E_i, E_j) = P(E_i|E_j)P(E_j)$. Since $P(E_i|E_j) = \frac{1}{2}$, because if one packet arrived to its correct buffer, we have only two other options. Therefore, $P(E_i, E_j) = \frac{1}{2} \times \frac{1}{3} = \frac{1}{6}$. If two sensor packets arrived in their correct buffer, then the last one has only one buffer left, so that $P(E_1, E_2, E_3) = \frac{1}{6}$.

B The probability that at least one packet arrived in its correct order is

$$\begin{aligned} P(E_1 \cup E_2 \cup E_3) &= P(E_1) + P(E_2) + P(E_3) - P(E_1 \cap E_2) \\ &\quad - P(E_1 \cap E_3) - P(E_2 \cap E_3) + (E_1 \cap E_2 \cap E_3) \\ &= \frac{1}{3} + \frac{1}{3} + \frac{1}{3} - \frac{1}{6} - \frac{1}{6} - \frac{1}{6} + \frac{1}{6} = \frac{2}{3}. \end{aligned}$$

C The probability that no packet arrives in its correct order is

$$1 - P(E_1 \cup E_2 \cup E_3) = 1 - \frac{2}{3} = \frac{1}{3}.$$

Let's develop a Monte Carlo simulation for this scenario to verify the accuracy of our analytical solutions.

The code to solve the previous example in *Scilab* can be written as following:

```
clear
N=1000;
x=grand(N, "prm", (1:3));
y=kron(ones(N,1),[1,2,3]);
z=abs(x-y);
// Probability to receive all packets in their correct
    order
P1=length(find(sum(z,2)==0))/N
// Probability that at least one packet is received
    correctly
P2=length(find(z(:,1)==0 | z(:,2)==0 | z(:,3)==0))/N
// Probability that no packet is received in its
    correct order
P3=length(find(z(:,1)~=0 & z(:,2)~=0 & z(:,3)~=0))/N
```

The upcoming section will solely elaborate on the new functions introduced in this code.

– In this code, we randomly generate 3-tuples [1, 2, 3] with equal probabilities. Each tuple represents the order in which sensor packets arrive at the receiver's buffers.

The first column corresponds to the first buffer, which should receive data from the first sensor, and likewise for the second and third columns. The numbers in each column indicate the sensor packet's order. For instance, if the first column displays the number 2, it signifies that the packet from the second sensor arrived first and was placed in the first buffer. When the Time Division Multiplexing (TDM) system operates correctly, we expect to see 1 in the first column, 2 in the second column, and 3 in the third column. To simulate the random arrival of packets, we generate random 3-tuples using a function called *grand*. This function uses a parameter called *prm* to create random permutations of integers from 1 to 3. We generate N such random 3-tuples. The equivalent function in Octave for this operation is *randperm*.

- To assess the randomness of the arrival sequence, we must contrast it with the ideal scenario, where each sensor packet arrives in its sequential order. To accomplish this comparison, we employ the Kronecker product function. This function essentially multiplies the vector $[1, 2, 3]$ with a vector of ones, repeated a number of times corresponding to the experiment's length.
- In the following line, we establish the matrix z as the absolute difference between the ideal sequential order and the randomly generated sequence. It's important to note that the sum of these differences should equal zero only when both sequences are identical. Subsequently, we proceeded to count the number of zeros in this matrix and divide the count by the total number of trials. This calculation provides us with the probability that all packets will arrive in the correct order. Executing the code in Scilab, we obtained a result of $P_1 = 0.165$. Notably, this result is very close to the analytical prediction of 0.166.
- Try to understand the next lines to evaluate the other requirements. It is also essential to run the code and check the results.
- Try to rewrite the code using the *Octave* programming language.

Example 7 (Cognitive Radio and Spectrum Sensing). The wireless spectrum faces considerable congestion, with nearly all valuable frequency bands already licensed, except for a few ISM bands. This has posed a significant challenge to the quest to introduce new wireless services within the confines of the licensed spectrum. One promising solution is known as cognitive radio. Cognitive radio's fundamental premise is to repurpose licensed bands under the crucial stipulation that "no harmful interference should affect the licensed users." In this context, the cognitive radio device is referred to as the "secondary user (SU)," while the licensed user is termed the "primary user (PU)". The SU is allowed to utilize these bands only when they are unoccupied, or there will be no potential for harmful interference. Achieving this objective can be realized through either prior knowledge of unoccupied bands (i. e., when the PU is not using them) during certain times or by employing spectrum sensing. In spectrum sensing, the SU actively monitors the spectrum and makes determinations regarding its availability. Various techniques are employed for spectrum sensing, yet no single technique is flawless. Spectrum sensing errors manifest in two primary forms: misdetection and false alarm. A misdetection error occurs when the SU fails to detect the presence of the PU, leading to unintended transmission and the possibility of harmful interference. Conversely, a false alarm error arises when the SU incorrectly identifies the PU as active when, in fact, it is in an idle state, potentially diminishing the usability and performance of SUs. These two types of errors are quantified using probabilities, specifically the probability of misdetection (denoted as P_M) and the probability of false alarm (denoted as P_{FA}). The goal is to minimize both P_M and P_{FA} as much as possible, although achieving absolute zero for both is often unattainable. Fine-tuning and optimizing the trade-off between P_M and P_{FA} can be accomplished through various compromise techniques.

Example 8. Cognitive radio utilizes the same spectrum as a weather monitoring radar system. The radar takes samples randomly during the day. However, we know that it occupies the spectrum only 3 % of the day (i. e., it is active for about 45 mins/day). The cognitive radio device senses the spectrum before transmission to ensure that the spectrum is white (i. e., the radar is idle). Transmitting during the radar's active phase could lead to interference and inaccuracies in the radar's measurements. The sensing method used by the cognitive radio has a detection accuracy of 98 % in detecting the radar signal. However, it has a false alarm probability of 20 %. At a particular time, the cognitive radio sensing alarm was positive (i. e., it assumes the radar is active); what is the probability that the radar is indeed active at that particular time?

Solution. Let's define the event A as the radar is active and the event B as the positive alarm has been received. We need to compute the probability that the radar is active given that a positive alarm has been received, i. e., $P(A|B)$. Using conditional probability:

$$P(A|B) = \frac{P(A \cap B)}{P(B)} = \frac{P(B|A)P(A)}{P(B)} \tag{2.15}$$

$$P(B) = P(B|A)P(A) + P(B|\bar{A})P(\bar{A}). \tag{2.16}$$

Where $\bar{A}$ is the complement of A, i. e., the radar is idle. We can attribute the likelihood of a positive spectrum sensing outcome to either the radar being active or the presence of a false alarm. It is given from the example that: $P(B|A) = 0.98$, $P(B|\bar{A}) = 0.20$, $P(A) = 0.03$, and $P(\bar{A}) = 1 - 0.03 = 0.97$. Substituting these values in the above equation we obtain:

$$P(A|B) = \frac{0.98 \times 0.03}{0.98 \times 0.03 + 0.20 \times 0.97} = 0.132. \tag{2.17}$$

The results highlight an interesting observation: when the sensing alarm registers as positive, it implies that the radar is active only 13 % of the time. This might seem misleading, as one might assume that the secondary user is mostly inactive. However, this low percentage of radar activity is due to the small window of time during which the radar is actually operational, and it is essential to maintain the safety of the primary user. Conversely, if we calculate the probability of a negative sensing alarm, meaning that the cognitive radio can transmit, we find it to be $P(\bar{B}) = 1 - P(B) = 0.78$, indicating that the secondary user has the opportunity to transmit for 78 % of the time. It's worth noting that with perfect sensing (zero false alarms and misdetections), the secondary user would ideally transmit 97 % of the time.

Example 9. Using Octave programming, create a Monte Carlo simulation code to replicate the previous example. Afterward, you can compare the simulation results with the analytical solution.

Solution. You can structure the code as follows. All the commands used are familiar, so you should find it straightforward to comprehend.

```
% Octave Code
clear all
N=1e8; % Number of repeating
Pt=0.03; % The probability that PU transmits
x=rand(N,1); %generating uniform random from 0 to 1
y=sign(Pt-x); %y=1 if PU transmit and -1 elsewhere
y1=y+rand(size(y)); % emulating the sensing errors due to false alarm
y2=y-rand(size(y)); % emulating the sensing errors due to misdetection
N1=length(find(y==1)); %Number of times where PU actually transmit
N2=length(find(y1>=-0.2))-N1; %Number of false alarm times
N3=length(find(y2>=0.02)); %Number of SU detects PU without false alarm
R=-ones(size(y)); %generating (-1) vector with same size of y
% filling R vector with ones where SU decides PU is active
R(find(y2>=0.02))=1;
R(find(y1>=-0.2))=1; % same as before but for false alarming
N4=length(find(R==1)); %The number where SU decides the PU is active
P1=N1/N4; % The probability that PU active given that positive alarm
```

Upon executing the code, we obtain a value of $P1 = 0.134$. Additional values can also be calculated using the code, such as the probability of a positive alarm, which is given by $\frac{N4}{N}$. In our Monte Carlo simulation, this probability is computed as 0.224. This outcome further validates our analytical solution.

Rewrite the previous code in Scilab.

Example 10. A blood test method is employed to detect a specific disease, boasting an accuracy of 99 %. However, it does yield a 1 % false-positive rate. Historical screening data reveals that only 1 % of the population is afflicted with this disease. The test was administered to a randomly chosen individual, yielding a positive result. Now, we want to determine the probability that this person is truly infected with the disease.

Solution. Let's define the following two events:
- Event A represents the blood test result returning as positive.
- Event B denotes the individual actually having the disease.

Consequently, our goal is to calculate $P(B|A)$. We can utilize Bayes's theorem to do so, as follows:

$$P(B|A) = \frac{P(A|B)P(B)}{P(A)} = \frac{0.99 \times 0.01}{0.99 \times 0.01 + 0.01 \times 0.99} = 0.5. \tag{2.18}$$

This result could be confusing because the accuracy of the blood test is 99 %. How can there be only 50 % chance that the person is actually infected with the disease after a positive test? To understand this result, let's assume that we have 1000 randomly selected persons. Since only 1 % of the population is infected with this disease, then on average, we have 10 persons infected in those 1000. If we make the blood test for all 1000 persons, since the accuracy is 99 %, therefore it will detect almost all infected people (who are 10). However, since the test yields 1 % false positive, it means that it will also show $0.01 \times 990 = 9.9 \approx 10$ healthy persons as infected! The total of people who have a positive test is about 20 persons; however, only half of them are actually infected.

Example 11. Consider a random number generator that generates $x \in \{1, 2, 3\}$. Generally, the probability of each number appearing next is $\frac{1}{3}$, except if the current generated number is 2. In this case, the probability that the next number will also be 2 is 80 %. The other two numbers will appear with equal probability (it is, of course, 0.1 each). Find:
- The probability that $x = 2$.
- The probability that $x = 1$ (it is the same as $x = 3$).
- Write Octave and Scilab codes to simulate this example and compare the results with the theoretical one.

Many real applications could be found to match such a dependent process. One example is stochastic channel modeling. We may model the wireless channels as three states. The states are high SNR, medium SNR, and low SNR. When the channel is at low SNR, then it might have a higher probability of staying in this state (it is called deep fading). In your area of research, you may think about a similar random process, where the next outcome depends on the previous one. A more general analysis will be given later under the topic of the Markov Process. Before you look at the answer, try it yourself first!

Answer. Since the next outcome of this experiment depends on the current outcome, let's define x_n as the outcome at time n. Therefore, and using (2.13),

$$P(x_n = 2) = P(x_n = 2|x_{n-1} = 1)P(x_{n-1} = 1) \\ + P(x_n = 2|x_{n-1} = 2)P(x_{n-1} = 2) + P(x_n = 2|x_{n-1} = 3)P(x_{n-1} = 3).$$

Based on the example, it's evident that:

$$P(x_n = 2|x_{n-1} = 1) = P(x_n = 2|x_{n-1} = 3) = \frac{1}{3}.$$

However,

$$P(x_n = 2|x_{n-1} = 2) = 0.8.$$

Furthermore, we know

$$P(x_{n-1} = 1) + P(x_{n-1} = 2) + P(x_{n-1} = 3) = 1.$$

Both the outcomes 1 and 3 have the same chance to appear, i. e., $P(x_{n-1} = 1) = P(x_{n-1} = 3)$. Let's define $P(x_{n-1} = 1) = y$, therefore,

$$P(x_{n-1} = 2) = 1 - 2y \rightarrow y = \frac{1 - P(x_{n-1} = 2)}{2}.$$

Substituting this equation in the above total probability equation we get

$$P(x_n = 2) = \frac{1}{3}y + 0.8P(x_n = 2) + \frac{1}{3}y,$$

substituting for y,

$$P(x_n = 2) = \frac{1 - P(x_{n-1} = 2)}{3} + 0.8P(x_n = 2) \rightarrow P(x_n = 2) = 0.625.$$

Finally, the probability that $P(x_n = 1) = P(x_n = 3) = \frac{1-0.625}{2} = 0.1875$.

The Octave code can be expressed as follows. Since all the commands have been previously discussed, you should be able to comprehend it.

```
clear all
N=100000; %Number of iterations
r=[1 2*ones(1,8) 3]; % 80 percent of number 2
x=unidrnd(3,1,N); % generating N random integers from
    1 to 3
for i=1:N,
y=x(i);
if y==2,
```

```
z=unidrnd(10,1);
x(i+1)=r(z);
end
end
P=length(find(x==2))/N %Probability x=2
P1=length(find(x==1))/N %Probability x=1
```

When running the Octave code provided above, the results yield $P = 0.62437$ and $P1 = 0.18509$, which validate our analytical solution.

To implement the same code in Scilab with only minor adjustments, refer to the code snippet below:

```
\\ Scilab code
clear
N=100000; //Number of iterations
r=[1 2*ones(1,8) 3]; // 80 percent of number 2
x=grand(1,N,'uin',1,3); // generating N random number
    from 1 to 3
for i=1:N,
y=x(i);
if y==2,
z=grand(1,'uin',1,10);
x(i+1)=r(z);
end
end
P=length(find(x==2))/N //Probability x=2
P1=length(find(x==1))/N //Probability x=1
```

Example 12. In Helsinki road conditions in March can be slippery on approximately 20 % of the days. When the roads are slippery, there is a high likelihood of heavy traffic (traffic jam) in the morning, with a probability of 0.7. Conversely, on non-slippery days, the probability of experiencing a traffic jam drops to 0.2. Furthermore, if it is slippery and there is a traffic jam, approximately 50 % of workers will arrive late at their workplace. In contrast, on non-slippery days without a traffic jam, an impressive 95 % of workers will arrive on time. In other scenarios, such as slippery roads without a traffic jam or non-slippery days with a traffic jam, only 65 % of workers will make it to work on time.

Now, for a specific day in March and a randomly selected worker, let's calculate the following probabilities:

- Determine the probability of a non-slippery road condition and a traffic jam occurring simultaneously, resulting in the worker arriving late.
- Calculate the overall probability of a worker arriving late, irrespective of the prevailing road conditions.

- Find the conditional probability that it was a slippery day when the worker arrived late to work.
- Conduct a simulation of this scenario using Scilab and Octave, and subsequently, compare the simulation results with the analytical calculations.

Solution. Let's start by defining our events as: R = {slippery road} $\Rightarrow \bar{R}$ = {not slippery}; J = {jamming road} $\Rightarrow \bar{J}$ = {no jam}; and L = {arriving late} $\Rightarrow \bar{L}$ = {arriving in time}. Based on the provided example, it is evident that during the month of March, $P(R) = \frac{1}{5}$, $P(J|R) = 0.70$, $P(J|\bar{R}) = 0.20$, $P(L|(R,J)) = 0.50$, $P(\bar{L}|(\bar{R},\bar{J})) = 0.95$, $P(\bar{L}|(R,\bar{J})) = 0.65$, and $P(\bar{L}|(\bar{R},J)) = 0.65$. These are all the pieces of information that we've extracted from the text. Now, let's proceed to solve the problem:

- The first requirement of the problem can be mathematically represented as follows: $P(\bar{R},J,L)$. We know from the conditional probability rules that: $P(\bar{R},J,L) = P(L|(\bar{R},J))P(\bar{R},J) = 0.35P(\bar{R},J)$. However, $P(\bar{R},J) = P(J|\bar{R})P(\bar{R}) = 0.2 \times (1 - \frac{1}{5}) = 0.16$. Therefore,

$$P(\bar{R},J,L) = 0.35 \times 0.16 = 0.06.$$

- The second requirement of the problem can be mathematically expressed as: $P(L)$. Using the law of total probability,

$$\begin{aligned} P(L) &= P(L|(R,J))P(R,J) + P(L|(R,\bar{J}))P(R,\bar{J}) \\ &\quad + P(L|(\bar{R},J))P(\bar{R},J) + P(L|(\bar{R},\bar{J}))P(\bar{R},\bar{J}). \end{aligned}$$

  Hence, using the corresponding values from the example text,

$$P(L) = 0.5P(R,J) + 0.35P(R,\bar{J}) + 0.35P(\bar{R},J) + 0.05P(\bar{R},\bar{J}).$$

  We can find the joint probability values using the role of conditional probability as: $P(R,J) = P(J|R)P(R) = 0.7 \times 0.2 = 0.14$; $P(R,\bar{J}) = P(\bar{J}|R)P(R) = 0.3 \times 0.2 = 0.06$, $P(\bar{R},J) = 0.16$ as computed in the previous requirement, and finally, $P(\bar{R},\bar{J}) = P(\bar{J}|\bar{R})P(\bar{R}) = 0.8 \times 0.8 = 0.64$. Substituting those values, we obtain: $P(L) = 0.5 \times 0.14 + 0.35 \times 0.06 + 0.35 \times 0.16 + 0.05 \times 0.64 = 0.18$.
- In this part we need to compute $P(R|L)$. This could be evaluated easily as:

$$P(R|L) = \frac{P(R,L)}{P(L)}.$$

  Therefore

$$\begin{aligned} P(R,L) &= P(R,L|J)P(J) + P(R,L|\bar{J})P(\bar{J}) \\ &= P(L|(R,J))P(R,J) + P(L|(R,\bar{J}))P(R,\bar{J}) \\ &= 0.50 \times 0.14 + 0.35 \times 0.06 = 0.09. \end{aligned}$$

  Substituting in the above formula we obtain: $P(R|L) = \frac{0.09}{0.18} = 0.50$.

In this example, there are numerous other scenarios to consider. For instance, you can explore the concept of $P(L|R)$ and its conceptual connection with $P(R|L)$?

- In this part, we show how to write a simple Octave code that can emulate this example.

```
% This code to simulate the Slippery-day  example
clear all
N=100000;
L=zeros(N,1);
R=(sign(0.2-rand(N,1))+1)/2;
Jp=(sign(R.*rand(N,1)-0.3)+1)/2;
J=Jp+(ones(size(R))-R).*(sign(0.2-rand(N,1))+1)/2;
I=R+J;
F1=find(I==2);
F2=find(I==0);
F3=find(I==1);
L(F1)=(sign(0.5-rand(length(F1),1))+1)/2;
L(F2)=(sign(0.05-rand(length(F2),1))+1)/2;
L(F3)=(sign(0.35-rand(length(F3),1))+1)/2;
T=[R J L];
P=T-kron([0 1 1],ones(N,1));
P1=length(find(sum(transpose(abs(P)))==0))/N
P2=length(find(L==1))/N
P3=sum(L.*R)/(P2*N)
```

The results are matching with the ones as $P_1 = 0.0568$, $P_2 = 0.1778$, and $P_3 = 0.5074$. When dealing with complex conditional relationships in problems, I tend to place more trust in my simulation results as opposed to my analytical solutions. I've encountered instances where discrepancies between the two became apparent, leading me to reevaluate my analytical work. To facilitate your understanding of the code, I'll provide some helpful pointers.

In the provided code, we execute the experiment 100,000 times (represented by N). We utilize the variable 'R,' which is a vector with N entries, consisting of zeros and ones. The probability of finding a one (1) in this vector is 20 %, and these ones are randomly distributed within R, signifying the days when road conditions are slippery. On slippery days (where $R = 1$), there is a 70 % chance of encountering a road jam, which is implemented in the code through the variable Jp, Jp is derived by element-wise multiplication of the 'R' vector with a uniformly distributed random number. When $R = 0$, the multiplication by zero yields a result of zero, indicating no jam. Conversely, where $R = 1$, there's a 70 % chance that 'Jp' will be set to 1, indicating a road jam.

The variable 'J' is employed in the code to identify jammed days, regardless of whether they were slippery or not. It is simply the sum of two vectors: 'Jp' (representing the probability of jam on slippery days) and another vector, which represents the probability of jam on non-slippery days.

Next, a new vector 'T' is defined, which is created by summing the random vector representing slippery days (where $R = 1$) and the random vector for jammed days (where $J = 1$). Since both vectors contain only 0 s and 1 s, the sum results in values of 0, 1, or 2. A value of 2 indicates a day that was both slippery and jammed, while 1 signifies a day that was either slippery or jammed. Finally, 0 indicates a day that was neither slippery nor jammed. With these hints, I hope you can navigate and comprehend the remaining portions of the code.

It's important to highlight that such interdependent events are prevalent in a wide array of real-life situations, and this underscores the importance of accounting for the dependence of random variables when formulating your simulation models. As demonstrated in the previous example of slippery road conditions, this concept can be applied to a broad spectrum of practical, real-world problems.

Now, let's consider an analogous problem in the realm of public health. Influenza, a highly contagious virus, tends to proliferate during the winter months, affecting millions of individuals annually. It poses significant risks, particularly to vulnerable groups such as the elderly, pregnant women, and individuals with underlying health conditions. Even for healthy individuals, influenza can result in a week of sick leave, impacting both companies and society as a whole. To mitigate these effects, many countries recommend proper vaccination against the influenza virus.

In this context, we can categorize people into three risk classes: high risk, medium risk, and low risk. Assuming a base probability of infection of 0.1 for all individuals without vaccination, this probability decreases to 0.02 for those who have received the recommended vaccine. The severity of infection varies with probabilities of 0.1, 0.02, and 0.001 for high, medium, and low risk classes, respectively. The proportions of these classes in the population are 0.1, 0.2, and 0.7, respectively.

To model this public health scenario, you can adapt the same simulation framework previously used for the slippery road example. You might even integrate real national data to make more accurate projections regarding the spread of the influenza virus within a given population.

Course instructor may ask the students to think about other different real applications and emulate the results.

2.1.2 Marginal probability

Suppose that the sample space $\mathbb{S} = \cup_{i=1}^{n} B_i = \cup_{i=1}^{m} A_i$. It means that it can be partitioned into two different families of disjoint sets $\{A_i\}$ and $\{B_i\}$. The marginal probability of event B_k is computed as

$$P(B_k) = \sum_{i=1}^{m} P(A_i \cap B_k). \tag{2.19}$$

This concept is illustrated in Figure 2.4.

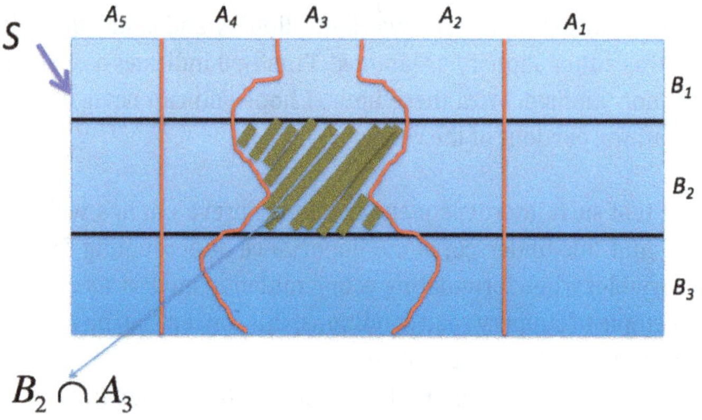

Figure 2.4: Venn diagram for marginal probability.

Example 13. Consider the scenario where we need to calculate the probability of a particular investment being profitable, disregarding the risk factor. In this context, we introduce a variable denoted as "R," which represents the level of risk associated with the investment based on certain project attributes. This variable can assume one of three levels: "High" (H), "Medium" (M), or "Low" (L), each signifying the degree of risk involved. The outcome of the investment can manifest as either "Good" (G), representing a favorable profit, or "Bad" (B), signifying losses or an unfavorable outcome. Figure 2.5 shows the conditional probabilities of risks and outcomes. If the probability of risks values are given as $P(H) = 0.3$, $P(M) = 0.5$, and $P(L) = 0.2$.

Profit \ Risk	High	Medium	Low
Bad	0.85	0.6	0.1
Good	0.15	0.4	0.9

Figure 2.5: Conditional profit-risk table.

Solution. Observe that the elements in the Table shown in Figure 2.5 are for the conditional probability, for example $P(B|H) = 0.85$. We should find the joint probability distribution in order to compute the marginal probability. This could be computed by multiplying the conditional probability by the probability of the risk level. For example, $P(B \cap H) = P(B|H)P(H) = 0.85 \times 0.3 = 0.255$. Applying the same for all elements in the table, we got the joint distribution table shown in Figure 2.6. The marginal probability that the investment is Good regardless of the risk factor is $P(G) = 0.425$.

Profit / Risk	High	Medium	Low	Marginal Probability
Bad	0.255	0.3	0.02	0.575
Good	0.045	0.2	0.18	0.425

Figure 2.6: Joint profit-risk table.

2.2 Random variables

It has been established that probability serves as a metric for quantifying the degree of uncertainty associated with the occurrence or nonoccurrence of specific events. We have explored various types of events and experiments, such as coin tosses, dice rolls, and the sequencing of sensor packets. In the realm of mathematics, continually dealing with such descriptive events can become mundane and confining. To address this, we employ random variables, which enable us to assign real numerical values to these descriptive events. In essence, random variables facilitate the transformation of the event space into the space of real numbers. As an illustrative example, consider the experiment involving three coin tosses, as we previously examined, where the sample space is as follows:

$$\mathbb{S} = \{TTT, TTH, THT, THH, HTT, HTH, HHT, HHH\}.$$

Now let's define the random variable x as the number of heads that appears out of the experiment, hence, $x \in \{0, 1, 2, 3\}$. Naturally, random variables can be defined in numerous alternative ways to suit specific interests and analytical needs. For instance, one might be concerned with the count of tails instead of heads. Alternatively, an interest may lie in the difference between the number of heads and the number of tails, which would yield random variable values within the set $x \in \{-3, -1, 1, 3\}$. In general, random variables can assume any value within the range of $-\infty < x < \infty$. Typically, we opt to define the random variable in a manner that best facilitates our analytical purposes.

Example 14. In the experiment of tossing a fair coin three times, we define the random variable x as the difference between the number of heads and the number of tails. Let's determine the probabilities associated with various x values and represent them graphically. Finally, find $P(x = 0)$, $P(x^2 \geq 4)$, and $P(x \leq \sqrt{3})$.

Solution. It is clear that $x \in \{-3, -1, 1, 3\}$. We can compute the probabilities as: $x = -3$ for the case where the outcome event is $\{(T, T, T)\}$. Therefore, $P(x = -3) = \frac{1}{8}$. The events which lead to $x = -1$ are $\{(T, T, H), (T, H, T), (H, T, T)\}$. Therefore, $P(x = -1) = \frac{3}{8}$.

The events which lead to $x = 1$ are $\{(T,H,H),(H,H,T),(H,T,H)\}$. Therefore, $P(x = 1) = \frac{3}{8}$. Finally, $x = 3$ for the event $\{(H,H,H)\}$, and hence, $P(x = 3) = \frac{1}{8}$. We can draw the probabilities as shown in Figure 2.7. This is called **probability mass function.**

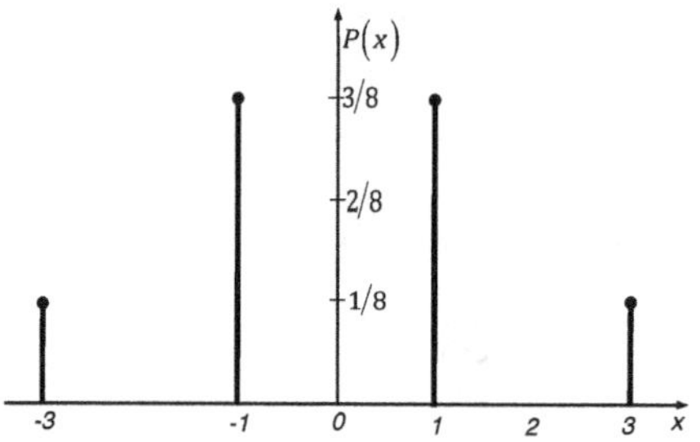

Figure 2.7: Probability mass function $P(x)$.

It is clear that $P(x = 0) = 0$, $P(x^2 \geq 4)$ is the same as $P(|x| \geq 2)$ and this will be achieved with $x = \pm 3$, therefore, $P(|x| \geq 2) = \frac{2}{8}$, and finally $x \leq \sqrt{3}$ is achieved at $x = \pm 1$. Therefore, $P(x \leq \sqrt{3}) = \frac{6}{8}$. This example shows the benefit of using random variables instead of descriptive events. We can use almost all kinds of mathematical operations with random variables.

In the previous example, the sample space consists of discrete and finite events, giving rise to discrete and finite random variables. This concept is referred to as the probability mass function (PMF), where the sum of all probabilities equals 1. However, it's important to note that random variables can also be discrete and infinitely large, as illustrated in the following example.

Example 15. In a particular experiment with only two distinct outcomes, denoted S_1 and S_2, the probability of observing S_1 is represented by p (naturally implying that the probability of observing S_2 is $1 - p$). We conduct the experiment repeatedly until we observe the outcome S_1. In this context, we define a random variable x as the number of experiment repetitions, with possible values for x being $1, 2, 3, \ldots \infty$. Our objective is to determine and visualize the probability mass function for this random variable. Additionally, try to prove that the summation of all probabilities equals 1.

Solution. After the initial observation, if the outcome is S_1, the experiment concludes with x taking the value of 1, and this event occurs with a probability of p. Therefore, $P(x = 1) = p$. On the other hand, if the first observation yields S_2 and S_1 subsequently appears in the second trial, x takes the value of 2. As both trials are assumed to be independent, the probability of this sequence is $P(x = 2) = p(1 - p)$. Similarly, the random variable $x = 3$ materializes when S_2 appears in the first two trials, followed by S_1 in the third trial. This means that $P(x = 3) = p(1 - p)^2$. In the general form,

$P(x = n) = p(1-p)^{n-1}$, $n = 1, 2, 3, \ldots \infty$. Figure 2.8 shows the probability mass function of this experiment.

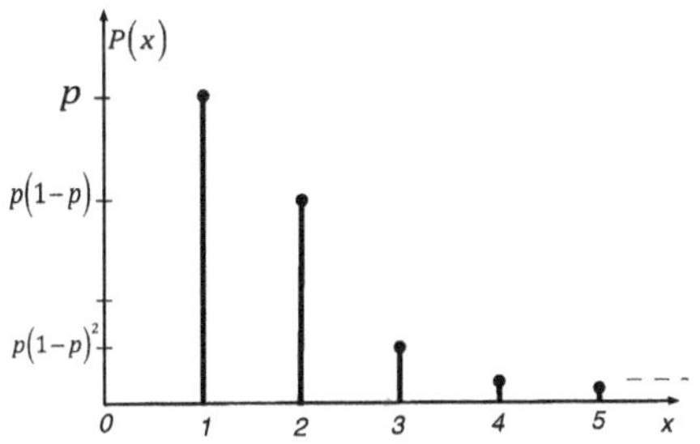

Figure 2.8: Probability mass function P(x).

The summation of all probabilities must be 1. It is possible to prove that as following:

$$A = \sum_{n=0}^{\infty} p(1-p)^n = p \sum_{n=0}^{\infty} (1-p)^n \tag{2.20}$$

$$\text{let's define } S = 1 + (1-p) + (1-p)^2 + (1-p)^3 + \cdots \tag{2.21}$$

$$\rightarrow S = 1 + (1-p)S \rightarrow S = \frac{1}{P} \tag{2.22}$$

$$A = p\frac{1}{p} = 1. \tag{2.23}$$

Random variables can also be continuous. Let's consider a simple example involving a random signal ranging from −2 to +2 volts, as illustrated in Figure 2.9. In this scenario, we encounter an infinite number of levels within the interval $(-2, 2)$. For any specific level, the probability is zero, meaning that $P(v = a_0) = 0$. For instance, if $a_0 = 0.4$, this pertains to the probability of the voltage being exactly 0.4 not, for example, 0.400000000000001. Nevertheless, $P(0.3999 \leq v \leq 0.4)$ may not be zero, depending on the underlying distribution. The reason lies in the continuous nature of the scenario, wherein between 0.3999 and 0.4 an infinite number of levels also exist. While in the discrete case, it's feasible to assign probabilities to each random variable, the challenge of assigning probabilities in the context of continuous random variables arises.

An approach for assigning probabilities to continuous random variables involves partitioning the range of the random signal into numerous small intervals denoted Δ_v and calculating the proportion of the signal that crosses each interval. Essentially, we determine the likelihood of each interval occurring, as depicted in Figure 2.10. This allows us to plot the magnitude levels of the random signal on the x-axis, with the y-axis representing the respective probabilities of occurrence. This representation is commonly referred to as the signal's histogram.

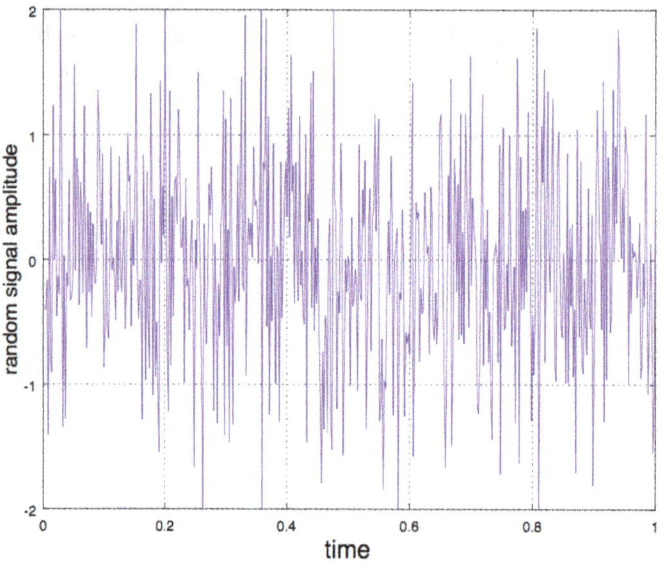

Figure 2.9: Continuous random signal.

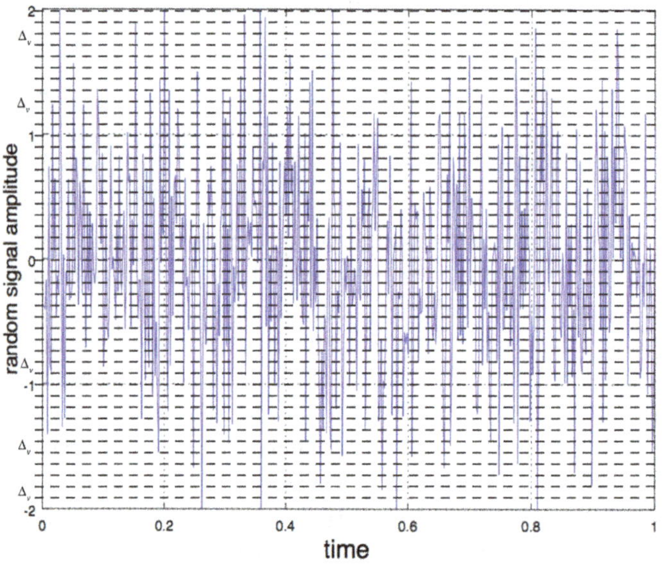

Figure 2.10: Continuous random signal with quantized amplitudes.

To illustrate this process step by step, we will discretize the signal amplitude into just four levels, as depicted in Figure 2.11. In this example, there are a total of 500 samples. As observed in the figure, the resulting probabilities are $P(-2 \le v < -1) = \frac{57}{500} = 0.114$, $P(-1 \le v < 0) = 0.386$, $P(0 \le v < 1) = 0.384$, $P(1 \le v < 2) = 0.116$.

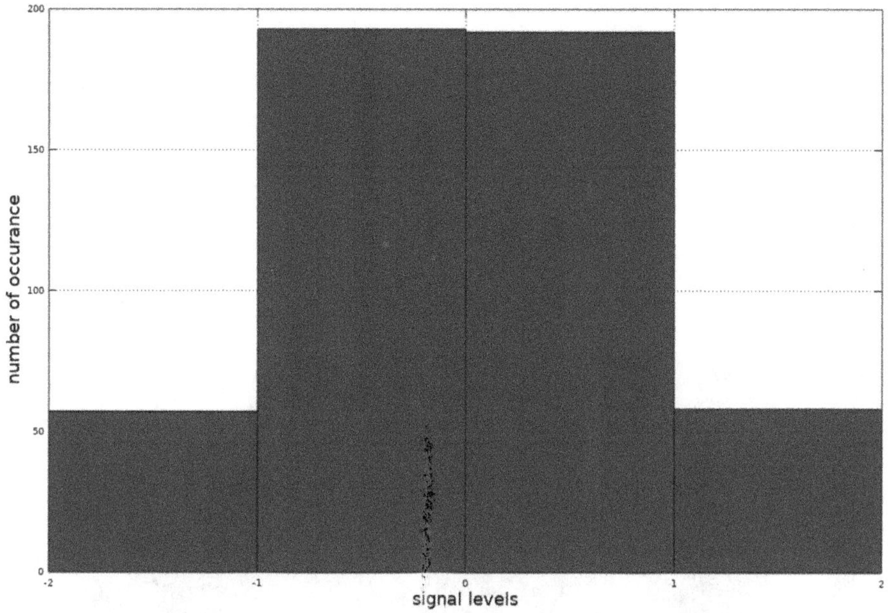

Figure 2.11: The histogram of the random signal example.

Notice that probabilities are defined for intervals, and the probability of each interval corresponds to the area of the respective rectangle. To enhance the precision of this measure of uncertainty, it's necessary to decrease the size of the intervals.

If we choose an interval size of ($\Delta_v = 0.1$) volt, we will have 40 levels, resulting in a more accurate histogram representation of a single realization of the random signal, as shown in Figure 2.12. As we further reduce the interval size, such as $(\Delta_v) \to 0$, we obtain a curve that encapsulates the dynamic behavior of the signal. This curve is known as the **probability density function** (PDF). Once more, the probability associated with each interval is determined by the area under the curve within that interval, as illustrated in Figure 2.13. In this figure, the probability that $P(-1 \le v < 0)$ is given by $\int_{-1}^{0} f_V(v)dv$. The function $f_V(v)$ is referred to as the *probability density function* and is used to describe the random variations of continuous random variables. The probability of events can be computed as the area under the curve corresponding to the event. For a random variable x to fall within event **A**, the probability is expressed as,

$$P(x \in \mathbf{A}) = \int_{x \in \mathbf{A}} f_X(x)dx. \tag{2.24}$$

The probability density function (PDF) is a fundamental concept in probability theory and stochastic processes. It encapsulates all the random fluctuations and their corresponding probabilities of occurrence. The PDF plays a crucial role in computing

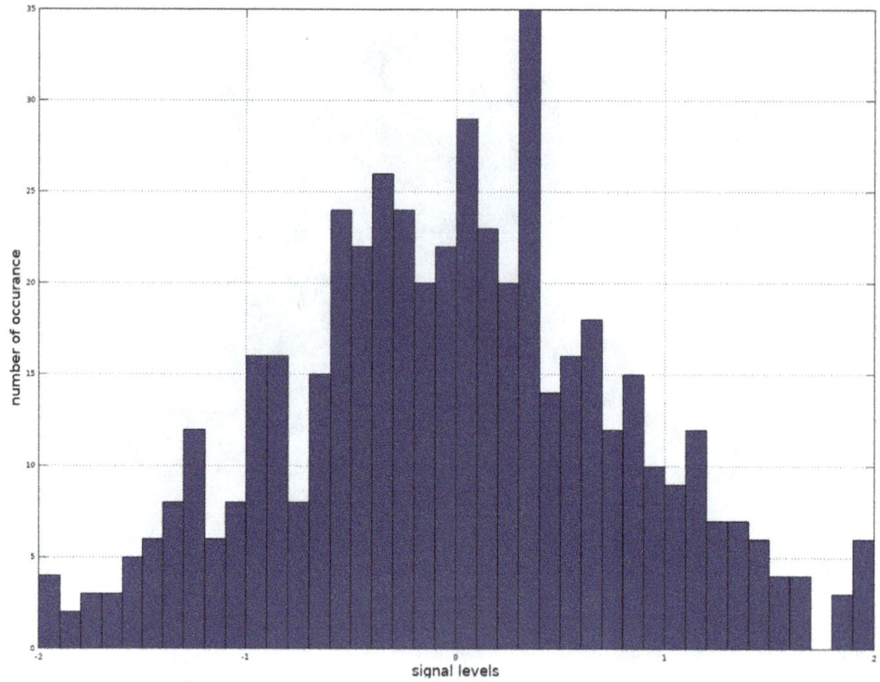

Figure 2.12: The histogram of the random signal example.

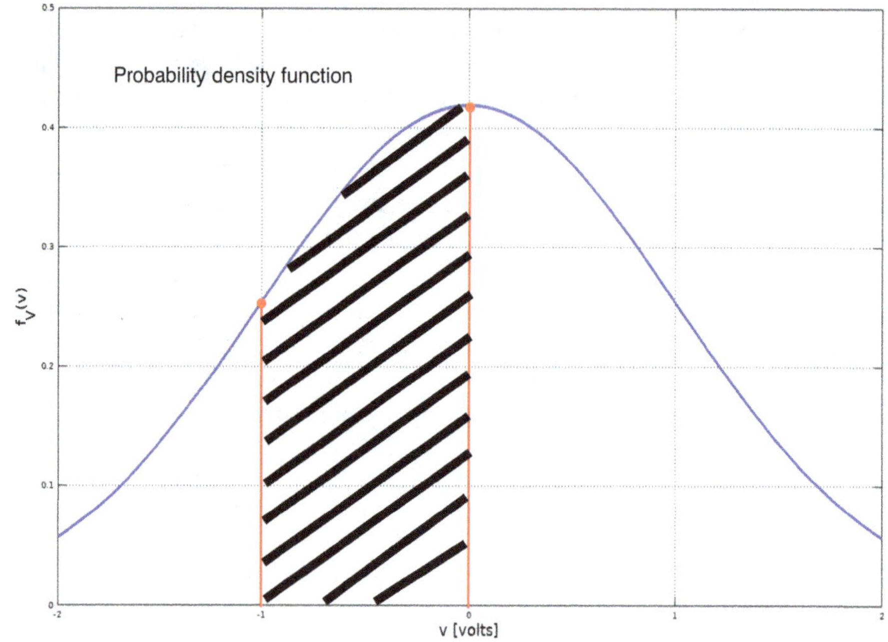

Figure 2.13: The histogram of the random signal example.

various statistical properties of a random variable, including measures such as the mean, median, mode, and more. Since the PDF quantifies the relative likelihood of events, it must be nonnegative, and its overall area under the curve equals 1. Mathematically,

$$f_X(x) \geq 0 \quad \forall x \tag{2.25}$$

$$\int_{-\infty}^{\infty} f_X(x)dx = 1. \tag{2.26}$$

The same principles apply to discrete random variables, but instead of integration we employ summation, as follows:

$$P(x_i) \geq 0 \quad \forall x_i \tag{2.27}$$

$$\sum_{i=-\infty}^{\infty} P(x_i) = 1. \tag{2.28}$$

An essential function derived from the probability distribution is the cumulative distribution function (CDF). The CDF is defined as the probability that a particular random variable is less than or equal to a specific value:

$$F_X(a) = P(x \leq a). \tag{2.29}$$

For discrete random variables, the CDF is given as:

$$F_X(a) = \sum_{\text{all } x_i \leq a} P(x_i). \tag{2.30}$$

For continuous random variables, the CDF is given by:

$$F_X(a) = \int_{-\infty}^{a} f_X(x)dx. \tag{2.31}$$

Random processes exhibit variability and are contingent upon the behavior of real systems. For instance, when flipping a coin, we encounter one of two possible states. Similarly, in the case of a random signal generator spanning the range of 0 to 5 volts, if we define our random variable as the signal level, it will also fall within the 0 to 5 range. Nevertheless, even in the latter scenario, the behavior of the random signal is governed by a probability model. Figure 2.14 illustrates the outcomes of two distinct models for the same random signal, both falling within the 0 to 5 volts range.

The subsequent section will delve into several key models for both discrete and continuous random variables.

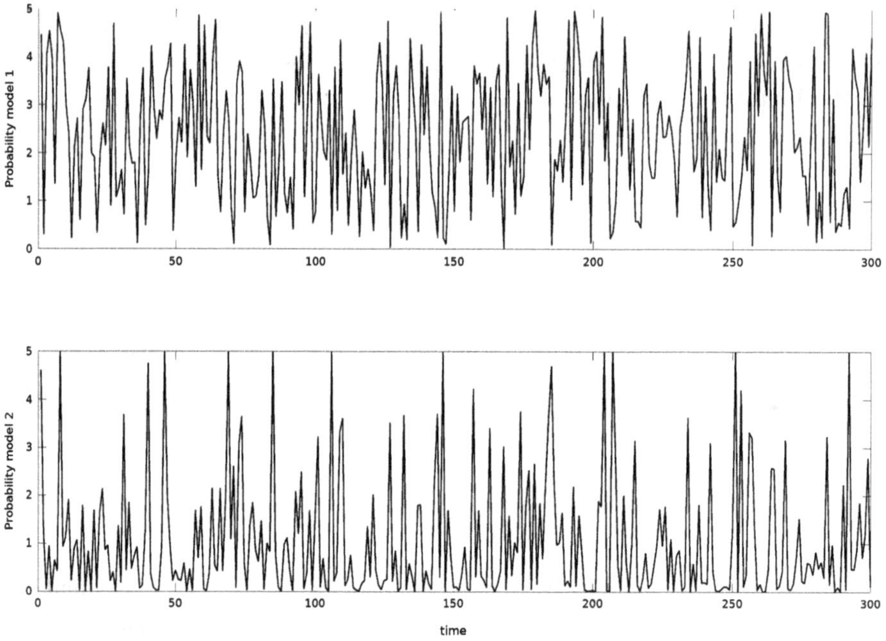

Figure 2.14: Random signals from 0 to 5 volts with different probability models.

2.3 Random variable models

In this section, we explore various important types of random variable models. Each of them effectively represents a specific experiment with uncertain outcomes.

2.3.1 Bernoulli random variable

The simplest type of random variable represents an uncertain experiment with only two possible outcomes. This could be a coin toss with *heads* or *tails*, a received binary bit with *One* or *Zero*, or generally *Success* or *Failure* outcomes. As previously explained, a random variable serves as the link between events and real numbers, allowing us to assign distinct numerical values to each outcome. For instance, we can define the random variable $X = 1$ for the event *Success* and $X = 0$ for the event *Failure*. The experiment is uncertain because we lack complete knowledge about the precise outcome. Therefore, we assign probabilities to each possible outcome, such as $P(1) = P(\{X = 1\}) = p$ and $P(0) = P(\{X = 0\}) = 1 - p$. The probability p associated with the *Success* outcome can be determined through theory or observation. For example, if we know that the experiment is fair with equally likely outcomes, then $p = 0.5$. However, if the experimental process is unknown, we may estimate p by observing the historical outcomes, contingent on whether the process is stationary or not. This aspect will be further discussed in the section on *Stochastic Processes*. The probability mass function of the Bernoulli random variable is illustrated in Figure 2.15.

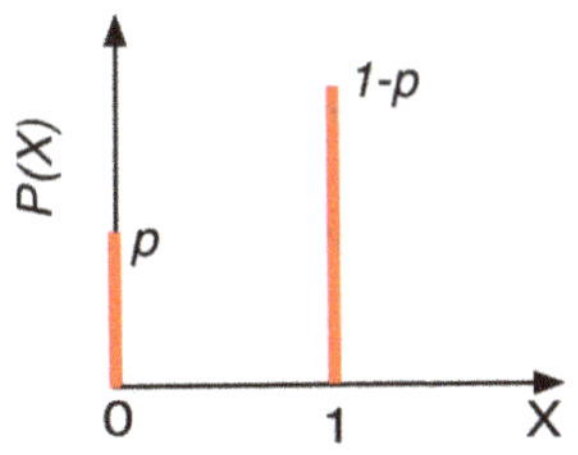

Figure 2.15: Probability mass function of Bernoulli RV.

2.3.2 Binomial random variable

In an abstract sense, a *binomial* random variable quantifies the count of *success* outcomes achieved when a Bernoulli experiment is conducted independently n times.

Mathematically, a binomial random variable Y is defined as:

$$Y = \sum_{i=1}^{n} X_i, \quad X_i \in \{0, 1\}. \tag{2.32}$$

It is clear that $Y \in \{0, 1, 2, \ldots, n\}$.

Example 16. Consider a football team wins a match with a score of 3:0. Historical data reveals that player *MM* has accounted for 20 % of all the team's goals. Given that we have no specific information about who scored the goals, what is the probability that Player *MM* has scored two goals?

Solution. We can readily frame this problem as a binomial one. We can define success (S), denoted as $X = 1$, when player MM scores the goal, and failure (F) $X = 0$ as another player scoring the goal Assuming independent events, the probability that player MM scores the first two goals (i. e., SSF) is $p^2(1 - p) = 0.2^2(0.8) = 0.032$. However, there are two other possible scenarios. Player MM might score the first and third goals, or the second and third goals. All possible combinations of scoring 2 goals out of 3 are SSF, SFS, FSS. Therefore, the probability that player MM scores two goals out of three is: $3 \times 0.032 = 0.096$.

The general formula of the probability of binomial random variable (i. e., to have k success out of n trails) is

$$P(Y = k) = P(k) = \binom{n}{k} p^k (1 - p)^{n-k}, \tag{2.33}$$

where $\binom{n}{k} = \frac{n!}{k!(n-k)!}$, and n! $= n(n - 1)(n - 2) \ldots 3 \times 2 \times 1$.

Example 17. Five unfair coins are tossed with a probability of heads of 0.7. Find:
- The probability that two heads are obtained.
- The probability that at least one heads is obtained.
- Draw roughly the probability mass function.
- Write an Octave Code to simulate this example and compare the results.

Solution. The probability that two heads are obtained can be found by applying (2.33) as $P(2) = \binom{5}{2}0.7^2(1-0.7)^3 = 0.132$.

The probability that heads is obtained at least once means $P(1 \cup 2 \cup 3 \cup 4 \cup 5) = 1 - P(0) = 1 - 0.3^5 = 0.9976$.

To draw the probability mass function we need to find the values of $P(k)\ \forall k = 0, \dots 5$ using (2.33). Figure 2.16 shows the probability mass function. Finally, a simple Octave code to simulate this example is given next. It gives the answers of 0.13186 and 0.99746, respectively.

```
clear all
N=100000; %Number of repeated experiments
x=(sign(0.7-rand(N,5))+1)/2; %Generating random matrix
    with 1s and 0s
y=sum(x,2); % computing the total number of heads in
    each experiment
P1=length(find(y==2))/N; % computing the prob. that
    number of heads =2
P2=1-length(find(y==0))/N; % At least one head appears
```

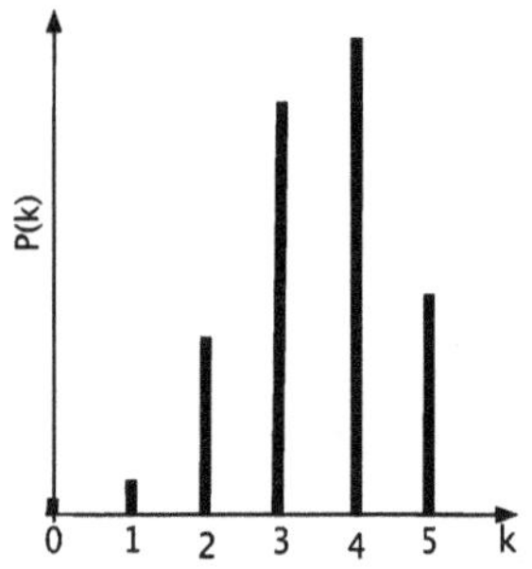

Figure 2.16: Probability mass function.

2.3.3 Poisson random variable

A Poisson random variable can be thought of as a special case of a binomial random variable, particularly when the probability of an event occurring is very low and the number of trials is exceedingly high. Let's derive the Poisson random variable from the binomial distribution. Consider a random occurrence of events over time (for instance, when your mobile phone rings). We know that, on average, λ events happen per unit of time. However, the exact timing of these events is unknown, and they occur independently. It's evident that the expected number of events within a specific time interval T will be λT. Now, what's the probability of precisely k events occurring during this time period of T? We can address this by breaking down the time interval T into extremely small time segments. These segments are so small that the probability of two independent events happening within a single interval is practically zero! Consequently, the number of intervals can be represented as $n = \frac{T}{\Delta t}$. Therefore, the probability of having precisely k events within the time interval of T follows the binomial distribution as follows:

$$P(k) = \binom{n}{k} p^k (1-p)^{n-k}, \tag{2.34}$$

where p is the probability that an event will occur within a single time interval Δt. It is clear that $p = \Delta t \lambda = \frac{\lambda T}{n}$. Since $\Delta t \to 0$, thus, $n \to \infty$ and $p \to 0$. Therefore, directly solving the binomial equation is not feasible. It is necessary to establish the limits of the Binomial equation. This equation can be expressed in the following form:

$$P(k) = \frac{n!}{k!(n-k)!} \left(\frac{\lambda T}{n}\right)^k \frac{(1-\frac{\lambda T}{n})^n}{(1-\frac{\lambda T}{n})^k} \tag{2.35}$$

$$\lim_{n\to\infty} \left(1 - \frac{\lambda T}{n}\right)^n = e^{-\lambda T} \tag{2.36}$$

$$P(k) = \frac{n(n-1)\dots(n-k+1)}{k!} \frac{(\lambda T)^k}{n^k} e^{-\lambda T} \tag{2.37}$$

$$P(k) = \frac{(\lambda T)^k}{k!} e^{-\lambda T}. \tag{2.38}$$

The final equation is referred to as the Poisson random variable. To aid in your comprehension of the mathematical derivations above, keep in mind that, when $n \to \infty$, then for finite k, $(1 - \frac{\lambda T}{n})^k = 1$ and

$$n(n-1)\dots(n-k+1) \approx n^k.$$

The Poisson random variable finds extensive applications across a wide array of fields, spanning technology, biology, finance, and even social systems, such as in the structural analysis of social networks. Researchers appreciate the Poisson formula for its elegant simplicity, often rendering problems solvable in a closed mathematical form. Nevertheless, in certain scenarios, the Poisson distribution may be deemed as an overly simplified model, as it presupposes independent event occurrences and a constant rate of events.

Example 18. At a traffic light intersection, the average arrival rate of cars is 170 cars per hour, following a Poisson distribution. What is the probability of ten cars arriving within a ten minute interval? Additionally, what is the probability that the number of cars falls between 25 and 35 in the same ten minute time frame?

Solution. Since the rate is give per hour, it would be convenient to use it as the time unit. Therefore, the observation period $T = 0.167$ hours, and $k = 10$. Therefore,

$$P(10) = \frac{(170 \times 0.167)^{10}}{10!} e^{-170\times0.167} = 0.0000455. \tag{2.39}$$

The probability that the number of cars will be between 25 and 35 is computed as

$$P(25 \leq k \leq 35) = e^{-\lambda T} \sum_{k=25}^{35} \frac{(\lambda T)^k}{k!} = 0.667. \tag{2.40}$$

Example 19. For a specific mobile phone brand, the probability of a battery explosion within an hour is 10^{-6}. If there are five passengers using the same mobile phone on an airplane for a 6-hour flight, what's the probability of at least one mobile phone exploding during the flight? Furthermore, what's the probability that exactly two mobiles will explode? Let's also explore these scenarios using the Poisson approximation.

Solution. The probability that each mobile will explode during the entire travel interval time is 6×10^{-6}. The probability that at least one mobile will explode during the flight is $1 - P(0) = 1 - (1 - 6 \times 10^{-6})^5 = 0.00003$. You should be able to complete the answers!

2.3.4 Hypergeometric random variable

In a total collection of N items, K of them are classified as type A, while the remainder, $N - K$, are classified as type B. If n items were drawn sequentially without replacement (or perhaps all at once), the probability that there are exactly $k \leq n$ items within those n items belonging to Type A is called hypergeometric distribution, and it is given by:

$$P(X = k) = \frac{\binom{K}{k}\binom{N-K}{n-k}}{\binom{N}{n}}, \tag{2.41}$$

where $\max(0, n - N + K) \leq k \leq \min(K, n)$.

Example 20. Consider a group of 100 individuals, 30 of whom are over the age of 60. From this group, a random sample of 20 persons has been chosen. Now we want to calculate the probability of that exactly 5 of them are older than 60. Write a Scilab code to emulate this example.

Answer. We have $N = 100$, $K = 30$, $n = 20$, and $k = 5$, then using (2.41): $P(X = 5) = \frac{\binom{30}{5}\binom{100-30}{20-5}}{\binom{100}{20}} = 0.1918$.

Before you study the code below, try to write the code yourself first. We try to keep it simple, but not necessarily efficient. In programming, we usually try to avoid using loops (like for, while, etc.), as they consume time and memory. However, in this code, we used the loop as shown next.

```
1   clear
2   N=1000; // number of experiment repetition
3   x=zeros(N,100);
4   y=grand(N,'prm',(1:100));
5   k=0; // counting variable
6   for i=1:N
7   x(i,y(i,1:70))=1;
8   z=sum(x(i,1:20));
9   if z==15,
10  k=k+1;
11  end
12  end
13  P=k/N // the required probability
```

Upon running the Scilab code, the obtained result was $P = 0.1904$. Remarkably, this outcome closely aligns with our theoretical expectation.

Now, let's explain how the code works. In *Line* 3, we assigned a zero matrix x with 100 columns and N rows. In *Line* 4, we generated a matrix y of the same dimension as x, but each row has numbers from 1 to 100 and is randomly allocated. The loop starts at *Line* 6. Because in this experiment, we already knew that there are exactly 30 persons over 70. We marked those who are under that age with a 1 and those over that age with a 0. Since x is already an all zeros matrix, we assign 70 ones in each row, as shown in *Line* 7. However, those ones are distributed randomly within the rows of x. Because the first 70 numbers of y are random numbers between 1 to 100, these values of y have been used as *indices* of x, which is marked with the number 1. In *Line* 8, we count how many ones we have in the first 20 samples. Because we are looking for the probability that we have exactly five people over the mentioned age, this is equivalent to 15 persons under the age within the sample of 20. If this condition has been achieved, then we increment the counter k by one. Finally, the probability is computed in *Line* 13.

I highly encourage you to experiment with each of the functions used in this code, as it's a valuable way to grasp the inner workings of the program. If you are using Octave, you should be able to reconfigure this code using Octave functions seamlessly.

Example 21. The shipments of canned food from several companies arrived at the port in one of the countries. Each shipment consists of 10,000 cans. Security information was received, that one of the exporting companies used prohibited health materials in 20 %

of its shipment. Inspecting all the cans from all the companies will require a significant amount of time and effort. What is the minimum number of cans that should be inspected from each shipment, so that the shipment containing the prohibited materials will appear in the sample at a percentage of not less than 90 %?

Solution. It is clear that we can model this problem with hypergeometric distribution. We have two classes, one containing clean cans with an average of 8000 and the second containing contaminated cans with an average of 2000. This is the case in one shipment, the other shipments all contain safe cans. We want to determine the contaminated shipment with a minimum effort. From (2.41), we want to determine the value of n in order to have $P(k \geq 1) = 0.9$. It means that the probability of finding at least one contaminated can is 90 %. Remember that $P(k \geq 1) = 1 - P(k = 0)$, in other words, $P(k = 1) = 0.1$. Substituting these values in (2.41)

$$\frac{\binom{2000}{0}\binom{8000}{n}}{\binom{10000}{n}} = 0.1.$$

The above equation could be simplified as:

$$\frac{8000! \times (10000 - n)!}{10000! \times (8000 - n)!} = 0.1.$$

I guess your calculator will not be able to handle the factorial of such huge numbers. But fortunately, it is not needed. We can try the first few numbers of n. Let's try what will be the probability with n=1! substituting in the previous equation we obtain

$$\frac{8000! \times (9999)!}{10000! \times (7999)!} = \frac{8000 \times 7999! \times (9999)!}{10000 \times 9999! \times (7999)!} = 0.8.$$

Certainly, a probability of 0.8 indicates that the likelihood of encountering at least one contaminated can is merely 0.2. Clearly, this is insufficient. Therefore, we ought to explore larger values of n. It can be readily demonstrated that for any given n, the equation remains as follows:

$$\prod_{k=0}^{k=n} \left(\frac{8000 - k}{10000 - k} \right). \tag{2.42}$$

After a few tries, we found that at $n = 11$, equation (2.42) leads to

$$\prod_{k=0}^{k=11} \left(\frac{8000 - k}{10000 - k} \right) = 0.086. \tag{2.43}$$

This implies that the likelihood of identifying a contaminated shipment stands at 91.4 %, based on an examination of just 11 randomly chosen cans out of a total of 10,000. To validate our analytical findings, the following code demonstrates a Monte Carlo simulation of this scenario using Octave. Upon executing this code, we determine the probability of discovery to be $P = 92\,\%$.

```
% Shipment Example
clear all
N=10000; %Number of Cans
p=0.2; % Percentage of contaminated Cans
x=sign(rand(N,1)-p); %x=-1 for Cont. Cans
s=0; %Initialization of counter
T=200; % Repeating the simulation
n=11; % Number of Samples
for k=1:T,
a=randperm(N); % generating random sorting
y=x(a(1:n)); % selecting n random samples
if sum(y)<n, % if there is any finding of contamination
s=s+1;
end
end
P=s/T % probability of finding
```

2.3.5 Uniform distribution random variable

In our previous discussions, we focused on discrete random variables. Now, let's explore the uniform distribution, which represents one of the simplest continuous random variables. A uniform random variable is characterized by its uniform distribution over a specific range, defined as:

$$f_X(x) = \begin{cases} \frac{1}{b-a} & a \le x \le b, \\ 0 & \text{elsewhere.} \end{cases} \tag{2.44}$$

The uniform distribution's magnitude should validate that the total area beneath the distribution curve equals one, as previously explained in (2.25). Figure 2.17 illustrates the characteristic shape of the uniform distribution function.

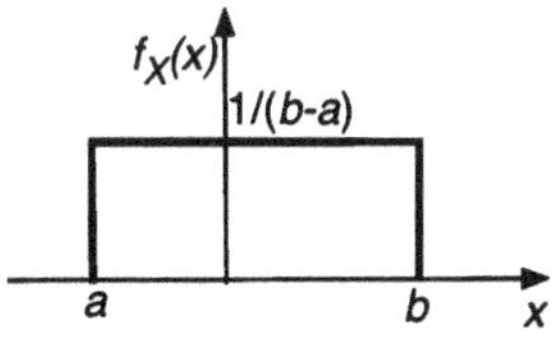

Figure 2.17: Uniform distribution function.

Example 22. The output voltage of a random generator falls within the range of −2 to 5 volts, following a uniform distribution. When you took a voltage measurement at a specific moment, consider the following evaluation:

- What is the probability that the measured voltage is less than 3 *volts*.
- What is the probability that the voltage is greater than 4 *volts*.
- What is the probability that the measured voltage is 2.5 *volts*.
- What is the probability that the measured voltage is greater than 6 *volts*.
- Write Octave code to emulate the example and compare the theoretical and simulated results.

Answers. We can just apply equation (2.24), hence,

- $P(X \leq 3) = \int_{-2}^{3} \frac{1}{7} dx = \frac{x}{7}\Big|_{-2}^{3} = \frac{5}{7}$.
- $P(X \geq 4) = \int_{4}^{5} \frac{1}{7} dx = \frac{x}{7}\Big|_{4}^{5} = \frac{1}{7}$.
- The probability that the measured value is exactly 2.5 *volts* is simply *zero*. Revise the discussion in the Section 2.2.
- The probability that the measured value is greater than 6 is also zero, because the maximum voltage value is 5.
- Computer-generated numbers are inherently discrete due to finite precision, but for practical purposes, they can often be treated as continuous. Most modern computers can store more than 14 floating-point digits. The command *rand* is a common tool in many software packages to generate uniform numbers within the range of 0 to 1. Consequently, it can be readily adapted to produce numbers within any desired uniform distribution range, as demonstrated in the following Octave code example.

```
clear all
N=10000; % number of generated samples
x=7*rand(N,1)-2; %generating random from -2 to 5
P1=length(find(x<3))/N %Pr(x<3)
P2=length(find(x>4))/N% Pr(x>4)
P3=length(find(x==2.5))/N % Pr(x=2.5)
P4=length(find(x>6))/N % Pr(x>6)
```

Compare the outcomes. They should exhibit a high degree of similarity. Experiment by decreasing and increasing the number of samples, denoted as N. How does this variation impact the accuracy of the results? Guidance on selecting the appropriate number of samples will be covered in more detail later.

Example 23. A transmitter sends a lengthy message, lasting 5 seconds, once every 1 minute. However, the transmission time within the minute is randomly and uniformly distributed. There's another transmitter nearby that follows the same pattern. If the messages from both transmitters coincide in time, it results in a collision, causing both transmitters to resend their messages later. Determine the probability of collision. Develop Octave code to simulate this scenario and compare the results with the theoretical prediction.

Solution. Let's assume that the transmission times of the first and second transmitters are denoted as t_x and t_y, respectively. Both of these variables follow a uniform distribution ranging from 0 to 60 seconds. A collision will occur if $|t_x - t_y| \le 5$ seconds, within the interval $0 \le t_x, t_y \le 60$. A graphical representation of this problem can be seen in Figure 2.18. The probability of collision can be calculated as the area of the shaded strip divided by the total area of the square:

$$P(|t_x - t_y| \le 5) = \frac{60^2 - 55^2}{60^2} \approx 0.1597. \tag{2.45}$$

To emulate this scenario, you can use the following Octave code (or a similar one in Scilab, after changing the comment sign):

```
clear all
N=100000;
tx=60*rand(1,N); %generating uniform RV from 0 to 60
ty=60*rand(1,N); %generating uniform RV from 0 to 60
z=abs(tx-ty); % finding the absolute time difference
P=length(find(z<=5))/N
```

Running this code gives the result of $P = 0.15920$.

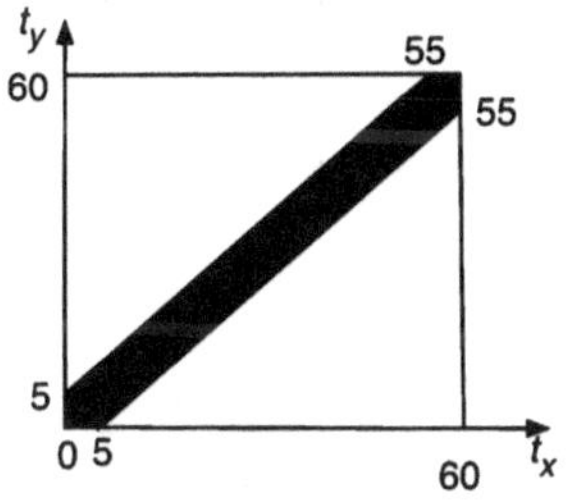

Figure 2.18: Example 23.

2.3.6 Exponential random variables

True to its name, the exponential distribution takes the form of an exponential function, defined as:

$$f_X(x) = \lambda e^{-\lambda x}, \quad x \ge 0. \tag{2.46}$$

Figure 2.19 illustrates the shape of the exponential distribution function for various values of λ. It is evident that the random variable x is positive and has a higher probability

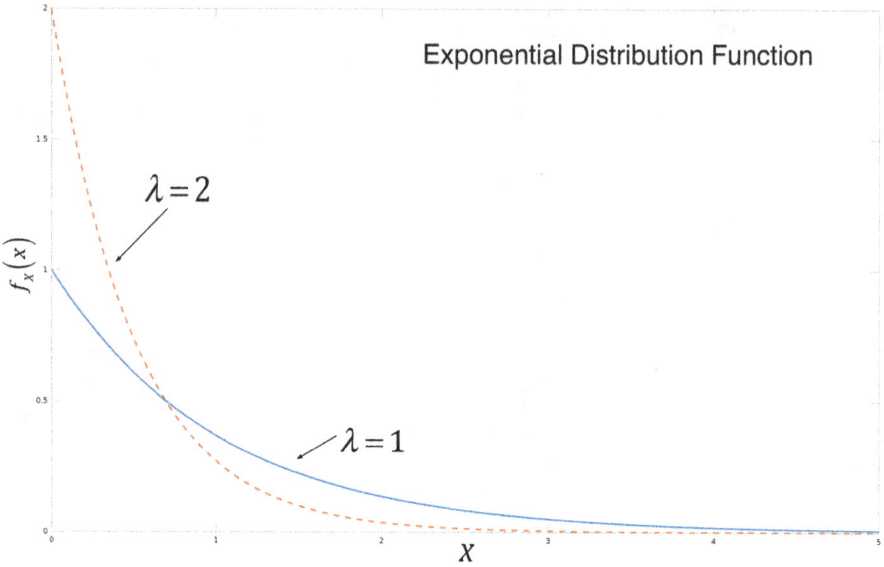

Figure 2.19: Exponential distribution function.

of being closer to zero. However, it can also have larger values as it extends towards infinity ∞.

In addition to its manageable and straightforward mathematical representation, the exponential distribution exhibits numerous applications and distinctive characteristics. One notable property of the exponential distribution is its memoryless property.

Example 24. Prove the memoryless characteristic of the exponential distribution.

Solution. Assume two variables $z > y$, the conditional probability that

$$P(x > z|x > y) = \frac{P(x > z \cap x > y)}{P(x > y)}.$$

However, the event $x > y$ is subset (i. e., included) of the event $x > z$, because $z > y$, hence, $P(x > z \cap x > y) = P(x > z)$. Furthermore, $P(x > z) = \lambda \int_z^{\infty} e^{-\lambda x} dx = e^{-\lambda z}$. Therefore,

$$P(x > z|x > y) = \frac{P(x > z)}{P(x > y)} = \frac{e^{-\lambda z}}{e^{-\lambda y}} = e^{-\lambda(z-y)}. \tag{2.47}$$

This proves the memoryless characteristic of the exponential distribution. For example, $P(x > 2|x > 1)$ gives identical result as $P(x > 1001|x > 1000)$. In an exponential probability density function, the uncertainty of future events is completely independent of their past occurrences. While this assumption can prove challenging to justify in certain practical scenarios, it often provides a valuable and reliable approximation.

It is established that the exponential distribution accurately models the distribution of waiting times between events in a Poisson distribution. In this context, the random variable is time, and the distribution can be expressed as follows

$$f_T(t) = \lambda e^{-\lambda t}, \quad t \geq 0. \tag{2.48}$$

Example 25. Derive the cumulative distribution function CDF of the empirical distribution.

Solution. As previously explained, the cumulative distribution function (CDF) of the exponential distribution can be readily derived as follows:

$$P(t \leq t_0) = \lambda \int_0^{t_0} e^{-\lambda t} dt = 1 - e^{-\lambda t_0}. \tag{2.49}$$

Example 26. A certain system comprises three components, as illustrated in Figure 2.20. The system will experience a failure if either S_1 or S_2 fails AND S_3 fails. Assuming that the random time to failure for each component follows an exponential distribution with respective average failure rates of $\lambda_1 = 0.001$/day, $\lambda_2 = 0.003$/day, and $\lambda_3 = 0.0005$/day, determine the probability that the system will operate without failure for at least the next 50 days. Additionally, provide an Octave code to simulate the system's behavior and verify the theoretical results.

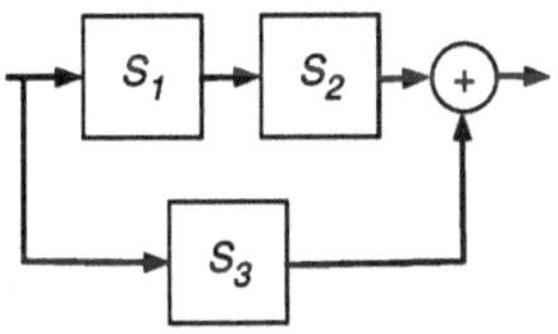

Figure 2.20: System configuration.

Solution. The probability that a device will remain operational until time t_0 implies that a failure will occur after t_0, expressed as $P(t \geq t_0) = \int_{t_0}^{\infty} \lambda e^{-\lambda t} dt$, which simplifies to $e^{-\lambda t_0}$. We can calculate the probability that each device will function for 50 days as follows: $P_1 = e^{-0.001\times 50} = 0.951$; $P_2 = e^{-0.003\times 50} = 0.861$; and $P_3 = e^{-0.0005\times 50} = 0.975$. For the whole system to work, we need both S_1 AND S_2 OR S_3 to work. Hence, the probability that it will work after 50 days is $P_1 \times P_2 + P_3 - P_1 P_2 P_3 = 0.995$.

Before we delve into how to simulate this example using a Monte Carlo simulation, it's important to note that Octave defines the exponential function as:

$$f_T(t) = \frac{1}{\lambda} e^{-\frac{t}{\lambda}}.$$

To align with the example, we need to reverse the value of λ. Additionally, we will present two different codes to solve the problem. One of these codes is significantly more efficient in terms of execution time.

Let's start with the first code:

```
1   clear all
2   N=10000; % number of experiments
3   lambda1=0.001;
4   lambda2=0.003;
5   lambda3=0.0005;
6   TTF1=exprnd(lambda1^-1,1,N);
7   TTF2=exprnd(lambda2^-1,1,N);
8   TTF3=exprnd(lambda3^-1,1,N);
9   k=0;
10  for i=1:N,
11  if TTF1(i)>50 && TTF2(i)>50 || TTF3(i)>50,
12  P1= k=k+1;
13  end
14  end
15  P= k/N
```

In the code snippet above, we employ the Octave command *exprnd* to generate random numbers following an exponential distribution with the specified λ. TTF stands for Time To Failure, and it represents the random time to failure for each device. We create a raw vector containing $N = 10{,}000$ random values, corresponding to the time to failure for each device, labeled as TTF1, TTF2, and TTF3. To calculate the probability that the entire system will function beyond 50 days, we need the failure time of $\{S_1$ AND $S_2\}$ OR S_3 to exceed 50 days. We check this condition for each entry in the code at *Line 11*. Running the provided code yields a result of approximately 0.9957, which aligns with our theoretical expectations. Nevertheless, this code is not particularly efficient due to its use of a *for* loop. We can enhance efficiency and conciseness by rewriting the code, as demonstrated in the following example:

```
clear all
N=10000; % number of experiments
lambda1=0.001;
lambda2=0.003;
lambda3=0.0005;
TTF1=exprnd(lambda1^-1,1,N);
TTF2=exprnd(lambda2^-1,1,N);
TTF3=exprnd(lambda3^-1,1,N);
TT=max(min(TTF1,TTF2),TTF3);
P=length(find(TT>50))/N
```

The idea here is that the system will work after 50 days if the minimum downtime of S_1 and S_2 is greater than 50 or if the downtime of S_3 is greater than 50. Therefore, we have calculated this operation time in *Line* 9. Consequently, we find the numbers where this time is greater than 50 in *Line* 10. The results of both codes are identical.

I employed the commands *tic* and *toc* to measure the execution time of both codes on my computer. The first code required 0.14309 seconds, while the second code needed only 0.00982 seconds. This signifies that the second code is approximately 15 times faster than the first code, even though both yield identical results. Although 0.14 seconds is a negligible amount of time, when simulating considerably larger problems, execution times can increase significantly. It's not uncommon to encounter simulation tasks that demand days of processing time. Hence, writing efficient code becomes crucial in many situations. As a general guideline, it's advisable to minimize the use of loops whenever possible.

Finally, if you are using Scilab, you can generate the exponentially distributed random numbers using the command *grand*(1, N, 'exp', 1/lambda).

Example 27. Let's revisit the transmitter collision case provided in Example 23. In this scenario, both transmitters send their packets when they have data to transmit. We assume that the time to send a packet for each transmitter follows an exponential distribution with a rate parameter of $\lambda = 2$ per minute. Additionally, the duration of each packet is 5 seconds for both transmitters. The question at hand is: What is the probability that both packets will interfere with each other within a long time of transmission window? To answer this question, we will solve the problem using a Monte Carlo simulation.

Solution.

```
% Exponential Collision
clear all
lambda=4/60; % The average transmission is 4 packets/minute
N=2; %number of Transmitters
M=100000; % number of transmissions
T=exprnd(lambda^-1,N,M); % Generating random times of transmission
D=T(1,:)-T(2,:); % time differences between transmissions
C=find(abs(D)<5); % the time difference within 5 seconds
P=length(C)/M; %Probability of collision
```

Running the code we find that the probability of collision is 0.284.

2.3.7 Erlang or Gamma distribution

The exponential distribution is often used to model the random waiting time until the next event takes place. If you're interested in the waiting time until a sequence of k consecutive events occurs, you can define it as follows:

$$T_k = \sum_{i=1}^{k} t_i,$$

where each t_i is a random variable with an exponential distribution. Hence, the distribution of T_k is called Erlang distribution and it has the following form:

$$f_{T_k}(t) = \frac{\lambda(\lambda t)^{k-1} e^{-\lambda t}}{(k-1)!} \quad t \geq 0. \tag{2.50}$$

The Erlang and Poisson distributions exhibit a unique complementarity: while the Poisson distribution quantifies the frequency of events within a given time frame, the Erlang distribution characterizes the duration of time it takes for a specific number of events to occur. The generalized form of the Erlang distribution, where the shape parameter k does not need to be an integer, is called the Gamma distribution. However, we may use the name Gamma distribution for integer shape parameters as well. One command to generate random numbers with Gamma distribution in Octave is *gamrnd* and in Scilab *grand(m, n, 'gam', ...)*.

Example 28. Three sensors play a pivotal role in monitoring a critical physical parameter during a specific chemical reaction. To ensure the integrity of the reaction, a comprehensive redundancy strategy is employed. At any given time, one sensor actively operates, while the remaining two are on standby. If one sensor encounters a malfunction, the second standby sensor seamlessly assumes the active role and continues to function until it too experiences failure. If the second sensor fails, the third sensor takes over. For each of these sensors, the failure rate has been quantified at $\lambda = 0.1$ per year. With this information at hand, our objective is to calculate the probability that all three sensors will fail within a 10-year time frame.

Solution. This example is a typical application of Gamma distribution. We may compute the probability that all sensors will fail within 10 years as:

$$\begin{aligned} P(t \leq 10) &= \frac{\lambda^3}{2!} \int_0^{10} t^2 e^{-\lambda t} dt \\ &= -\frac{\lambda^3}{2} \left[e^{-\lambda t} \left(\frac{2}{\lambda^3} + \frac{2t}{\lambda^2} + \frac{t^2}{\lambda} \right) \right]_{t=0}^{10} \\ &= -\left[e^{-0.1t} \left(1 + 0.1t + \frac{0.01t^2}{2} \right) \right]_{t=0}^{10} \end{aligned}$$

$$= 0.0803.$$

Next we show simple Octave code to emulate the same problem. The result obtained from the simulation is 0.0804.

```
clear all
lambda = 0.1;
N= 100000;
x=exprnd(lambda^-1,3,N);
y=sum(x);
Ne=length(find(y<=10));
P=Ne/N
```

Example 29. In the previous example, system reliability was enhanced by incorporating a redundancy strategy involving two standby sensors. Should one sensor experience a malfunction, the next one seamlessly assumes its role. Consequently, the cumulative uptime can be described using an Erlang distribution, which accounts for the sum of all random failure times. Another alternative is to use all sensors simultaneously to maintain system operation. In this scenario, the system's downtime corresponds to the longest failure duration among the sensors. If there are N sensors and only one is required to maintain system functionality, the system's failure time can be calculated as follows:

$$T_{\max} = \max_{i=1,\dots N} t_i,$$

where t_i is an exponential random variable representing the failure time of the ith sensor. Determine the probability density function of $T_{\max}$ and then proceed to recalculate the probability of system failure within a 10-year time frame, considering the parallel operation of sensors, as demonstrated in the previous example.

Solution. Since $T_{\max} \geq t_i$, $\forall i$, hence:

$$P(T_{\max} \leq a) = P(t_1 \leq a, t_2 \leq a, \dots, t_N \leq a).$$

Assuming all sensors are independent,

$$P(T_{\max} \leq a) = P(t_1 \leq a).P(t_2 \leq a)\dots P(t_N \leq a).$$

Therefore, the CDF function of $T_{\max}$, and from equation (2.49) is given by

$$F_{T_{\max}}(a) = (1 - e^{-\lambda a})^N.$$

The probability density function of $T_{\max}$ is the derivative of the CDF function, and it is easily derived as:

$$f_{T_{\max}}(t) = N\lambda e^{-\lambda t}(1 - e^{-\lambda t})^{N-1}.$$

For $N = 3$ as in the previous example, the probability density function is

$$f_{T_{\max}}(t) = 3\lambda e^{-\lambda t}(1 - e^{-\lambda t})^2.$$

The probability that all sensors will fail during a 10-years period for $\lambda = 0.1$/year is computed as follows:

$$\begin{aligned}\int_0^{10} f_{T_{\max}}(t)dt &= 3\lambda \int_0^{10} e^{-\lambda t}(1 - e^{-\lambda t})^2 dt \\ &= (-3e^{-\lambda t} + 3e^{-2\lambda t} - e^{-3\lambda t})_{t=0}^{10} \\ &= 0.2526.\end{aligned}$$

It is more likely to experience failure over a 10-year operating period than the previous standby scenario. Next, we provide Octave code to replicate this scenario, and the outcome of the simulation indicates a probability of $P = 0.2531$.

```
clear all
lambda = 0.1;
N= 100000;
x=exprnd(lambda^-1,3,N);
y=max(x);
Ne=length(find(y<=10));
P=Ne/N
```

2.3.8 Normal (or Gaussian) distribution

Without a doubt, the *Normal Distribution* stands as the paramount probability density function across a multitude of scientific disciplines. Its significance can be attributed to several compelling reasons, which underscore its pivotal role, such as

- The aggregation of numerous independent and identically distributed random variables leads to the emergence of a normal distribution, a phenomenon known as the central limit theorem. This theorem, which is fundamental in statistics, serves as the foundation for various other statistical principles. In practical applications, such as telecommunications, the cumulative effect of diverse physical behaviors like thermal noise, gunshot noise, and interference can be accurately represented by a normal distribution.
- Many natural processes can be effectively approximated by the normal distribution, providing a suitable representation for their inherent uncertainties.
- Various distributions can be closely approximated by the normal distribution. For instance, a binomial distribution with a substantially large n and a p value that's not

extremely close to 0 or 1 can be well approximated using the normal distribution. Similarly, when dealing with a Poisson distribution with a sizable λ it can also be accurately approximated as a normal distribution.

We say that a random variable X has a normal distribution with mean (average) μ and variance σ^2 if it has the following probability density function:

$$f_X(x) = \frac{1}{\sqrt{2\pi\sigma^2}} e^{-\frac{(x-\mu)^2}{2\sigma^2}}. \tag{2.51}$$

Figure 2.21 shows the shape of a normal distribution with the mean $\mu = 1$ and two different values of the variance.

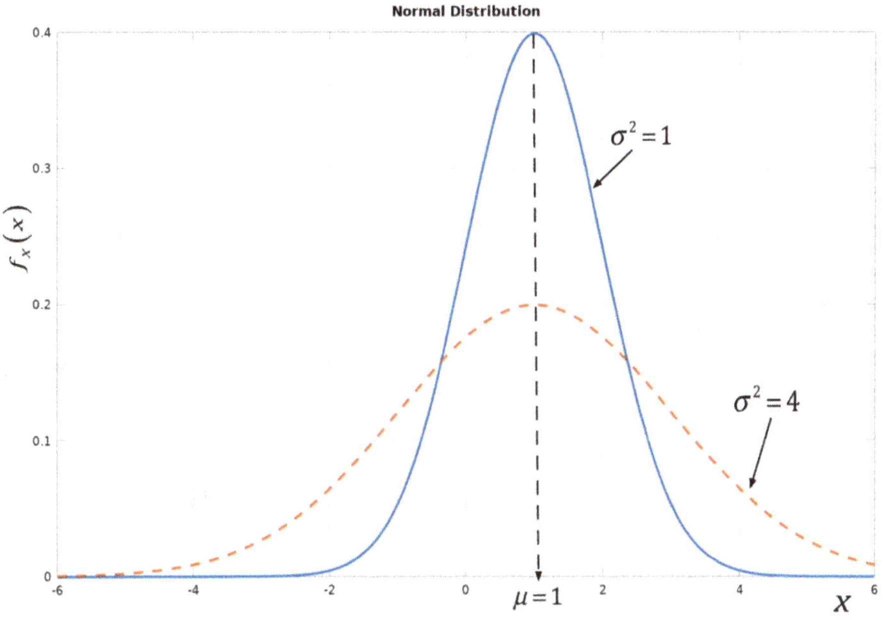

Figure 2.21: Normal distribution function.

There is a minor issue here because it's impossible to represent the integral of the normal distribution with a closed-form mathematical expression, which typically involves elementary functions such as polynomials, trigonometric, logarithmic, and exponential functions. Consequently, it is not possible to express the probability and the cumulative distribution function (CDF) in terms of elementary functions. Nevertheless, there are specialized functions designed specifically to address this challenge, such as the *error function*.

Example 30. Express the CDF of the normal distribution using the *error function*.

Answer. The CDF $F_X(x_0)$ can be calculated as:

$$P(x \leq x_0) = \frac{1}{\sqrt{2\pi\sigma^2}} \int_{-\infty}^{x_0} e^{-\frac{(x-\mu)^2}{2\sigma^2}} dx.$$

Let's define $t = \frac{(x-\mu)}{\sqrt{2}\sigma} \Rightarrow dt = \frac{1}{\sqrt{2}\sigma} dx$, and change the variables in the previous integration to obtain:

$$\begin{aligned} P(x \leq x_0) &= P\left(t \leq \frac{x_0 - \mu}{\sqrt{2}\sigma}\right) = \frac{1}{\sqrt{\pi}} \int_{-\infty}^{\frac{x_0-\mu}{\sqrt{2}\sigma}} e^{-t^2} dt \\ &= \frac{1}{2} + \frac{1}{2}\operatorname{erf}\left(\frac{x_0 - \mu}{\sqrt{2}\sigma}\right), \end{aligned} \tag{2.52}$$

where erf(x) is called the error function and it is given by

$$\operatorname{erf}(x) = \frac{2}{\sqrt{\pi}} \int_0^x e^{-t^2} dt. \tag{2.53}$$

The values of the error function can be calculated using numerical methods, and are additionally available in tabular form.

There is another form to compute the probability of a normal distribution, known as the *Q-function*. It is defined as $Q(x_0) = P(x \geq x_0)$. It is widely used to compute the probability of errors in digital communications.

Example 31. Derive the formula for the *Q-function*.

Solution. For a normal distribution with zero mean and unit variance, the *Q-function* is defined as:

$$Q(x_0) = \frac{1}{\sqrt{2\pi}} \int_{x_0}^{\infty} e^{-\frac{x^2}{2}} dx. \tag{2.54}$$

For a normal distribution with arbitrary mean and variance, the probability of $P(x \geq x_0)$ could be derived in terms of the *Q-function* as:

$$P(x \geq x_0) = Q\left(\frac{x_0 - \mu}{\sigma}\right). \tag{2.55}$$

You can find the values of $Q(x)$ in tables, calculators, and various software packages.

The connection between the Q-function and the error function can be readily derived as

$$Q(x_0) = \frac{1}{2} - \frac{1}{2}\text{erf}\left(\frac{x_0}{\sqrt{2}}\right). \tag{2.56}$$

In both *Octave* and *Scilab* you may use the command erf(x) to compute the *error function*. The *Q-function* could be computed using the relation (2.56). You may also find the *Q-function* in the *Communications* package under *Octave* as *qfunc*(x). Figure 2.22 shows the CDF of the Gaussian distribution. An intriguing characteristic of the Gaussian distribution is that its mean, median, and mode are all equal. The significance of this property will become apparent when we introduce some concepts of estimation theory later on.

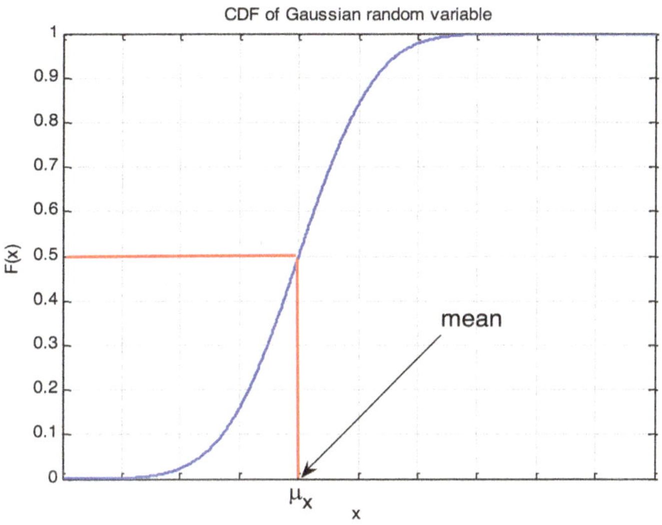

Figure 2.22: CDF of Gaussian distribution function.

Example 32. The random signal $x(t)$ has a normal distribution with mean = 1 and variance = 4. At a given time, we measured the signal and analytically determined the following probabilities:
- $P(x \geq 2.5)$;
- $P(x \leq 5)$;
- $P(1 \leq x \leq 3)$.

Using Monte Carlo simulations, write a code in Octave to compute the above probabilities.

Answer. We will solve this example with using *Q-function*. You can use the *error function* as well. Let $z = \frac{x-\mu_x}{\sigma_x} = \frac{x-1}{2}$, hence,
- $P(x \geq 2.5) = P(z \geq \frac{2.5-1}{2}) = Q(0.75) = 0.22$.

- $P(x \leq 5) = 1 - P(x \geq 5) = 1 - Q(\frac{5-1}{2}) = 1 - 0.02 = 0.98.$
- $P(1 \leq x \leq 3) = P(x \geq 1) - P(x \geq 3) = Q(0) - Q(1) = 0.5 - 0.158 = 0.342.$

Finally, we may evaluate the problem by simulations, as shown in the next code (for *Octave*). The three probabilities are evaluated as P_1, P_2, and P_3 respectively. The simulation results rounded to 4 digits were $P_1 = 0.2269$, $P_2 = 0.9769$, and $P_3 = 0.3431$. They align with the theoretical calculations.

```
clear all
N=100000; % number of experiments
x=2*randn(N,1)+1;
P1=length(find(x>=2.5))/N;
P2=length(find(x<5))/N;
y=x(find(x>=1));
P3=length(find(y<=3))/N;
```

The code is clear and straightforward. However, it may be beneficial to provide a more detailed explanation of P_3. As $P_3 = P(1 \leq x \leq 3)$, we created a new vector y containing all values of x that are greater than or equal to 1. Subsequently, we counted the number of elements in the vector y that are less than 3.

2.3.9 Rayleigh distribution

The Rayleigh distribution is important in the field of wireless communications, where it is utilized to model the amplitude of received signals in multipath fading channels. Furthermore, the Rayleigh distribution has a wide range of applications in diverse fields, including its use in modeling financial data, characterizing particle sizes in materials, and representing the variability of wind speeds in environmental science.

Suppose that a random variable Y is defined as:

$$Y = \sqrt{X_1^2 + X_2^2}, \tag{2.57}$$

where $X_i \; \forall i = 1, 2$ are independent and identically distributed (i.i.d) Gaussian random variables with zero mean and variance σ^2. The probability density function PDF of Y is called the Rayleigh distribution and it is given by:

$$f_Y(y) = \frac{y}{\sigma^2} e^{-y^2/(2\sigma^2)}, \quad y \geq 0. \tag{2.58}$$

You can generate a $m \times n$ matrix of random numbers with Rayleigh distribution using the σ parameter in Octave and Scilab with the commands:

$$y = \sigma \sqrt{\text{randn}(m, n).^2 + \text{randn}(m, n).^2}\,.$$

Nevertheless, the *Octave* has a dedicated function in the statistics package to generate random numbers with Rayleigh distribution as *raylrnd(sigma, m, n)*. Figure 2.23 shows the Rayleigh probability density function for different σ values. The following code shows how to make such a plot in Octave.

```
% Draw the Rayleigh PDF for different Sigma values
%On Octave
x=0:.01:6;
sig=[0.5;1;2];
f=x./sig.*exp(-x.^2./(2*sig.^2));
plot(x,f(1,:),"linewidth",3,x,f(2,:),"linewidth",3,x,
    f(3,:),"linewidth",3);
text(0.75, 0.5,['\leftarrow \sigma=0.5'],"fontsize",13);
text(1.5, 0.5,['\leftarrow \sigma=1'],"fontsize",13);
text(3.0, 0.5,['\leftarrow \sigma=2'],"fontsize",13);
grid
xlabel("x","fontsize",14)
ylabel("f_X(x)","fontsize",14)
title('The PDF of Rayleigh RV for different \sigma
    values')
```

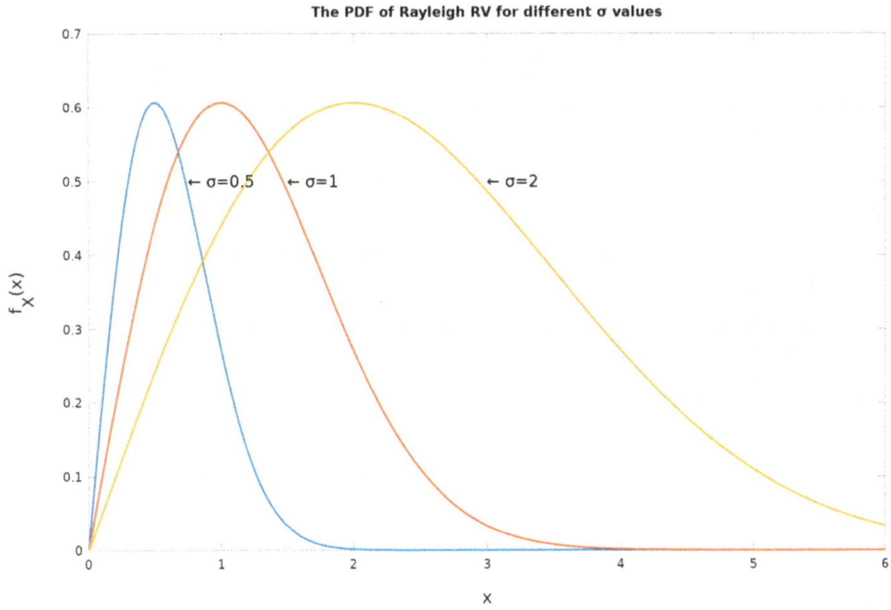

Figure 2.23: Rayleigh distribution function.

It is a straightforward to derive the cumulative distribution function $F(y_0) = P(y \le y_0)$ of Rayleigh random variables as:

$$P(y \le y_0) = \int_0^{y_0} \frac{y}{\sigma^2} e^{-y^2/(2\sigma^2)} dy$$
$$= 1 - e^{-\frac{y_0^2}{2\sigma^2}} \quad y_0 \ge 0. \tag{2.59}$$

Example 33. Most wind turbines are designed to operate at their full capacity when exposed in wind speeds ranging from 50 to 60 km/h. However, when wind speeds exceed 100 km/h, it becomes necessary to halt the operation of most turbines to prevent potential damage. The variability of wind speeds can be effectively represented by a Rayleigh distribution. For a specific month, it is established that the Rayleigh distribution parameter, denoted by σ, is equal to 40. Now, let's determine the percentage of time during that month that the turbines will be stopped.

Solution. The wind turbines should be stopped when the speed $y > 100$. The probability of this event is $P(y > 100) = 1 - P(y \le 100)$, and from equation (2.59),

$$P(y > 100) = e^{-\frac{100^2}{2\times 40^2}} = 0.044.$$

In other words, out of a total of 720 hours in that month, the wind turbines are expected to stop for about 32 hours.

Next, we will present a straightforward simulation code that generates Rayleigh random numbers to model wind speed. This code allows us to determine the percentage of time when the wind speed exceeds 100 km/h.

```
clear all
Sigma = 40;
N= 100000;
x=raylrnd(Sigma,1,N);
y0=100;
Ne=length(find(x>100));
P=Ne/N
```

2.3.10 The Rician distribution

Rician or Rice-distributed random variables hold significant relevance in different fields such as representing noise that occurs in magnetic resonance imaging (MRI), in economics and finance, and in wireless communications. In wireless communications, Rician distribution is instrumental in representing the amplitude of received signals in multipath fading channels, especially when a strong path (line of sight) is present. These random variables share an identical form with the Rayleigh distribution (2.57), with the distinction that they incorporate nonzero mean values for the normally distributed random variables X_1 and X_2. The probability density function for this distribution is defined as follows:

$$f_Y(y) = \frac{y}{\sigma^2} e^{\frac{-(y^2+a^2)}{2\sigma^2}} I_0\left(\frac{ya}{\sigma^2}\right), \quad y \geq 0, \tag{2.60}$$

where $I_0(x)$ is the modified Bessel function of the first kind with order zero.

Example 34. In the single-coil system, the noisy distribution in MRI images can be modeled as a Rician distribution. Rician noise makes image-based quantitative measurements difficult because it degrades the image quality considerably. What is the probability that the random noise is less than the shape parameter $K = \frac{a^2}{2\sigma^2}$?

Solution. The probability can be found as

$$\begin{aligned} P(y \leq K) &= \frac{1}{\sigma^2} \int_0^K y e^{-\frac{(y^2+a^2)}{2\sigma^2}} I_0\left(\frac{ya^2}{\sigma^2}\right) dy \\ &= \frac{e^{-K}}{\sigma^2} \int_0^K y e^{-\frac{y^2}{2\sigma^2}} I_0(2yK) dy. \end{aligned}$$

Unfortunately, the formula presented above lacks a straightforward solution using elementary functions. While numerical techniques can be employed to tackle this challenge. The function that handles this integration is called the Marcum Q-function with order one as $F(a_0) = 1 - Q_1(\frac{a}{\sigma}, \frac{a_0}{\sigma})$. An alternative approach is to employ Monte Carlo simulation for its evaluation as shown in the next code for $a = 1$ and $\sigma = 2$. With these given values the probability was 0.00179.

```
% Set parameters for the Rician distribution
alpha = 1;
sigma = 2;  % Ricean factor
N=100000; %Number of samples
theta = 2*pi*rand(1);
x=sigma*randn(2,N) + alpha*[cos(theta);sin(theta)];
% Generate Rician random numbers
rician = (x(1,:).^2+x(2,:).^2).^0.5 ; % Generate N
    samples
K=alpha^2/(2*sigma^2);
Ne=length(find(rician<=K));
P=Ne/N
```

2.3.11 Chi-square (χ^2) distribution

Numerous probability distributions have their origin in the normal distribution. Let's consider a random variable, denoted as Y, which is defined as follows:

$$Y = \sum_{k=1}^{n} X_k^2, \tag{2.61}$$

where X_i $\forall i = 1, 2, \ldots, n$ are independent and identically distributed (i.i.d) Gaussian random variables with zero mean and unit variance. The probability density function PDF of Y is called Chi-square with n degrees of freedom and it is given by:

$$f_Y(y) = \frac{1}{2^{n/2}\Gamma(\frac{n}{2})} y^{\frac{n}{2}-1} e^{-y/2}, \quad y \geq 0, \tag{2.62}$$

where $\Gamma(x)$ is known as the *Gamma function* and it is given by

$$\Gamma(x) = \int_0^{\infty} t^{x-1} e^{-t} dt, \quad x > 0. \tag{2.63}$$

We can generate a $M \times N$ matrix of random numbers with Chi-square distribution and with n degrees of freedom and unit variance on Octave and Scilab with the commands: *chi2rnd(n, M, N)* and *grand(M, N, 'chi', n)* respectively. *Chi-Square distribution* has several important applications in statistics and hypothesis testing. Furthermore, it has many other applications in reliability analysis, image processing, biostatistics, risk management, and social science. The CDF of the *Chi-Square distribution* can be computed as:

$$P(y \leq y_0) = \frac{1}{2^{n/2}\Gamma(\frac{n}{2})} \int_0^{y_0} y^{n/2-1} e^{-y/2} dy. \tag{2.64}$$

Again, there is no closed-form solution for this integration. However, it can be evaluated numerically or through simulation. Figure 2.24 shows the Chi-square probability density function for different degrees of freedom.

2.3.12 Other distributions

There are a multitude of distribution functions employed in various academic disciplines. Many of these distributions can be seen as alternative representations of relationships that derive from the normal distribution of random variables. For instance, the distribution of $y = \frac{X_1}{X_2}$, where X_1 and X_2 are independent normal distributions, is known

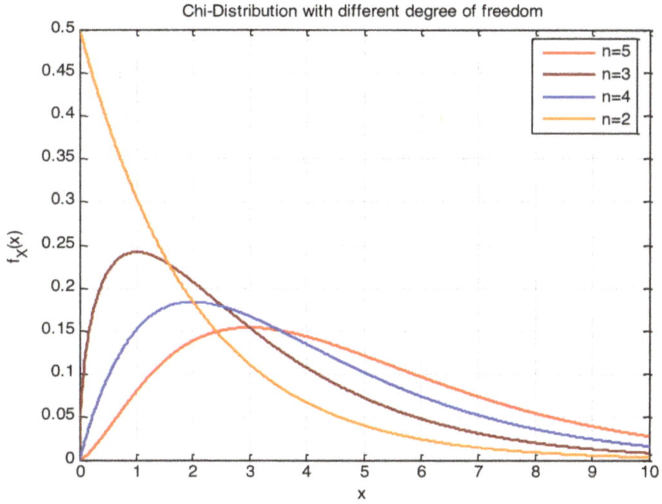

Figure 2.24: Chi-Square Probability density function.

as the *Cauchy distribution*. Another example is the log-normal distribution $y = 10^{\frac{X}{10}}$, which arises when X is a normally distributed random variable. The log-normal distribution finds extensive use in modeling the uncertainty in wave propagation associated with shadowing effects, such as those caused by buildings, trees, and other large obstacles in wireless communication channels.

Each probability density function (PDF) typically holds a unique significance in modeling the uncertainty of specific behaviors or observations. For instance, the 'Weibull distribution' is commonly applied to model expected lifetimes in reliability theory and data traffic flow. In the realm of statistics, essential PDFs such as the 'empirical distribution' and the 't-student distribution' play a crucial role.

Example 35. In this example, we illustrate the process of deriving the probability density function for a straightforward problem. Imagine a circular white paper with a radius of R. Now, let's say you randomly mark a point on this paper with a pencil without looking at it. The question at hand is: What is the probability distribution of the mark's location, characterized by its distance from the center (r) and the angle (θ) it makes with respect to the center?

Solution. We can derive the probability density function (PDF) simply by first computing the cumulative density function (CDF), i. e., $P(r \leq r_0)$. The derivative of the CDF function gives the PDF. The location of the point on the paper could be specified by two parameters, the distance from the center and the angle it makes with the x-axis. In terms of rotation around the circle, the point could be on any angle, i. e., the distribution of angles should be uniform from 0 to 2π. For the distance from the center, $P(r \leq r_0)$ could be computed as the ratio of the area of the circle $r \leq r_0$ to the total area of the circle as:

$$P(r \le r_0) = \frac{\pi r_0^2}{\pi R^2} = \frac{r_0^2}{R^2}. \tag{2.65}$$

Finally, the PDF of the distance is the derivative of the CDF as

$$\begin{aligned} f_r(x) &= \left.\frac{dP(r \le r_0)}{dr_0}\right|_{r_0=x} \\ &= \frac{2x}{R^2} \quad 0 \le x \le R. \end{aligned} \tag{2.66}$$

Example 36. In the previous example, considering a circle with a radius of 20 cm, what is the probability that the randomly marked point falls within the region defined by a radius between 17 cm and 19 cm, and an angle between $\frac{\pi}{4}$ and $\frac{\pi}{3}$?

Solution. The probability of being in the ring between 17 cm and 19 cm is

$$P(17 \le x \le 19) = \int_{17}^{19} \frac{2x}{R^2} dx = \left.\frac{x^2}{R^2}\right|_{17}^{19} = \frac{72}{400} = 0.18.$$

The probability of being in the angle between $\frac{\pi}{4}$ and $\frac{\pi}{3}$ is $\frac{1}{24}$ (because it is assumed uniform distributed between 0 to 2π). Finally, and since x and θ are independent, then

$$P(17 \le x \le 19, \pi/4 \le \theta \le \pi/3) = \frac{0.18}{24} = 0.0075.$$

Write a simulation code to emulate the above example and compare the results.

While many documented probability density functions are readily available in software packages such as Matlab, Octave, Scilab, R, Python, Julia, and others, enabling the swift generation of random numbers following these distributions, there are instances where a custom probability distribution is needed for a specific purpose, and it might not be readily available in existing software packages. In such cases, there exist straightforward methods for generating random numbers that conform to the desired probability distribution. We know that the cumulative distribution function is given as

$$F(x_0) = P(x \le x_0) = \int_{-\infty}^{x_0} f_X(x)dx = u, \tag{2.67}$$

where $0 \le u \le 1$. If we generate u as uniform random numbers from 0 to 1, we can subsequently obtain the corresponding random numbers x using the inverse transform method, i. e., $x = F^{-1}(u)$, where $F^{-1}(.)$ is the inverse CDF function.

Example 37. We have derived the probability density function (PDF) as

$$f_X(x) = \frac{1}{4\sqrt{x}} \quad 0 \le x \le 4.$$

This distribution is nonstandard and not readily available in common software packages. Nevertheless, there is a requirement to generate random numbers that conform to this specific distribution.

Solution. First we derive the cumulative distribution function as

$$F(x_0) = \int_0^{x_0} \frac{1}{4\sqrt{x}} dx = \frac{\sqrt{x_0}}{2}.$$

Hence, it is possible to generate the random samples that follow the given PDF distribution by computing the inverse function as $x = 4u^2$, where u are uniformly distributed numbers from 0 to 1.

The following Octave code generates 10000 numbers of this distribution. It also shows the histogram of the generated random numbers. It is compatible with the original PDF as shown in Figure 2.25.

```
% Generating random numbers
N=10000;
u=rand(N,1);
x=4*u.^2;
hist(x,200)
xlabel("x","fontsize",14)
ylabel("f_X(x)","fontsize",14)
grid
```

Unfortunately, finding the inverse of the cumulative distribution function (CDF) is not always a straightforward task. In fact, in many cases, it requires the use of iterative numerical methods o find a solution. However, there exists another highly efficient and versatile method for generating a sequence of random numbers corresponding to an arbitrary PDF. It is known as the *acceptance-rejection method*. The method is based on selecting another probability density function, say $v_X(x)$, that can generate its random numbers easily (for example, it is implemented already in the software package or the inverse of its CDF is straightforward). However, this $v_X(x)$ should be close to the PDF we want to generate the random sequence from, say $f_X(x)$. The selected PDF should achieve the bound $\frac{f_X(x)}{v_X(x)} \leq m \; \forall x$, where the scalar $m > 0$ and of course it should be close to 1. Therefore,

$$0 \leq \frac{f_X(x)}{m \times v_X(x)} \leq 1.$$

To accomplish this, we initiate the process by generating two random numbers, V from the distribution $v_X(x)$, and U from a uniform distribution spanning from 0 to 1. Next, we compare the ratio $\frac{f_X(V)}{m \times v_X(V)}$ with U. If this ratio exceeds U, we accept V as a random number generated in accordance with the distribution $f_X(x)$. However, if it falls short

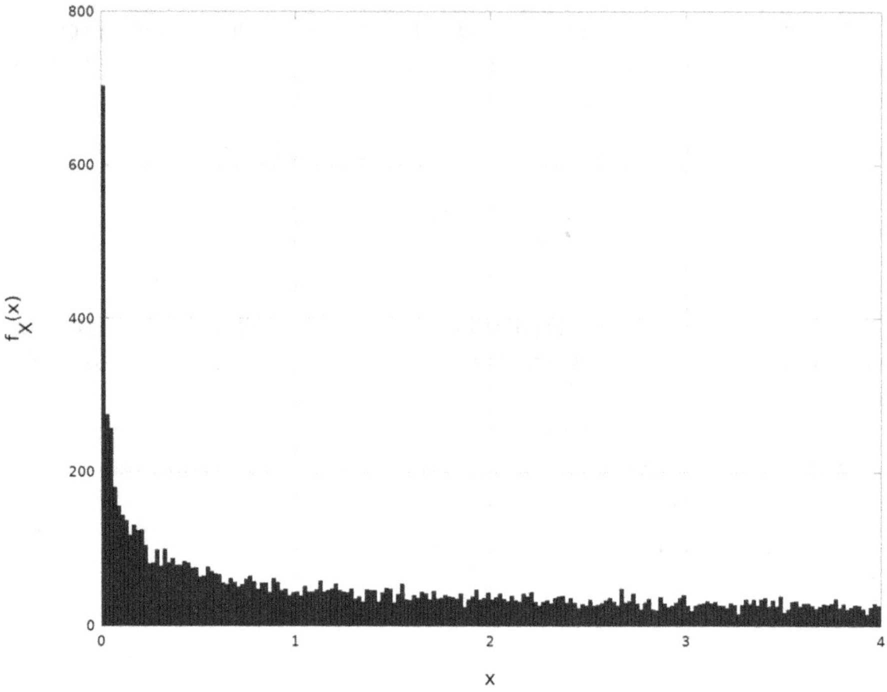

Figure 2.25: Histogram of random numbers.

of U, we reject V and initiate the process again. The validation of this method's efficacy is a straightforward process and can be found in numerous textbooks.

Example 38. Random number generation follows the probability density function given by

$$f_X(x) = \frac{2.48xe^{-x}}{x+1}.$$

Solution. For simplicity let's consider $v_X(x) = e^{-x}$, i. e., an exponential distribution with $\lambda = 1$ as shown in eq. (2.46). Therefore, $\frac{f_X(x)}{v_X(x)} = \frac{2.48x}{x+1} \leq 2.48$, so we can apply the acceptance-rejection method as shown in the next Octave code.

```
clear all
N=10000; % number of generated randoms
i=1;
while i<=N,
U=rand(1);
V=exprnd(1);
g=V/(V+1);
if g>U,
```

```
f(i)=V;% vector with the required distribution
i=i+1;
end
end
r=0:.01:10;
y=2.48*r.*exp(-r)./(r+1); % the PDF
[n,x]=hist(f,100,1); %Histogram
n  = n / diff(x)(1);
bar(x,n,0.5);
hold on;
plot(r, y, 'r','linewidth', 3);
hold off
```

You may compare the histogram of the generated random numbers with the required PDF function as shown in Figure 2.26. There are a few other methods that could be used to generate random numbers for nonstandard distributions.

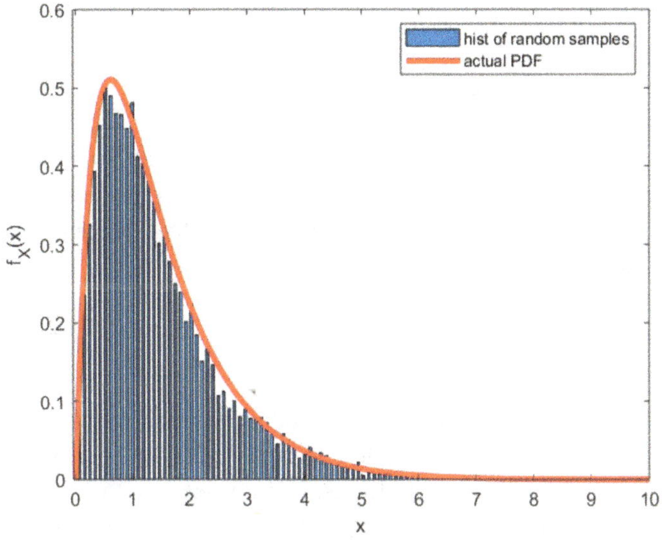

Figure 2.26: Histogram of generated random samples.

2.4 Expectations of random variables

To introduce the concepts of expectation theory, let's examine a scenario involving a class of 8 students, each receiving a final grade on a 5-point scale. These grades are represented by the vector $X = (0, 1, 1, 2, 3, 4, 4, 5)$. If we randomly select a student from this class, we can calculate the probabilities associated with different grade outcomes. The probability of a student receiving a grade of 0, denoted as $P(x = 0)$, is $\frac{1}{8}$, since there is only one occurrence of grade 0 in the vector. Similarly, the probability of a student receiving a grade of 1, denoted as $P(x = 1)$, is $\frac{1}{4}$ (or $\frac{2}{8}$), because grade 1 appears twice in the vector of all outcomes X. Similarly, the other probabilities are: $P(x = 2) = P(x = 3) = P(x = 5) = \frac{1}{8}$, and $P(x = 4) = \frac{1}{4}$. If you're tasked with calculating the average of the grades provided above, simply add them together and then divide the total by the number of students as

$$E[x] = \frac{0+1+1+2+3+4+4+5}{8} = 2.5,$$

where $E[.]$ is known as the expectation operator. The same equation above could be reformulated as:

$$E[x] = 0 \times \frac{1}{8} + 1 \times \frac{2}{8} + 2 \times \frac{1}{8} + 3 \times \frac{1}{8} + 4 \times \frac{2}{8} + 5 \times \frac{1}{8} = 2.5.$$

This implies that the average is a weighted sum of the grades, factoring in their associated probabilities or likelihood of occurrence. In mathematical terms, we can encapsulate the previous discussion as follows:

$$E[x] = \sum_{\forall i} x_i P(x = x_i). \tag{2.68}$$

Consequently, the average value of any discrete random variable can be determined through the aforementioned weighted sum. We commonly represent the average value of a random variable X as $\mu_X = E[x]$. Moreover, we can calculate the average of any function $g(x)$ applied to discrete random variables as follows:

$$E[g(x)] = \sum_{\forall i} g(x_i) P(x = x_i). \tag{2.69}$$

As an example, consider the case where $g(x) = x^2$. In this scenario, we are calculating the average of squared values. If the variable x represents electrical current or voltage, then x^2 can symbolize normalized power. Hence, $E[x^2]$ corresponds to the mean normalized power of the signal. Applying this concept to the previous example of course grades, we can compute the average of squared grade values as follows:

$$E[x^2] = 0 \times \frac{1}{8} + 1 \times \frac{2}{8} + 4 \times \frac{1}{8} + 9 \times \frac{1}{8} + 16 \times \frac{2}{8} + 25 \times \frac{1}{8} = 9.$$

In fact, as previously demonstrated, the average of the grades in this class is 2.5, which provides valuable insight into the performance of the students. It allows comparisons to be made between different classes to determine which one is performing better. However, the average alone does not convey information about the spread or variability of the grades. For instance, if all students received exactly 2.5 points, the average will remain unchanged. Likewise, if half of the students scored a perfect 5 while the other half received zeros, the average will still remain the same.

To quantify the dispersion or variability around the mean, one can compute the variance of the random variable. The variance of random variables is defined by

$$\sigma_X^2 = E[(x - \mu_X)^2] = E[x^2] - \mu_X^2. \tag{2.70}$$

In the previous grading example, the variance is calculated as $\sigma_X^2 = 9 - 2.5^2 = 2.75$. However, if all students have a score of 2.5 points, then $\sigma_X^2 = 0$, indicating no deviation from the mean. In general, a higher variance implies greater scattering around the mean, signifying a wider range of values relative to the average. The square root of the variance is referred to as the *standard deviation* and is denoted as σ_X. The subscript X is added to distinguish between different variables, as we may be dealing with multiple random variables later, such as σ_X^2, μ_Y, and so forth. For continuous random variables, the summation is replaced by integration, and the probability mass function is replaced by the probability density function, as follows:

$$E[g(x)] = \int_{-\infty}^{\infty} g(x) f_X(x) dx, \tag{2.71}$$

where $g(x)$ represents a function of the random variable x. When $g(x) = x$, the expectation gives the mean (average) value, i. e.,

$$\mu_X = E[x] = \int_{-\infty}^{\infty} x f_X(x) dx. \tag{2.72}$$

The variance is computed with $g(x) = (x - \mu_X)^2$, i. e.,

$$\sigma_X^2 = E[(x - \mu_X)^2] = \int_{-\infty}^{\infty} (x - \mu_X)^2 f_X(x) dx. \tag{2.73}$$

The expectation is a linear operator, i. e., $E[\alpha x + \beta y] = \alpha E[x] + \beta E[y]$, where x and y are random variables and α and β are any deterministic numbers.

Example 39. A discrete random signal generator generates four voltage levels as $x \in \{-2, 0, +3, +5\}$ volts, with probabilities of 0.2, 0.1, 0.5, and 0.2 respectively. Compute:
- The average voltage (DC), i. e., $E[x]$.
- The average total normalized power ($R = 1$), i. e., $E[x^2]$.
- The average power in the AC, i. e., variance.
- The average of the signal $y = x\cos(\pi x)$.

Answers.
- $E[x] = -2 \times 0.2 + 0 \times 0.1 + 3 \times 0.5 + 5 \times 0.2 = 2.1$ volts.

- $E[x^2] = 4 \times 0.2 + 0 \times 0.1 + 9 \times 0.5 + 25 \times 0.2 = 10.3$ watts.
- $E[(x - \mu_X)^2] = E[x^2] - \mu_X^2 = 10.3 - 2.1^2 = 5.89$ watts.
- $E[y] = E[x\cos(\pi x)] = -2\cos(-2\pi) \times 0.2 + 0\cos(0\pi) \times 0.1 + 3\cos(3\pi) \times 0.5 + 5\cos(5\pi) \times 0.2 = -2.9$.

Figure 2.27 shows a screenshot of an emulation code written on *Scilab* and the results obtained. It is clear that the simulation results match the theoretical ones very well.

```
// This Scilab code to emulate Example 2.31
clear
N=10000;
x=rand(N,1);
y=zeros(N,1);
y(find(x<0.2))=-2;
y(find(x>0.2 & x<0.3))=0;
y(find(x>0.3 & x<0.8))=3;
y(find(x>0.8))=5;
Average=mean(y)
Ave_Square=mean(y.^2)
Variance=variance(y)
MeS=mean(y.*cos(y*%pi)) // You should add % to pi
```

```
Scilab 6.0.0 Console

--> Average
 Average  =

   2.0999

--> Ave_Square
 Ave_Square  =

   10.3277

--> Variance
 Variance  =

   5.9187119

--> MeS
 MeS  =

  -2.9059
```

Figure 2.27: Scilab emulation code of Example 39.

Example 40. Calculate the mean and variance for a set of random integers following a Poisson distribution.

Solution. The random variable x with Poisson probability mass function is given as $P(x = k) = \frac{(\lambda T)^k}{k!} e^{-\lambda T}$. Hence, we can compute the mean as:

$$E[x] = \sum_{k=0}^{\infty} k\, P(x = k) \tag{2.74}$$

$$= e^{-\lambda T} \sum_{k=0}^{\infty} k \frac{(\lambda T)^k}{k!} \tag{2.75}$$

$$\because e^{\lambda T} = \sum_{k=0}^{\infty} \frac{(\lambda T)^k}{k!} \tag{2.76}$$

$$\therefore \frac{d}{d\lambda T} e^{\lambda T} = \sum_{k=0}^{\infty} k \frac{(\lambda T)^{k-1}}{k!} = e^{\lambda T} \tag{2.77}$$

$$\rightarrow \sum_{k=0}^{\infty} k \frac{(\lambda T)^k}{k!} = \lambda T e^{\lambda T} \tag{2.78}$$

$$\rightarrow E[x] = \lambda T e^{\lambda T} e^{-\lambda T} = \lambda T. \tag{2.79}$$

In a similar way, we can determine $E[x^2]$, as

$$E[x^2] = \sum_{k=0}^{\infty} k^2\, P(x = k) \tag{2.80}$$

$$= e^{-\lambda T} \sum_{k=0}^{\infty} k^2 \frac{(\lambda T)^k}{k!} \tag{2.81}$$

$$\because \frac{d^2}{d(\lambda T)^2} e^{\lambda T} = \sum_{k=0}^{\infty} k(k-1) \frac{(\lambda T)^{k-2}}{k!} = e^{\lambda T} \tag{2.82}$$

$$\rightarrow \sum_{k=0}^{\infty} k^2 \frac{(\lambda T)^k}{k!} = \lambda T(\lambda T + 1) e^{\lambda T} \tag{2.83}$$

$$\rightarrow E[x^2] = (\lambda T)^2 + \lambda T. \tag{2.84}$$

The variance of the Poisson distribution is $\sigma^2 = E[x^2] - (E[x])^2 = (\lambda T)^2 + \lambda T - (\lambda T)^2 = \lambda T$. The variance has equal to the mean.

Example 41. Determine the relationship between the parameter of a Rayleigh distribution random variable and its mean, variance, and median.

Solution. Random variables that follow a Rayleigh distribution are of a continuous nature, requiring integration rather than summation for computation. Therefore, the mean is calculated as follows:

$$E[x] = \int_0^{\infty} x \frac{x}{\sigma^2} e^{-x^2/(2\sigma^2)} dx. \tag{2.85}$$

Using integration by parts we get

$$E[x] = \left[x \int \frac{x}{\sigma^2} e^{-x^2/(2\sigma^2)} dx - \int \int \frac{x}{\sigma^2} e^{-x^2/(2\sigma^2)} \right]_0^{\infty} \tag{2.86}$$

$$\because \int \frac{x}{\sigma^2} e^{-x^2/(2\sigma^2)} dx = -e^{-x^2/(2\sigma^2)} \tag{2.87}$$

$$\therefore E[x] = \left[-xe^{-x^2/(2\sigma^2)} + \int e^{-x^2/(2\sigma^2)} dx\right]_0^\infty \tag{2.88}$$

$$= \int_0^\infty e^{-x^2/(2\sigma^2)} dx = \sqrt{\frac{\pi}{2}}\sigma. \tag{2.89}$$

In the last equation we used the well-known formula of normal distribution, that is:

$$\frac{1}{\sqrt{2\pi}\sigma} \int_0^\infty e^{-x^2/(2\sigma^2)} dx = \frac{1}{2}.$$

To calculate the variance, we can start by computing $E[x^2]$ using the same integration by parts method as follows:

$$E[x^2] = \left[-x^2 e^{-x^2/(2\sigma^2)} + 2\int xe^{-x^2/(2\sigma^2)} dx\right]_0^\infty$$

$$= 2\int_0^\infty xe^{-x^2/(2\sigma^2)} dx = 2\sigma^2.$$

Hence, the variance of Rayleigh distributed random variables is:

$$\sigma_X^2 = 2\sigma^2 - \frac{\pi}{2}\sigma^2 = \left(2 - \frac{\pi}{2}\right)\sigma^2. \tag{2.90}$$

Remember that σ^2 here is not the variance of the Rayleigh random variable, but it is the variance of the normal distributed random variables as explained before in Subsection 2.3.9. The median of a random variable, denoted as x_{med}, represents the point at which the cumulative distribution function reaches 0.5. This signifies that there is an equal probability of the variable being greater than or less than the median, each with a 50 % chance. The concept of the median is important in statistics and can be computed as follows: $P(x \leq x_{\text{med}}) = 0.5$.

In the case of Rayleigh distribution random variables, we may compute the general form of the median as:

$$F_X(x_{\text{med}}) = P(x \leq x_{\text{med}}) = \int_0^{x_{\text{med}}} \frac{x}{\sigma^2} e^{-x^2/(2\sigma^2)} dx = \frac{1}{2}$$

$$= 1 - e^{-x_{\text{med}}^2/(2\sigma^2)}$$

$$\therefore x_{\text{med}} = \sigma\sqrt{2\ln(2)}. \tag{2.91}$$

In the previous example, we demonstrated how to easily compute certain statistical parameters of the Rayleigh distribution using basic high-school mathematics. However, it's important to acknowledge that not all statistical problems are as straightforward. For instance, when you're tasked with finding the expected value of $E_X[e^{1/(1+x)}|\cos(2\pi x)|]$, where x follows a Rayleigh distribution, the mathematical derivations can become highly challenging. In fact, in many instances, closed-form solutions simply don't exist, meaning that the solution cannot be expressed using elementary functions. As a result, computer simulations are becoming indispensable tools in numerous scientific disciplines.

Example 42. Let's validate our previous derivations with Octave code. To do this, generate 10,000 random numbers following a Rayleigh distribution with a parameter of $\sigma = 3$. Calculate the mean, variance, and median of this dataset using the same simulation approach. Additionally, use the simulator to find:

$$E_X\left[e^{1/(1+x)}\left|\cos(2\pi x)\right|\right].$$

Solution. For $\sigma = 3$, from the derived parameters in the previous example, $\mu_X = 3.7599$, $E_X[x^2] = 18$, Var $= 3.8628$, and the median $x_{\text{med}} = 3.5322$.

The simulation code could easily be written as shown next:

```
clear all
N=10000;
x=3*randn(N,2);%generating two Normal distributed
    vectors
y=sqrt(x(:,1).^2+x(:,2).^2); %generating Rayleigh
Average=mean(y)
AverSq=mean(y.^2)
Variance=var(y)
Med=median(y)
EqMean=mean(exp(1./(1+y)).*abs(cos(2*pi*y)))
```

After executing the code, the following results were obtained:
- Average = 3.7767;
- Average of Squares = 18.112;
- Variance = 3.8681;
- Median = 3.5561;
- EqMean = 0.82974.

It's evident that the simulation results closely match the theoretical values. Notably, the expectation of the equation mentioned earlier was not analytically derived due to its perceived complexity. However, it proved to be relatively straightforward to compute through simulation. Alternatively, numerical integration could be employed to determine it. Nevertheless, it's important to emphasize that numerical integration is fundamentally different from Monte Carlo simulation.

In many cases, we can have confidence in simulation results, provided that we utilize a sufficient number of trials and assume the existence of a solution. The next example will illustrate the importance of this last condition.

Example 43. A random variable Y is termed as having a standard *Cauchy* distribution when its probability density function is defined as:

$$f_Y(y) = \frac{1}{\pi(y^2+1)}. \tag{2.92}$$

The standard *Cauchy* distribution can be generated using the formula: $y = \frac{x_1}{x_2}$, where x_1 and x_2 represent two independent random variables following a normal distribution with zero mean and unit variance. Let's calculate the mean and variance of the standard *Cauchy* distribution.

Solution. We can determine the mean as

$$E_Y[y] = \int_{-\infty}^{\infty} \frac{y}{\pi(y^2+1)} dy = \frac{1}{2\pi} \ln(y^2+1)\Big|_{-\infty}^{\infty} = NaN.$$

The results demonstrate that the standard Cauchy distribution lacks a specific mean. As a consequence, it also does not possess a well-defined variance. However, a noteworthy point to consider is that if you employ simulation to generate random variables following the Cauchy distribution and calculate the mean, the computer will provide you with numerical results. Nevertheless, it's essential to be careful. Upon executing the same code repeatedly, you will observe varying outcomes each time. These results can vary significantly, with the mean taking on both large and small, positive and negative values. This variability serves as a clear indicator that the mean is not converging to any particular fixed value. To establish the convergence of the mean, it is advisable to repeat the execution of your averaging simulation code multiple times. As an example, Scilab was used to generate 10,000 random numbers following the Cauchy distribution. The average value was computed three times, resulting in the following outcomes: −5.1591, 0.72205, and −13.776. Clearly, these results do not converge to a specific constant value.

Example 44. Determine the mean, variance, mode, and median of the random distribution as illustrated in the Example 35.

Solution. Since the probability density function of the random distance from the point to the center of the circle is given by: $f_r(x) = \frac{2r}{R^2}$ and $0 \le x \le R$. The mean is calculated as:

$$E[x] = \frac{2}{R^2} \int_0^R x^2 dx = \frac{2}{R^2} \left[\frac{x^3}{3}\right]_0^R = \frac{2}{3} R. \tag{2.93}$$

The mean square value is:

$$E[x^2] = \frac{2}{R^2}\int_0^R x^3 dx = \frac{2}{R^2}\left[\frac{x^4}{4}\right]_0^R = \frac{1}{2}R^2. \tag{2.94}$$

The mode is defined as the value where the probability density function is maximum, i. e., $x_{\text{mode}} = \text{argmax}[f_r(x)] \rightarrow x_{\text{mode}} = R$.

Finally, the median is computed as

$$P(x \leq x_{\text{med}}) = \frac{2}{R^2}\int_0^{x_{\text{med}}} x dx = \frac{x_{\text{med}}^2}{R^2} = \frac{1}{2} \rightarrow x_{\text{med}} = \frac{R}{\sqrt{2}}. \tag{2.95}$$

In the preceding examples, we've explored methods for calculating expectations of random variables, in both discrete and continuous scenarios. This underscores the significance of having, or at the very least estimating, the probability density (or mass) function that characterizes the uncertainty associated with the problem at hand.

When dealing with independent random variables, denoted as x and y, we have the ability to disentangle the expectation operation as:

$$E_{XY}[xy] = E_X[x]E_Y[y] \tag{2.96}$$

$$E_{XY}[g_1(x)g_2(y)] = E_X[g_1(x)]E_Y[g_2(y)] \tag{2.97}$$

$$E_{XY}[g(x,y)] = E_Y[E_X[g(x,y)]]. \tag{2.98}$$

Example 45. Assume three independent random variables x, y, and z, where x has a normal distribution with $\mu_X = 2$ and $\sigma^2 = 4$; y has Rayleigh distribution with parameter $\sigma = 1$; and z is uniformly distributed from 0 to 1. Find the mean and variance of the random variable $g = xy + x\sin(\pi z)$.

Solution. With a direct application of the expectation rules

$$\begin{aligned} E[g] &= E_{XYZ}[xy + x\sin(\pi z)] \\ &= E_{XY}[xy] + E_{XZ}[x\sin(\pi z)] \\ &= E_X[x]E_Y[y] + E_X[x]E_Z[\sin(\pi z)]. \end{aligned}$$

The expected values of x are given by: $E_X[x] = \mu_X = 2$, $E_X[x^2] = \sigma_X^2 + \mu_X^2 = 8$, moreover,

$$E_Z[\sin(\pi z)] = \int_0^1 \sin(\pi z)dz = -\frac{1}{\pi}\cos(\pi z)\Big|_0^1 = \frac{2}{\pi}.$$

Furthermore, $E[y] = \sqrt{\frac{\pi}{2}}$, $E[y^2] = 2$. Therefore,

$$\mu_g = E[g] = 2 \times \sqrt{\frac{\pi}{2}} + 2 \times \frac{2}{\pi} = 3.7799.$$

The variance of the random variable g is computed as $\sigma_g^2 = E[g^2] - \mu_g^2$

$$\begin{aligned} E[g^2] &= E_{XYZ}[x^2y^2 + 2x^2y\sin(\pi z) + x^2\sin^2(\pi z)] \\ &= E_X[x^2]E_Y[y^2] + 2E_X[x^2]E_Y[y]E_Z[\sin(\pi z)] \\ &\quad + E_X[x^2]E_Z[\sin^2(\pi z)]. \end{aligned}$$

Since, $\sin(\pi z)^2 = \frac{1}{2}(1 - \cos(2\pi z))$, then,

$$\begin{aligned} E[g^2] &= 8 \times 2 + 2 \times 8 \times \sqrt{\frac{\pi}{2}} \times \frac{2}{\pi} + 8 \times \frac{1}{2} = 32.766 \\ \sigma_g^2 &= E[g^2] - \mu_g^2 = 32.766 - 3.7799^2 = 18.478. \end{aligned}$$

The following code shows how to simulate this example in Octave. The results were very close to the theoretical ones.

```
clear all
N=100000;
x=2*randn(N,1)+2; %generating Normal distributed
w=randn(N,2);
y=sqrt(w(:,1).^2+w(:,2).^2); %generating Rayleigh
z=rand(N,1); %generating uniform distribution 0-1
g=x.*y+x.*sin(pi*z);
Average=mean(g)
AverSq=mean(g.^2)
Variance=var(g)
```

It is crucial to bear in mind that the separation of multiplied random variables is only feasible when they exhibit independence.

Example 46. For a uniform random variable x between 1 and 4, what is the average value of x^{-1}?

Solution. The expected value can be easily found as:

$$E\left[\frac{1}{x}\right] = \frac{1}{3}\int_1^4 \frac{1}{x}dx = \frac{1}{3}\ln(x)\Big|_1^4 = \frac{\ln(4)}{3}. \tag{2.99}$$

It is very important to note that $E[\frac{1}{x}] \neq \frac{1}{E[x]}$.

2.4.1 Conditional expectations

In estimation theory, the concept of conditional expectation is frequently encountered.

$$E_{X|Y}[X|Y]. \tag{2.100}$$

If X and Y are independent, then the value of Y will not affect the expectation of X, i. e., $E_{X|Y}[X|Y] = E_X[X]$. However, if they are dependent and Y is a random variable; therefore, $E_{X|Y}[X|Y]$ is generally a random variable in Y. However, we could determine the general expectation of X by taking the expectation of the expectation as:

$$E_X[X] = E_Y[E_{X|Y}[X|Y]]. \tag{2.101}$$

Actually, this is the method to find the expectations in the case of dependent random variables. In the case that $X \in \Omega$, where Ω is the sample space which is partitioned by n disjoint events $A_1, A_2, \dots A_n$ so that $\Omega = \cup_{i=1}^{n} A_i$, hence,

$$E[X] = \sum_{i=1}^{n} E[X|A_i]P(A_i). \tag{2.102}$$

Example 47. Assume X is a random variable with an exponential distribution whose parameter λ is a random variable such as $\lambda \in \{1, 5\}$ with equal probability. Find $E[X]$. Validate your result by simulation.

Answer. Using eq. (2.102) $E[X] = E[X|\lambda = 1] \times \frac{1}{2} + E[X|\lambda = 5] \times \frac{1}{2} = \frac{6}{10}$.

The expectation of dependent random variables is discussed later.

2.5 Multivariate random variables

In many cases, we encounter multiple random variables that may exhibit interdependencies or correlations. These random variables can follow either a purely discrete, a purely continuous, or a mixed distribution. When considering a pair of joint random variables, denoted as x and y, the joint distribution is typically represented as follows: $f_{XY}(x,y)$. Generally, the joint distribution of a vector of n random variables $\mathbf{X} = [x_1, x_2, \dots, x_n]^T$ is represented as $f_{\mathbf{X}}(x_1, x_2, \dots, x_n)$. Next, we will explore the properties of joint distributions involving two random variables, denoted as x and y. It's worth noting that all of the results we discuss can be extended to cases involving n random variables.

In the context of joint distributions for two random variables, it is possible to derive the distribution of one random variable by averaging over the other, as demonstrated by the following expression:

$$f_Y(y) = \int_{-\infty}^{\infty} f_{XY}(x,y)dx. \tag{2.103}$$

As we know, for a single random variable, the total area under its probability density function must equal 1. This fundamental principle extends to joint distribution functions involving any number of variables. For instance, in the case of two joint random variables, the total area under their two-dimensional probability density function must be equal to 1, as expressed by:

$$\int_{-\infty}^{\infty}\int_{-\infty}^{\infty} f_{XY}(x,y)dxdy = 1. \tag{2.104}$$

The cumulative distribution function for the joint random variables is defined in the same way as a single RV PDF. For two RVs, the CDF is defined as:

$$F_{XY}(x_0,y_0) = P(x \leq x_0, y \leq y_0) = \int_{-\infty}^{y_0}\int_{-\infty}^{x_0} f_{XY}(x,y)dxdy. \tag{2.105}$$

Hence, we can find the joint PDF from the joint CDF as

$$f_{XY}(x,y) = \frac{\partial^2 F_{XY}(x,y)}{\partial x \partial y}. \tag{2.106}$$

The conditional probability density function is very similar for the case if discrete random variables given by eq. (2.11). For any two continuous joint random variables, it is always true that:

$$f_{X|Y}(x|y) = \frac{f_{XY}(x,y)}{f_Y(y)} \tag{2.107}$$

$$f_{Y|X}(y|x) = \frac{f_{XY}(x,y)}{f_X(x)}. \tag{2.108}$$

As discussed before, in the case of independent random variables, we can say that $f_{X|Y}(x|y) = f_X(x)$, because y has no impact on the distribution of x. Therefore, for independent x and y, it is always valid to say

$$f_{XY}(x,y) = f_X(x)f_Y(y). \tag{2.109}$$

The expectation formula for joint multiple random variables closely resembles that of a single random variable. It involves averaging a given function $g(x,y)$, and it can be expressed as:

$$E_{XY}[g(x,y)] = \int_{-\infty}^{\infty}\int_{-\infty}^{\infty} g(x,y)f_{XY}(x,y)dxdy. \tag{2.110}$$

In general, for any two random variables x and y, with joint distribution $f_{XY}(x,y)$, we can state that:

- The random variables x and y are orthogonal if $E[xy] = 0$.

- They are independent if $E[xy] = E[x]E[y]$.
- They are uncorrelated if their covariance $E[(x - \mu_X)(y - \mu_Y)] = 0$.

Example 48. A joint probability density function given by $f_{XY}(x,y) = Ae^{-xy}$, where both $x, y \geq 1$. Find:

- The value of A.
- The probability that $x \leq 2$.
- The probability $P_{XY}(x \geq 5, 2 \leq y \leq 3)$.
- The probability density functions $f_X(x)$ and $f_Y(y)$.
- The conditional probability of $f_{X|Y}(x|y)$.
- Are x and y independent?
- Find the average of x and the average of y.
- Find the average of xy.
- Find the joint CDF function.
- Find the covariance, $\text{cov}(x, y)$.

Solution.

- According to the fundamental principle of probability density functions (PDFs), it is a well-established concept that the total integrated area under the probability distribution for two-dimensional random variables must equal 1. In other words,

$$A \int_1^\infty \int_1^\infty e^{-xy} dxdy = 1.$$

We can approach the integration process in two stages: initially, we integrate with respect to the variable x while holding y constant, and subsequently, we perform the integration with respect to the variable y. The first integration with respect to x leads to

$$-A \int_1^\infty \left[\frac{e^{-xy}}{y} dy \right]_{x=1}^\infty .$$

Let's complete the integration before we take the limits of x. Let $u = xy \to du = xdy$. Substituting into the previous equation we get

$$-A \int_x^\infty \left[\frac{e^{-u}}{u} du \right]_{x=1}^\infty .$$

The integration formula of $\int_x^\infty \frac{e^{-u}}{u} du$ cannot be implemented in the form of elementary functions. This integration is very well known as *Exponential Integral*. Its values are tabulated and also implemented in most of numerical packages, for example, in

Octave it is implemented as *expint(.)* and usually denoted as $E_1(x)$. Finally, we substitute the integration limits as

$$-A(E_1(\infty) - E_1(1)) = -A(0 - 0.2913) = 1 \rightarrow A = 4.558.$$

- The second requirement is to compute the probability:

$$P(x \leq 2) = 4.558 \int_1^{\infty} \left[\int_1^2 e^{-xy} dx \right] dy$$
$$= -4.558 \int_1^{\infty} \left[\frac{e^{-xy}}{y} \right]_1^2 dy$$
$$-4.558(E_1(2) - E_1(1))$$
$$= -4.558(0.0489 - 0.2193) = 0.777.$$

- The joint probability is computed as:

$$P_{XY}(x \geq 5, 2 \leq y \leq 3) = 4.558 \int_2^3 \left[\int_5^{\infty} e^{-xy} dx \right] dy$$
$$= 4.558 \int_2^3 \frac{e^{-5y}}{y} dy$$
$$= 4.558(E_1(10) - E_1(15)) = 1.886 \times 10^{-5}.$$

- To find the separate distributions of x and y,

$$f_X(x) = 4.558 \int_1^{\infty} e^{-xy} dy = 4.558 \frac{e^{-x}}{x}$$

and

$$f_Y(y) = 4.558 \int_1^{\infty} e^{-xy} dx = 4.558 \frac{e^{-y}}{y}.$$

- The conditional probability density function is computed as

$$f_{X|Y}(x|y) = \frac{f_{XY}(x,y)}{f_Y(y)} = ye^{-y(x-1)}.$$

- It is clear that x and y are dependent. One simple checking test is to see if $f_{XY}(x,y)$ equals $f_X(x)f_Y(y)$ or not. It is clear that

$$f_X(x)f_Y(y) = 4.558^2\frac{e^{-(x+y)}}{xy} \neq 4.558e^{-xy}.$$

– The average of x is computed as

$$E_X[x] = \int_1^{\infty} xf_X(x)dx$$
$$= 4.558\int_1^{\infty} e^{-x}dx = 4.558e^{-1} = 1.6768.$$

The average of y will be identical because both have the same PDFs. Hence, $E_Y[y] = 1.6768$.

– The average of xy is computed as $E_{XY}[xy]$. We may simplify the computation by first finding the average of x at a given value of y and then averaging the result over y. It can be done as

$$E_{XY}[xy] = E_Y[E_X[x|y]] = \int_1^{\infty} y\left[\int_1^{\infty} xe^{-xy}dx\right]dy$$
$$= 4.558\int_1^{\infty} y\left[-\frac{e^{-xy}}{y}\left(x+\frac{1}{y}\right)\Big|_1^{\infty}\right]dy$$
$$= 4.558\int_1^{\infty}\left[e^{-y}\left(1+\frac{1}{y}\right)\right]dy$$
$$= 4.558(e^{-1} + E_1(1)) = 2.6767.$$

– The joint CDF function as stated in equation (2.105) as:

$$F_{XY}(x_0, y_0) = 4.558\int_1^{y_0}\int_1^{x_0} e^{-xy}dxdy$$
$$= 4.558\int_1^{y_0}\frac{e^{-xy}}{y}\Big|_1^{x_0}dy$$
$$= 4.558(E_1(1) - E_1(y_0) - E_1(x_0) + E_1(x_0y_0)).$$

– The covariance

$$E[(x-\mu_X)(y-\mu_Y)] = E[xy] - \mu_X\mu_Y$$
$$= 2.6767 - 1.6768 \times 1.6768 = -0.1346.$$

Example 49. Let's consider a random variable x that follows a normal distribution with a variance of 4. However, the mean of this distribution undergoes random fluctuations between two possible values: 0 with a probability of 0.3 and 3. Now, let's determine

- The mean and variance of x.
- The Probability $P(x \leq 1)$.

Solution.

- The mean of the random variable will be

$$E_{\mu}[E_X[x]] = E_{\mu}[\mu_X] = 0.3 \times 0 + 0.7 \times 3 = 2.1.$$

Calculating the variance in this scenario can be somewhat tricky. One might assume that since the variance of a normal distribution is typically independent of the mean (i. e., $\sigma^2 = 4$ in this case), it should remain constant. However, this assumption doesn't hold when the mean is subject to random fluctuations, as the mean square value also varies. To compute the variance under these conditions, we can use the following formula:

$$\begin{aligned}\sigma_X^2 &= E[x^2] - (E[x])^2 \\ E[x^2] &= (4+0) \times 0.3 + (4+3^2) \times 0.7 = 10.3 \\ \therefore \sigma_X^2 &= 10.3 - 2.1^2 = 5.89.\end{aligned}$$

- The probability value is computed as

$$P(x \leq 1|\mu) = 1 - Q\left(\frac{1-\mu_X}{2}\right).$$

This probability term is inherently stochastic, exhibiting two potential values contingent on the mean. Consequently, we can determine the mean probability as follows:

$$P(x \leq 1) = \left[1 - Q\left(\frac{1-0}{2}\right)\right] \times 0.3 + \left[1 - Q\left(\frac{1-3}{2}\right)\right] \times 0.7 = 0.3185.$$

- Next we show how to validate the theoretical results using simulation in Octave

```
clear all
N=10000;
mu=3*(sign(rand(N,1)-0.3)+1)/2;
x=2*randn(N,1)+mu;
Mean=mean(x)
Var=var(x)
P=length(find(x<1))/N
```

The simulation results are: Mean = 2.1211, Var = 5.9390, and $P = 0.31390$.

Example 50. For a random variable $y = \frac{\alpha}{x}$, where x is a uniform distribution from 1 to 2, and α is random variable with two values as $\alpha \in \{-4, 6\}$ with probabilities 0.3 and 0.7, respectively. Find $E[y]$.

Solution. We can find the expectation as

$$\begin{aligned} E[y] &= E[y|\alpha = -4]P(\alpha = -4) + E[y|\alpha = 6]P(\alpha = 6) \\ &= -4E\left[\frac{1}{x}\right] \times 0.3 + 6E\left[\frac{1}{x}\right] \times 0.7 \\ &= 3\ln(2). \end{aligned}$$

Below is a simple code to validate this result.

```
clear all
N=10000;
x=1+rand(N,1);
u=rand(N,1);
z=6*ones(N,1);
z(find(u<=0.3))=-4;
y=z./x;
mu=mean(y);
```

Example 51. Imagine a wireless communication system where the channel can be represented by two distinct states: Excellent and Bad. The probability of encountering the Excellent state is 0.2. A packet is ready to be transmitted from the transmitter. If the channel is in the Excellent state, the packet will be successfully delivered within 0.01 seconds. However, if the channel is in the Bad state, the transmission process will still take 0.01 seconds, but the receiver will fail to decode the packet correctly. This error triggers a negative acknowledgement message from the receiver to the transmitter, requesting that the packet be retransmitted. The retransmission process adds an additional 0.02 seconds to the total transmission time. Determine the average time required for the packet to be correctly decoded.

Solution. Assume that T is a random variable representing the time required to deliver the packet correctly. The average time can be expressed as:

$$\begin{aligned} E[T] &= E[T|\text{Excellent}]P(\text{Excellent}) + E[T|\text{Bad}]P(\text{Bad}) \\ E[T|\text{Excellent}] &= 0.01 \\ E[T|\text{Bad}] &= 0.01 + 0.02 + E[T] \\ \therefore E[T] &= 0.01 \times 0.2 + (0.03 + E[T]) \times 0.8 \\ \Rightarrow E[T] &= 0.13\,\text{seconds}. \end{aligned}$$

Example 52. Replicate the preceding example by simulating the average duration of successful data transfer in a wireless network. Compare the outcomes of the simulation with the theoretical findings. Construct the cumulative distribution function that represents the successful transmission time. From the CDF plot, find the minimum successful transmission time that achieves at least 80 %.

Solution. The code is shown below in the Octave programming tool. The average time in the simulation was around 0.131, which is very close to the theoretical results. From the CDF plot we can see that 0.2 seconds is the minimum time that can achieved at least 80 % of the transmission.

```
% Required successful transmission time
Pe=0.2; %probability of excellent channel
Pb=1-Pe; % probability of bad channel
Tb=0.01; % Packet duration
Tr=0.02; %Re-transmission time
T=0; %initial time
N=10000; % Number of experiments
n=1;
% Tt = total time before successful transmission
while (n<=N),
x=rand(1);
if x <= Pe,
Tt(n) = T+Tb;
n=n+1;
T=0;
else
T=T+Tb+Tr;
end
end
meanT=mean(Tt)
set(cdfplot(Tt), 'Linewidth',2, 'Color', 'r');
title('Plot of CDF', 'fontsize',14)
xlabel('Successful Transmission Time,
    T_t','fontsize',14);
ylabel('CDF  F(T_t)','fontsize',14);
```

The output result is shown in Figure 2.28.

One of the most intriguing joint probability distributions is the *Multivariate Normal Distribution*, which characterizes the collective distribution of n random variables, $\mathbf{X} = [x_1, x_2, \ldots, x_n]^T$, where $[.]^T$ refers to the *Transpose* operation. The multivariate (or multidimensional) normal distribution is represented as:

$$f_{\mathbf{X}}(x_1, x_2, \ldots, x_n) = \frac{1}{(2\pi)^{n/2} \det(\mathbf{R}_{XX})^{0.5}} e^{-\frac{1}{2}(\mathbf{X}-\mu_{\mathbf{X}})^T \mathbf{R}_{XX}^{-1} (\mathbf{X}-\mu_{\mathbf{X}})}. \tag{2.111}$$

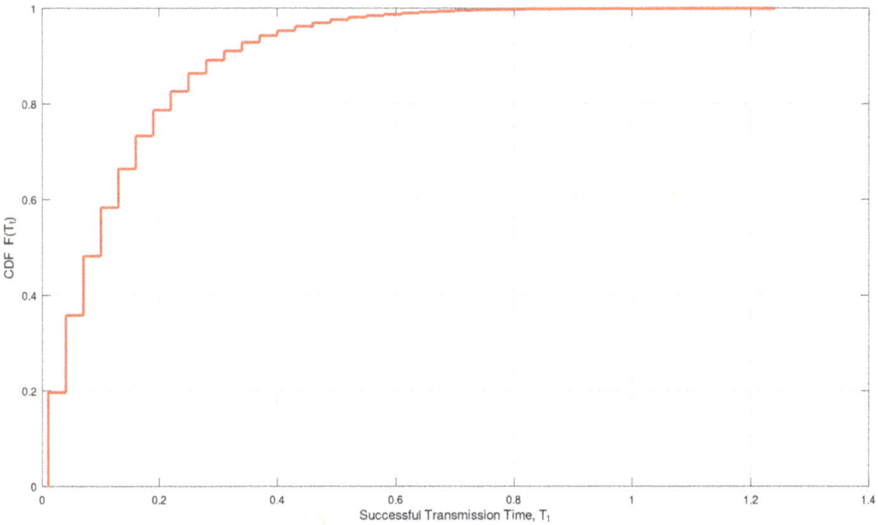

Figure 2.28: Cumulative distribution function of the successful transmission time.

Where $\mu_{\mathbf{X}} = [\mu_{x1}, \mu_{x2}, \ldots, \mu_{xn}]^T$, $\det(\mathbf{A})$ is the determinant of the matrix $\mathbf{A}$, and $\mathbf{R}_{XX}$ is the *covariance matrix* and it is given by:

$$\mathbf{R}_{XX} = E_{\mathbf{X}}[(\mathbf{X} - \mu_{\mathbf{X}})^T(\mathbf{X} - \mu_{\mathbf{X}})] \tag{2.112}$$
$$= \begin{bmatrix} \sigma_{x_1}^2 & \mathrm{cov}(x_1, x_2) & \ldots & \mathrm{cov}(x_1, x_n) \\ \mathrm{cov}(x_2, x_1) & \sigma_{x_2}^2 & \ldots & \mathrm{cov}(x_2, x_n) \\ \vdots & \ldots & \ddots & \vdots \\ \mathrm{cov}(x_n, x_1) & \mathrm{cov}(x_n, x_2) & \ldots & \sigma_{x_n}^2 \end{bmatrix}.$$

Where $\mathrm{cov}(x_i, x_j) = E[(x_i - \mu_{x_i})(x_j - \mu_{x_j})]$, and for the cases where $i = j$, the covariance is identical as the variance $\sigma_{x_i}^2$.

In case of uncorrelated random variables x_i and x_j for all $i \neq j$, the covariance $\mathrm{cov}(x_i, x_j) = 0$. Therefore, the covariance matrix $\mathbf{R}_{XX}$ is diagonal matrix as:

$$\mathbf{R}_{XX} = \begin{bmatrix} \sigma_{x_1}^2 & 0 & \ldots & 0 \\ 0 & \sigma_{x_2}^2 & \ldots & 0 \\ \vdots & \ldots & \ddots & \vdots \\ 0 & 0 & \ldots & \sigma_{x_n}^2 \end{bmatrix}. \tag{2.113}$$

In this case of uncorrelated random variables, the multivariate normal distribution (2.111) becomes

$$f_{\mathbf{X}}(x_1, x_2, \ldots, x_n) = f_{X_1}(x_1) f_{X_2}(x_2) \ldots f_{X_n}(x_n), \tag{2.114}$$

where

$$f_{X_i}(x_i) = \frac{1}{\sqrt{2\pi\sigma_{X_1}^2}} e^{-(x_i-\mu_{x_i})^2/(2\sigma_{x_i}^2)}.$$

In Octave, we can generate multivariate normal distribution using the command *mvnrnd*, and in Scilab using the command *Y = grand(n, 'mn', Mean, Cov)*.

Example 53. In this example we show how to generate three joint normally distributed random variables with

$$\mathbf{R}_{XX} = \begin{bmatrix} 1 & 2 & -2 \\ 2 & 2 & 1 \\ -2 & 1 & 5 \end{bmatrix}$$

and mean vector as $\mu_{\mathbf{X}} = [0, 1, 2]$. We will use Octave and Scilab respectively.

```
%Octave Code
clear all
N=200; %number of generated RVs
Cov=[1 0 -1; 0 2 1; -1 1 5]; %Covariance Matrix
Mean=[0;1;2]; %mean values
X=mvnrnd(Mean,Cov,N);
```

```
// Scilab Code
clear
N=200; //number of generated RVs
Cov=[1 0 -1; 0 2 1; -1 1 5]; // Covariance Matrix
Mean=[0;1;2]; // Mean vector
X=grand(N,'mn',Mean,Cov);
```

Example 54. Assume there are two base stations, one located at the origin $(0, 0)$ and the other one is located at $(0.5, 0.5)$ in unite square area. Assume one mobile has been randomly selected at (x, y) where x and y are uniformly distributed [0,1]. What is the probability that the mobile will be closer to the base station at the origin?

Solution. The probability can be computed as

$$\begin{aligned} P_r &= P(\sqrt{x^2 + y^2} \le \sqrt{(x-0.5)^2 + (y-0.5)^2}) \\ &= P(x^2 + y^2 \le (x-0.5)^2 + (y-0.5)^2) \\ &= P(x + y \le 0.5) = P(x \le 0.5 - y) = \int_0^{0.5}\int_0^{0.5-y} dxdy \\ &= \int_0^{0.5} (0.5 - y)dy = (0.5y - 0.5y^2)_0^{0.5} = \frac{1}{8}. \end{aligned}$$

2.6 Functions of random variables

When exploring uncertainties within systems, a common challenge arises: determining the probability density function of a derived random variable y when we know the probability density function of a source random variable x, and this relationship is defined by $y = g(x)$. In this context, the function $g(.)$ can take on various forms, including general functions and, as we delve into random processes, even differential equations. Moreover, this concept extends to multidimensional random variables, where $\mathbf{Y} = \mathbf{g}(\mathbf{X})$, where $\mathbf{X}$ represents an $n \times 1$ vector of independent random variables and $\mathbf{Y}$ represents an $m \times 1$ vector of dependent random variables. Nevertheless, we commence our exploration with individual random variables and subsequently extend our findings to broader contexts. Given the mapping of the random variable x to y through the function $g(.)$, it follows that y is also a random variable. It becomes evident that the probability of x falling within the interval x_1 to x_2 is equivalent to the probability of y falling within the interval defined by $g(x_1)$ to $g(x_2)$. This fact could be represented as

$$\int_{x_1}^{x_2} f_X(x)dx = \int_{y_1}^{y_2} f_Y(y)dy.$$

For a very small interval $\Delta x = x_2 - x_2$, the integration can be approximated as

$$f_X(x)\big|_{x_1}\Delta x = f_Y(y)\big|_{y_1}\Delta y.$$

Therefore,

$$f_Y(y) = f_X(g^{-1}(y))\frac{\Delta x}{\Delta y},$$

where $g^{-1}(y)$ is the inverse function, because the right side of the previous equation should be a function in the random variable y. Since $\lim_{\Delta x\to 0}\frac{\Delta y}{\Delta x} = \frac{dy}{dx}$ and $f_Y(y) \geq 0$, then:

$$f_Y(y) = \frac{f_X(g^{-1}(y))}{|\frac{dy}{dx}|}. \tag{2.115}$$

If the function $g(x)$ is not monotone, i. e., the same value of y may occur for different values of x as the case of $y = x^2$, which implies that $x_1 = +\sqrt{y}$ OR $x_2 = -\sqrt{y}$. Therefore, since they are mutually exclusive events, the distribution is computed as:

$$f_Y(y) = \frac{f_X(g^{-1}(y))}{|\frac{dy}{dx}|}\Bigg|_{x_1} + \frac{f_X(g^{-1}(y))}{|\frac{dy}{dx}|}\Bigg|_{x_2} + \cdots \tag{2.116}$$

Example 55. If the random variable x has a normal distribution with mean μ_X and variance σ_X^2, find the probability density function of y for the following five cases:

- Linear operation: $y = \alpha x + \beta$.
- Square value: $y = x^2$. If x is a random voltage or current, y represents the normalized power.
- Absolute value: $y = |x|$. It is the result of perfect voltage rectifiers (transforms AC to DC).
- Exponential function: $y = e^x$.
- Rational function: $y = \frac{1}{x}$.

Solutions. This problem can be effectively addressed by utilizing equation (2.115). Given that x follows a normal distribution, we can express its probability density function as:

$$f_X(x) = \frac{1}{\sqrt{2\pi\sigma_X^2}} e^{(x-\mu_X)^2/(2\sigma_X^2)}.$$

By directly applying eq. (2.115), we can make use of this expression to resolve the problem at hand as follows

- For the linear operation

$$\begin{aligned} y &= \alpha x + \beta \\ \frac{dy}{dx} &= \alpha \\ x = g^{-1}(y) &= \frac{1}{\alpha}(y - \beta) \\ \therefore f_Y(y) &= \frac{1}{\sqrt{2\pi\sigma_X^2\alpha^2}} e^{(y-\mu_X-\beta)^2/(2\alpha^2\sigma_X^2)}. \end{aligned}$$

  It's evident that the random variable y also follows a normal distribution, albeit with distinct parameters for its mean and variance, which are characterized as follows: $\mu_Y = \mu_X + \beta$ and $\sigma_Y^2 = \sigma_X^2\alpha^2$, respectively. This result demonstrates that performing any linear operation on a normal distribution will preserve its distribution type. The only aspects that can be altered are the mean and the variance, while the underlying distribution remains unchanged.
- Next for $y = x^2$,

$$\begin{aligned} \frac{dy}{dx} &= 2x \\ x = g^{-1}(y) &= \pm\sqrt{y} \\ \therefore f_Y(y) &= \frac{1}{2\sqrt{2\pi\sigma_X^2 y}} \left(e^{(\sqrt{y}-\mu_X)^2/(2\sigma_X^2)} + e^{(-\sqrt{y}-\mu_X)^2/(2\sigma_X^2)}\right). \end{aligned}$$

  It is evident that $y \geq 0$. As a special case when $\mu_X = 0$, the above distribution can simplified as

$$f_Y(y) = \frac{1}{\sqrt{2\pi\sigma_X^2 y}} e^{y/(2\sigma_X^2)}, \quad y \geq 0.$$

- For the full rectifier case, i. e., the absolute value: $y = |x|$.

$$\frac{dy}{dx} = \begin{cases} 1 & \text{if } x \geq 0 \\ -1 & \text{if } x < 0 \end{cases}$$
$$x = g^{-1}(y) = \pm y$$
$$\therefore f_Y(y) = \frac{1}{\sqrt{2\pi\sigma_X^2}} e^{(y-\mu_X)^2/(2\sigma_X^2)} + \frac{1}{\sqrt{2\pi\sigma_X^2}} e^{(y+\mu_X)^2/(2\sigma_X^2)}$$
$$= \frac{1}{\sqrt{2\pi\sigma_X^2}} [e^{(y-\mu_X)^2/(2\sigma_X^2)} + e^{(y+\mu_X)^2/(2\sigma_X^2)}] \quad y \geq 0.$$

The distribution $f_Y(y)$ referred to in this context is commonly known as the folded normal distribution, which finds significant and versatile applications in various domains of physics.

- For the case $y = e^x$

$$\frac{dy}{dx} = e^x$$
$$x = g^{-1}(y) = \ln(y)$$
$$\therefore f_Y(y) = \frac{1}{\sqrt{2\pi\sigma_X^2 y^2}} e^{(\ln(y)-\mu_X)^2/(2\sigma_X^2)} \quad y > 0.$$

This distribution $f_Y(y)$ is well known as *lognormal distribution.*

- For the last case, $y = \frac{1}{x}$

$$\frac{dy}{dx} = -\frac{1}{x^2}$$
$$x = g^{-1}(y) = \frac{1}{y}$$
$$\therefore f_Y(y) = \frac{1}{2\sqrt{\pi\sigma_X^2 y^4}} e^{(1/y-\mu_X)^2/(2\sigma_X^2)}.$$

In the case of a multidimensional joint distribution function $f_{\mathbf{X}}(x_1, x_2, \ldots, x_n)$ with general invertible and differentiable functions $y_1 = g_1(x_1, x_2, \ldots, x_n), y_2 = g_2(x_1, x_2, \ldots, x_n)$, $\ldots y_n = g_n(x_1, x_2, \ldots, x_n)$, we may find the joint distribution of the random vector $\mathbf{y} = [y_1, y_2, \ldots, y_n]$ as:

$$f_{\mathbf{Y}}(\mathbf{y}) = f_{\mathbf{X}}(g_1^{-1}(\mathbf{y}), g_2^{-1}(\mathbf{y}), \ldots, g_n^{-1}(\mathbf{y}))|\det(\mathbf{J}(\mathbf{y}))|^{-1}. \tag{2.117}$$

Where $\mathbf{J}(\mathbf{y})$ is the Jacobian matrix and it is given by:

$$\mathbf{J}(\mathbf{Y}) = \begin{bmatrix} \frac{\partial g_1}{\partial x_1} & \frac{\partial g_1}{\partial x_2} & \cdots & \frac{\partial g_1}{\partial x_n} \\ \frac{\partial g_2}{\partial x_1} & \frac{\partial g_2}{\partial x_2} & \cdots & \frac{\partial g_2}{\partial x_n} \\ \vdots & \vdots & \vdots & \vdots \\ \frac{\partial g_n}{\partial x_1} & \frac{\partial g_n}{\partial x_2} & \cdots & \frac{\partial g_n}{\partial x_n} \end{bmatrix}. \tag{2.118}$$

Example 56. In this example, we delve into handling functions with multiple random variables. Let's consider two independent random variables, x_1 and x_2, where x_1 follows a zero-mean normal distribution with unit variance, and x_2 follows an exponential distribution with $\lambda = 1$. Our focus lies on two specific random variables, namely:

$$y_1 = g_1(x_1, x_2) = \frac{x_1}{x_1 + x_2}$$
$$y_2 = g_2(x_1, x_2) = x_1 + x_2.$$

Find the joint distribution of $\mathbf{Y}$, i. e., $f_{\mathbf{Y}}(y_1, y_2)$.

Solution. As outlined in the case of a single random variable, the initial step is to determine the inverse functions. In this straightforward example involving two equations, our approach involves solving them concurrently to ascertain the solution as:

$$x_1 = y_1 y_2$$
$$x_2 = y_2 - y_1 y_2.$$

Since x_1 and x_2 are independent, therefore,

$$f_{\mathbf{X}}(x_1, x_2) = f_{X_1}(x_1) f_{X_2}(x_2) = \frac{1}{\sqrt{2\pi}} e^{-x_1^2/2} e^{-x_2}.$$

Applying in eq. (2.117) we obtain:

$$f_{\mathbf{Y}}(y_1, y_2) = \frac{1}{\sqrt{2\pi}} e^{-y_1^2 y_2^2/2} e^{-(y_2 - y_1 y_2)} \left|\det(\mathbf{J}(\mathbf{y}))\right|^{-1}.$$

The Jacobian matrix $\mathbf{J}(\mathbf{y})$ is computed as:

$$\mathbf{J}(\mathbf{x}) = \begin{bmatrix} \frac{\partial g_1}{\partial x_1} & \frac{\partial g_1}{\partial x_2} \\ \frac{\partial g_2}{\partial x_1} & \frac{\partial g_2}{\partial x_2} \end{bmatrix} = \begin{bmatrix} \frac{x_2}{(x_1+x_2)^2} & \frac{-x_1}{(x_1+x_2)^2} \\ 1 & 1 \end{bmatrix}.$$

Finally, the determinant of the Jacobian matrix is

$$\det(\mathbf{J}(\mathbf{x})) = \frac{1}{x_1 + x_2}.$$

Let's represent the determinant result as a function of the random variables y_1 and y_2. Given that $y_2 = x_1 + x_2$, substituting this into the joint distribution function of $\mathbf{Y}$ yields our conclusive result:

$$f_{\mathbf{Y}}(y_1, y_2) = \frac{1}{\sqrt{2\pi}} e^{-y_1^2 y_2^2/2} e^{-(y_2 - y_1 y_2)} |y_2|. \tag{2.119}$$

It's evident that y_1 and y_2 are dependent random variables. Could you provide a proof of this observation?

Example 57. In this example, we derive the Rayleigh distribution by determining the probability density function of $y = \sqrt{x_1^2 + x_2^2}$, where x_1 and x_2 are zero-mean, independent, and identically distributed as normal random variables with variance σ^2. While there are simpler methods to find the Rayleigh distribution, we provide a detailed explanation here for the sake of better understanding and learning.

Solution. We're dealing with a single function with two variables here. To utilize the method outlined in eq. (2.117), which requires two functions, we introduce the following pair of functions:

$$y_1 = \sqrt{x_1^2 + x_2^2}$$
$$y_2 = x_1.$$

Observe that $y_1 \geq y_2$. After obtaining $f_{\mathbf{Y}}(y_1, y_2)$, we can derive the desired distribution $f_{Y_1}(y_1)$ by integrating over the variable y_2 as:

$$f_{Y_1}(y_1) = \int_{-\infty}^{\infty} f_{\mathbf{Y}}(y_1, y_2) dy_2.$$

First we start by deriving the inverse functions as:

$$\begin{aligned} x_1 &= y_2 \\ x_2 &= \pm\sqrt{y_1^2 - y_2^2}. \end{aligned} \tag{2.120}$$

It is straightforward to find the differentiation as:

$$\begin{aligned} \frac{\partial g_1}{\partial x_1} &= \frac{x_1}{\sqrt{x_1^2 + x_2^2}} \\ \frac{\partial g_1}{\partial x_2} &= \frac{x_2}{\sqrt{x_1^2 + x_2^2}} \\ \frac{\partial g_2}{\partial x_1} &= 1 \\ \frac{\partial g_2}{\partial x_2} &= 0. \end{aligned} \tag{2.121}$$

The Jacobian matrix is then constructed as

$$\mathbf{J}(\mathbf{X}) = \begin{bmatrix} \frac{\partial g_1}{\partial x_1} & \frac{\partial g_1}{\partial x_2} \\ \frac{\partial g_2}{\partial x_1} & \frac{\partial g_2}{\partial x_2} \end{bmatrix} = \begin{bmatrix} \frac{x_1}{\sqrt{x_1^2+x_2^2}} & \frac{x_2}{\sqrt{x_1^2+x_2^2}} \\ 1 & 0 \end{bmatrix}$$

with the determinant value as

$$\det(\mathbf{J}(\mathbf{x})) = -\frac{x_2}{\sqrt{x_1^2 + x_2^2}}.$$

In terms of random variables $\mathbf{y}$, the determinant is

$$\mathbf{J}(\mathbf{g}^{-1}(\mathbf{y})) = \mp\sqrt{1 - (y_2/y_1)^2}.$$

Substitute these results in eq. (2.117) we obtain the joint probability distribution

$$f_{\mathbf{Y}}(y_1, y_2) = \frac{1}{2\pi\sigma^2} e^{-y_2^2/2\sigma^2} e^{-(y_1^2-y_2^2)/2\sigma^2} \frac{1}{\sqrt{1 - (y_2/y_1)^2}}.$$

With a few straightforward manipulations, it can be expressed as:

$$f_{\mathbf{Y}}(y_1, y_2) = \frac{1}{2\pi\sigma^2} e^{-y_1^2/2\sigma^2} \frac{1}{\sqrt{1 - (y_2/y_1)^2}}.$$

Finally, the required distribution is computed by integrating over y_2 as:

$$f_{Y_1}(y_1) = \frac{1}{2\pi\sigma^2} e^{-y_1^2/2\sigma^2} \int_{-\infty}^{\infty} \frac{1}{\sqrt{1 - (y_2/y_1)^2}} dy_2.$$

By making a change of variables as $u = \frac{y_2}{y_1}$, then $du = \frac{1}{y_1} dy_2$, and since $y_1 \geq y_2$, then $|u| \leq 1$, therefore,

$$f_{Y_1}(y_1) = \frac{y_1}{2\pi\sigma^2} e^{-y_1^2/2\sigma^2} \int_{-1}^{1} \frac{1}{\sqrt{1 - u^2}} du = \frac{y_1}{2\sigma^2} e^{-y_1^2/2\sigma^2}.$$

The result is the well-known form of the Rayleigh distribution.

We often encounter scenarios that involve the summation of various random variables, expressed as $y = x_1 + x_2 + \cdots$. Armed with the joint probability density function $f_{\mathbf{X}}(x_1, x_2, \ldots, x_n)$, we can determine the distribution of y. Let's begin with a simpler case of just two random variables, where $y = x_1 + x_2$, and then extend our findings to a more general context. Employing the same approach as previously, we define $y_1 = x_1 + x_2$ and $y_2 = x_1$. The determinant of the Jacobian matrix is

$$\mathbf{J}(\mathbf{X}) = \begin{bmatrix} 1 & 1 \\ 1 & 0 \end{bmatrix} \Rightarrow \det(\mathbf{J}(\mathbf{X})) = -1.$$

Applying these results in eq. (2.117), we obtain

$$f_{\mathbf{Y}}(y_1, y_2) = f_{\mathbf{X}}(y_2, y_1 - y_2).$$

To find the distribution of y_1, we integrate over y_2 as:

$$f_{Y_1}(y_1) = \int_{-\infty}^{\infty} f_{\mathbf{X}}(y_2, y_1 - y_2) dy_2.$$

If x_1 and x_2 are independent random variables, then

$$f_{\mathbf{X}}(x_1, x_2) = f_{X_1}(x_1) f_{X_2}(x_2).$$

Applied for the distribution of y_1 we obtain the following important result:

$$f_{Y_1}(y_1) = \int_{-\infty}^{\infty} f_{X_1}(y_2) f_{X_2}(y_1 - y_2) dy_2. \tag{2.122}$$

This form of integration is widely known in linear systems theory as the *convolution* integral. In the realm of linear systems, the output signal in the time domain is the convolution integral of the input signal and the system's impulse response function. Typically, performing the convolution integral in closed form is challenging, and even in straightforward cases, it can be a bit of a puzzle. To visualize, consider that for each value of y_1, the function f_{X_2} is mirrored horizontally and shifted to the left toward $-\infty$. This mirrored function is then shifted across to $+\infty$ while multiplying by f_{X_1}, and the area under the resulting curve is computed. Numerous insightful visual explanations of the convolution integral can be found in the classical linear systems literature. It is common to use an asterisk symbol to represent the convolution integral, for example, in the case of random variables above, it is represented as $f_{Y_1}(y_1) = f_{X_1}(y_1) * f_{X_2}(y_1)$. For a general n random variable, the distribution for their summation is given by

$$f_Y(y) = f_{X_1}(y) * f_{X_2}(y) * \cdots * f_{X_n}(y). \tag{2.123}$$

Where $f_Y(y)$ is called the nth fold convolution of all probability density functions $f_{X_i}(x_i)$, for $i = 1, \ldots, n$.

Example 58. Determine the probability density function for the random variable y, defined as the sum of two independent random variables, x_1 and x_2. The variable x_1 follows a uniform distribution ranging from -1 to 5, while x_2 follows an exponential distribution with $\lambda = 1$.

Solution. We may use the convolution integral in eq. (2.122). Here, we set $y_1 = x_1 + x_2$, and $y_2 = x_1$. The distribution $f_{X_1}(y_1)$ is a rectangular function from −1 to 5, and $f_{X_2}(y_1 - y_2) = e^{-(y_1-y_2)}$. We should perform the integration over two periods, the first period for $-1 \le y_1 \le 5$

$$\begin{aligned} f_{Y_1}(y_1) &= \int_{-1}^{y_1} \frac{1}{6} e^{-y_1+y_2} dy_2 \\ &= \frac{e^{-y_1}}{6} \int_{-1}^{y_1} e^{y_2} dy_2 \\ &= \frac{1}{6}(1 - e^{-y_1-1}). \end{aligned} \tag{2.124}$$

For $y_1 > 5$, there is no contribution from x_1 and all generated random numbers will be from the exponential distribution, i. e.,

$$\begin{aligned} f_{Y_1}(y_1) &= \int_{-1}^{5} \frac{1}{6} e^{-y_1+y_2} dy_2 \\ &= \frac{e^{-y_1}}{6} \int_{-1}^{5} e^{y_2} dy_2 \\ &= 24.67 e^{-y_1}. \end{aligned} \tag{2.125}$$

$$f_{Y_1}(y_1) = \begin{cases} 0 & \text{if } y_1 < -1, \\ \frac{1}{6}(1 - e^{-y_1-1}) & \text{if } -1 \le y_1 < 5, \\ 24.67 e^{-y_1} & \text{if } y_1 \ge 5. \end{cases} \tag{2.126}$$

Example 59. In Example 58, determine the cumulative distribution function that corresponds to the resulting probability density function. Additionally, demonstrate the proof that the total area under the probability density function equals 1.

Solution. The CDF is given by $F(x_0) = P(x \le x_0)$. We have replaced y_1 with x for convenience. From equation (2.126)

$$P(x \le x_0) = \int_{-\infty}^{x_0} f(x) dx \tag{2.127}$$

$$= \begin{cases} \frac{1}{6}(x_0 + e^{-x_0-1}) & \text{if } -1 \le x_0 < 5, \\ 0.834 + 24.67(e^{-5} + e^{-x_0}) & \text{if } x_0 \ge 5. \end{cases}$$

From the above it is clear that $\lim_{x_0 \to \infty} F(x_0) = 1$, so the total area under the probability density function is 1.

Example 60. In Example 58, compute the following items analytically:
- The probabilities $P_1(y \le 1)$ and $P_2(5 \le y \le 6)$.
- The mean and variance of y.
- The expected value of $\ln(1 + y)$, i. e., $E[\ln(1 + y)]$.

Write a simulation code in Scilab to confirm the previous analytical results.

Solution. We may use the derived PDF of y to find all items as:
- The solutions of the first item is:

$$P_1(y \le 1) = \frac{1}{6}\int_{-1}^{1} 1 - e^{-y-1} dy = \frac{2 + e^{-1}(e^{-1} - e^{1})}{6} = 0.1892$$

$$P_2(5 \le y \le 6) = \frac{[e^5 - e^{-1}]}{6}\int_{5}^{6} e^{-y} dy = 0.1051.$$

- The mean could be easily found by

$$E[y] = E[x_1 + x_2] = E[x_1] + E[x_2] = 2 + 1 = 3.$$

However, let's evaluate the mean again using the derived probability density function as:

$$E[y] = \int_{-1}^{5} y\left(\frac{1 - e^{-y-1}}{6}\right)dy + \int_{5}^{\infty} y\left(\frac{e^{-y}}{6}[e^5 - e^{-1}]\right)dy = 3.$$

The mean square value based on the derived PDF is:

$$E[y^2] = \int_{-1}^{5} y^2\left(\frac{1 - e^{-y-1}}{6}\right)dy + \int_{5}^{\infty} y^2\left(\frac{e^{-y}}{6}[e^5 - e^{-1}]\right)dy = 13.$$

Hence the variance is $13 - 9 = 4$.
- The expected value of the logarithmic function is computed as

$$E[\ln(1 + y)] = \frac{1}{6}\int_{-1}^{5} \ln(1 + y)(1 - e^{-(y+1)})dy + \frac{e^5 - e^{-1}}{6}\int_{5}^{\infty} \ln(1 + y)e^{-y} dy.$$

We may solve this integration by parts with defining $u = 1 + y$, therefore,

$$\int \ln(u)du = u\ln(u) - \int \frac{u}{u}du = u\ln(u) - u.$$

In the same way,

$$\int \ln(u)e^{-u}du = -e^{-u}\ln(u) + \int e^{-u}\frac{1}{u}du,$$

where $\int e^{-u}\frac{1}{u}du$ is the exponential integral explained before. Finally,

$$\begin{aligned}\int \ln(u)e^{-u+1}du &= e^{1}\int \ln(u)e^{-u}du \\ &= -e^{-u+1}\ln(u) + e^{1}\int e^{-u}\frac{1}{u}du.\end{aligned}$$

By substituting the integration limits and consulting tables to determine the corresponding exponential integral values, we can deduce that $E[\ln(1 + y)] = 1.2107$.

The code to evaluate the same example in Scilab is shown next:

```
//Scilab Code
clear
N=10000;
x1=6*rand(1,N)-1;
x2=grand(1,N,'exp',1);
y=x1+x2;
P1=length(find(y<1))/N
P2=length(find(y>0 & y<1))/N
mu=mean(y)
Var=variance(y)
F=mean(log(1+y))
```

The results after running the code are as follows: $P_1 = 0.1889$, $P_2 = 0.1084$, $mu = 3.003$, Var = 3.993, and $F = 1.2121$. The results closely match the analytical values. However, you may notice a difference in the effort required to obtain these results. By simulation, it is quite straightforward and simple to find. In fact, the analytical analysis for this example is also manageable; however, there are many other cases in practice where the mathematical solution can be complicated or infeasible.

Example 61. Find the general form of the probability density function of $y = x_1x_2$ as function in the joint distribution $f_{\mathbf{X}}(x_1, x_2)$.

Solution. Let's define $y_1 = x_1$, therefore,

$$f_{\mathbf{Y}}(y, y_1) = f_{\mathbf{X}}\left(y_1, \frac{y}{y_1}\right)|\det(\mathbf{J}(Y))|^{-1}.$$

The Jacobian matrix is

$$\mathbf{J}(\mathbf{X}) = \begin{bmatrix} x_2 & x_1 \\ 1 & 0 \end{bmatrix} \Rightarrow \det(\mathbf{J}(\mathbf{X})) = -x_1.$$

Therefore, as a function in y_1, $|\det(\mathbf{J}(Y))|^{-1} = |\frac{1}{y_1}|$.
Hence, the distribution of y is:

$$f_Y(y) = \int_{-\infty}^{\infty} \frac{1}{|y_1|} f_{\mathbf{X}}\left(y_1, \frac{y}{y_1}\right) dy_1.$$

If x_1 and x_2 are independent, then the distribution of y becomes

$$f_Y(y) = \int_{-\infty}^{\infty} \frac{1}{|y_1|} f_{X_1}(y_1) f_{X_2}\left(\frac{y}{y_1}\right) dy_1. \tag{2.128}$$

Example 62. Find the general form of the probability density function of $y = \max(x_1, x_2)$ as function in the joint distribution $f_{\mathbf{X}}(x_1, x_2)$.

Solution. It is unfeasible to employ the conventional techniques for discovering this distribution due to the nonsmooth and nondifferentiable nature of the function max(.). Nonetheless, there is an alternative straightforward approach to determining the distribution. By constructing the cumulative distribution function (CDF) of y and subsequently differentiating it, we can obtain the probability density function (PDF), as outlined in (2.106). The CDF of y is

$$F_Y(y_0) = P(y \le y_0) = P(\max(x_1, x_2) \le y_0). \tag{2.129}$$

If the greater value among the random variables is less than or equal to y_0, then the other variable must also be less than or equal to y_0. Hence,

$$F_Y(y_0) = P(x_1 \le y_0, x_2 \le y_0) = \int_{-\infty}^{y_0} \int_{-\infty}^{y_0} f_{\mathbf{X}}(x_1, x_2) dx_1 dx_2. \tag{2.130}$$

In the case of independent random variables,

$$F_Y(y_0) = P(x_1 \le y_0) P(x_2 \le y_0) = F_{X_1}(y_0) F_{X_2}(y_0). \tag{2.131}$$

The probability density function of y is then computed by:

$$f_Y(y) = \left.\frac{dF_Y(y_0)}{dy_0}\right|_{y_0 = y}. \tag{2.132}$$

Therefore, for independent random variables, we have interesting results that

$$f_Y(y) = f_{X_1}(y)F_{X_2}(y) + f_{X_2}(y)F_{X_1}(y). \tag{2.133}$$

Example 63. Two independent random variables x_1 and x_2, where x_1 has a uniform distribution from 0 to 1, and x_2 has an exponential distribution with $\lambda = 1$. For the random variable $y = \max(x_1, x_2)$, find

- The probability density function of y.
- The probabilities of $P(y \geq 1)$ and $P(y \leq 1)$.
- The mean and variance of y.
- Prove the results by simulation.

Solution.

- We may find the probability density function using the result in (2.133) as

$$f_Y(y) = \begin{cases} 1 - e^{-y} + ye^{-y} & \text{if } 0 \leq y < 1, \\ e^{-y} & \text{if } y \geq 1. \end{cases} \tag{2.134}$$

- Based on the probability density function given above we can compute the probabilities as

$$P(y \geq 1) = \int_1^{\infty} f_Y(y)dy = \int_1^{\infty} e^{-y}dy = e^{-1} = 0.3679$$

$$P(y \leq 1) = \int_0^1 f_Y(y)dy = \int_0^1 (1 - e^{-y} + ye^{-y})dy = 0.6321.$$

- The mean is computed based on the derived probability density function as:

$$\begin{aligned} E[y] &= \int_0^{\infty} yf_Y(y)dy = \int_0^1 yf_Y(y)dy + \int_1^{\infty} yf_Y(y)dy \\ &= \int_0^1 y(1 - e^{-y})dy + \int_0^1 y^2e^{-y}dy + \int_1^{\infty} ye^{-y}dy = 1.1322. \end{aligned}$$

In a similar manner,

$$\begin{aligned} E[y^2] &= \int_0^1 y^2(1 - e^{-y})dy + \int_0^1 y^3e^{-y}dy + \int_1^{\infty} y^2e^{-y}dy \\ &= 2.126 \Rightarrow \sigma^2 = 2.126 - 1.1322^2 = 0.8441. \end{aligned}$$

- We can easily write a short code to simulate this example as shown next:

```
%Octave Code
clear all
N=10000;
x1=rand(N,1);
x2=exprnd(1,N,1);
y=max(x1,x2);
P1=length(find(y>1))/N
P2=length(find(y<1))/N
mu=mean(y)
Var=var(y)
```

The results obtained after running this code on Octave are: $P_1 = 0.36800$, $P_2 = 0.63200$, $mu = 1.1237$, and Var $= 0.82486$.

Now, Let's now strengthen our grasp of the transition from single to multidimensional random variables and construct the corresponding joint distributions.

Example 64. Two independent random variables x_1 and x_2, where x_1 has a uniform distribution from 0 to 1, and x_2 also has a uniform distribution, but between -1 to 2. For the random variable $y = \max(x_1, x_2)$, find

- The probability density function of y.
- Draw the probability density function and compare with histogram.
- Generate random numbers representing y.

Solution. As shown in the previous example,

$$F_Y(y_0) = P(x_1 \le y_0)P(x_2 \le y_0) = F_{X_1}(y_0)F_{X_2}(y_0),$$

where

$$\begin{aligned} P(x_1 \le y_0) &= y_0, \quad 0 \le y_0 \le 1, \\ P(x_2 \le y_0) &= \frac{y_0 + 1}{3}, \quad -1 \le y_0 \le 2. \end{aligned}$$

Since $y = \max(x_1, x_2)$, therefore y will be always positive. Hence,

$$F_Y(y_0) = \begin{cases} 0 & y_0 < 0, \\ y_0 \frac{y_0+1}{3} & 0 \le y_0 < 1, \\ \frac{y_0+1}{3} & 1 \le y_0 \le 2. \end{cases}$$

Therefore the probability density function becomes

$$f_Y(y) = \begin{cases} 0 & y < 0, \\ \frac{2y+1}{3} & 0 \le y < 1, \\ \frac{1}{3} & 1 \le y \le 2. \end{cases}$$

Example 65. Find the general form of the probability density function of $y = \min(x_1, x_2)$ as function in the joint distribution $f_{\mathbf{X}}(x_1, x_2)$.

Solution. By constructing the cumulative distribution function (CDF) of y and subsequently differentiating it, we can obtain the probability density function (PDF), as outlined in (2.106). The CDF of y is

$$P(y \leq y_0) = P\big(\min(x_1, x_2) \leq y_0\big) = 1 - P\big(\min(x_1, x_2) \geq y_0\big). \tag{2.135}$$

If the minimum value of the random variables is greater than or equal to y_0, then the other variable must also be greater than or equal to y_0. Hence,

$$F_Y(y_0) = 1 - P(x_1 \geq y_0, x_2 \geq y_0) = 1 - \int_{y_0}^{\infty}\int_{y_0}^{\infty} f_{\mathbf{X}}(x_1, x_2)dx_1 dx_2. \tag{2.136}$$

In the case of independent random variables,

$$\begin{aligned} F_Y(y_0) &= 1 - \big((1 - P(x_1 \leq y_0))(1 - P(x_2 \leq y_0))\big) \\ &= 1 - \big((1 - F_{X_1}(y_0))(1 - F_{X_2}(y_0))\big). \end{aligned}$$

The probability density function of y is then computed as shown before:

$$f_Y(y) = \left.\frac{dF_Y(y_0)}{dy_0}\right|_{y_0 = y}.$$

This leads to:

$$f_Y(y) = f_{X_1}(y)(1 - F_{X_2}(y)) + f_{X_2}(y)(1 - F_{X_1}(y)). \tag{2.137}$$

2.7 Characteristic function

The analysis of the stochastic behavior of additive random variables often involves dealing with convolution integrals of their probability density functions, which can be challenging to solve analytically. However, drawing from the principles of linear systems, we understand that the *Fourier Transform* of a convolution integral equates to the multiplication of the *Fourier Transforms* of its individual components. This property greatly streamlines the analysis of linear systems, as will be discussed in more detail in Chapter 3. The fundamental concept behind the characteristic function is to obtain the Fourier transform of the probability density function, specifically with a positive exponent. This characteristic function uniquely defines the probability density function and finds extensive applications in probability theory. For a continuous random variable x, the characteristic functions are as follows:

$$\varphi_X(t) = E[e^{jXt}] = \int_{-\infty}^{\infty} e^{jxt} f_X(x)dx. \tag{2.138}$$

It is also possible to determine the probability density function from the characteristic function using the inverse Fourier transform as follows

$$f_X(x) = \frac{1}{2\pi} \int_{-\infty}^{\infty} e^{-jxt} \varphi_X(t)dt. \tag{2.139}$$

For discrete random variables x_i, the characteristic function is the Z-transformation of the probability mass function, such as:

$$\varphi_X(t) = E[e^{jXt}] = \sum_{x_i \in R_X} e^{jx_i t} p_X(x_i). \tag{2.140}$$

The probability mass function can also be derived from the characteristic function by applying the inverse Z-transformation.

Assume the random variable

$$y = \sum_{i=1}^{n} x_i,$$

where x_i are independent random variables with probability density function $f_{X_i}(x_i)$. We have already seen that the probability density function of y is the nth-fold convolution integral of the probability density functions which can be very complicated to solve. However, let's compute the characteristic function of y as:

$$\varphi_Y(t) = E[e^{jyt}] = E[e^{jt\sum_{i=1}^{n} x_i}]. \tag{2.141}$$

Since x_i are independent random variables,

$$E[e^{jt\sum_{i=1}^{n} x_i}] = \prod_{i=1}^{n} E[e^{jtx_i}]$$

$$\Rightarrow \varphi_Y(t) = \prod_{i=1}^{n} \varphi_{X_i}(t). \tag{2.142}$$

Hence, the characteristic function of the random variable y is the product of the characteristic functions of the individual random variables x_i. Many well-known probability distributions possess closed-form characteristic functions, which facilitates a straightforward assessment of the random behavior associated with the summation of these variables.

2.7.1 Moment generating function

To ascertain the moments of random variables, it is essential to have the probability density function. The ith moment of the random variable x is computed by:

$$E[x^i] = \int_{-\infty}^{\infty} x^i f_X(x)dx. \quad (2.143)$$

However, it is also possible to generate all moments using the characteristic function. If we differentiate the characteristic function in eq. (2.138) with respect to t we obtain:

$$\frac{d\varphi(t)}{dt} = \int_{-\infty}^{\infty} jxe^{jxt} f_X(x)dx. \quad (2.144)$$

At $t = 0$, the first differentiation becomes:

$$\left.\frac{d\varphi(t)}{dt}\right|_{t=0} = j\int_{-\infty}^{\infty} xf_X(x)dx = jE[x] \Rightarrow E[x] = \left.\frac{d\varphi(t)}{jdt}\right|_{t=0}. \quad (2.145)$$

You can find the second moment by double differentiating the characteristic function as:

$$\left.\frac{d^2\varphi(t)}{dt^2}\right|_{t=0} = -\int_{-\infty}^{\infty} x^2 f_X(x)dx \quad (2.146)$$

$$\Rightarrow E[x^2] = -\left.\frac{d^2\varphi(t)}{dt^2}\right|_{t=0}. \quad (2.147)$$

All moments could be found in the same manner, therefore, the ith moment is:

$$E[x^i] = \left.\frac{d^i\varphi(t)}{j^i dt^i}\right|_{t=0}. \quad (2.148)$$

The Taylor expansion of the exponential function around 0 is given by:

$$e^{jxt} = 1 + jxt + \frac{(jxt)^2}{2!} + \frac{(jxt)^2}{3!} + \cdots$$

$$\Rightarrow \varphi_X(t) = \sum_{i=0}^{\infty} \frac{(jt)^i}{i!} E[x^i]. \quad (2.149)$$

Next, we offer a few examples to derive the characteristic function of some well-known probability mass and density functions. Some references define the moment generating function as:

$$MG(t) = \varphi(-jt), \tag{2.150}$$

which means simply getting rid of the complex symbol j.

Example 66. Derive the characteristic function of the binomial distribution and use it to find the first two moments and the variance.

Solution. The binomial distribution defines the mass function of

$$y = \sum_{i=1}^{n} x_i \tag{2.151}$$

with independent $x_i \in \{0, 1\}$, and $P(x_i = 1) = p$.

Apply the characteristic function algorithm as in eq. (2.140):

$$\varphi_Y(t) = E[e^{jyt}] = E[e^{jt\sum_{i=1}^{n} x_i}] = E\left[\prod_{i=1}^{n} e^{jtx_i}\right]. \tag{2.152}$$

Since all x_i are independent, therefore

$$\varphi_Y(t) = \prod_{i=1}^{n} E[e^{jtx_i}] = \prod_{i=1}^{n} (e^{jt}p + e^{j0t}(1-p)) = (e^{jt}p - p + 1)^n. \tag{2.153}$$

The characteristic function can be employed to determine any moment. For instance, the first moment is

$$E[y] = \left.\frac{d\varphi(t)}{jdt}\right|_{t=0} = \frac{n}{j}(e^{jt}p - p + 1)^{n-1} e^{jt}\Big|_{(t=0)} jp = np. \tag{2.154}$$

The second moment is:

$$\begin{aligned} E[y^2] = -\left.\frac{d^2\varphi(t)}{dt^2}\right|_{t=0} &= [n(n-1)(e^{jt}p - p + 1)^{n-2} e^{2jt} p^2 \\ &+ n(e^{jt}p - p + 1)^{n-1} e^{jt}]_{t=0} p = n(n-1)p^2 + np. \end{aligned}$$

The variance of the binomial distribution is computed as usual:

$$\sigma_Y = E[y^2] - (E[y])^2 = np(1-p). \tag{2.155}$$

Example 67. Derive the characteristic function of the uniform distribution X from a to b.

Solution. Similarly, we use eq. (2.138) as

$$\varphi_X(t) = E[e^{jXt}] = \frac{1}{b-a} \int_a^b e^{jxt} dx = \frac{e^{jbt} - e^{jat}}{jt(b-a)}. \tag{2.156}$$

Example 68. Derive the characteristic function of the exponential distribution X.

Solution. With the reference of eq. (2.138):

$$\varphi_X(t) = E[e^{jXt}] = \int_0^{\infty} \lambda e^{jxt} e^{-\lambda x} dx = \frac{\lambda}{\lambda - jt}. \tag{2.157}$$

One of the most important characteristic functions is the one that represents the normal distribution. It can be derived as:

$$\varphi_X(t) = E[e^{jXt}] = \frac{1}{\sqrt{2\pi\sigma^2}} \int_{-\infty}^{\infty} e^{jxt} e^{-(x-\mu)^2/2\sigma^2} dx \tag{2.158}$$

$$= \frac{1}{\sqrt{2\pi\sigma^2}} \int_{-\infty}^{\infty} e^{-(x^2-2\mu x+\mu^2-2j\sigma^2 tx)/2\sigma^2} dx \tag{2.159}$$

$$= \frac{1}{\sqrt{2\pi\sigma^2}} \int_{-\infty}^{\infty} e^{-(x^2-2(\mu+2j\sigma^2 t)x+\mu^2)/2\sigma^2} dx. \tag{2.160}$$

Since

$$x^2 - 2(\mu + 2j\sigma^2 t)x + \mu^2 = (x - (\mu + 2j\sigma^2 t))^2 - \sigma^2 t(2j\mu - \sigma^2 t). \tag{2.161}$$

Therefore,

$$\varphi_X(t) = e^{(j\mu t - \sigma^2 t^2/2)} \frac{1}{\sqrt{2\pi\sigma^2}} \int_{-\infty}^{\infty} e^{(x-(\mu+2j\sigma^2 t))^2/2\sigma^2} dx \tag{2.162}$$

$$\because \int_{-\infty}^{\infty} e^{(x-(\mu+2j\sigma^2 t))^2/2\sigma^2} dx = 1 \tag{2.163}$$

$$\therefore \varphi_X(t) = e^{(j\mu t - \sigma^2 t^2/2)}. \tag{2.164}$$

Hence, the characteristic function of a zero-mean normal distribution is $\varphi_X(t) = e^{-\sigma^2 t^2/2}$. We may generalize the characteristic function of the multivariate normal distribution given in eq. (2.111) as:

$$\varphi_{\mathbf{X}}(\mathbf{t}) = e^{(j\mathbf{t}^T \mu_X - \mathbf{t}^T \mathbf{R}_{XX} \mathbf{t}/2)}, \tag{2.165}$$

where $\mathbf{t} = [t_1, t_2, \ldots, t_n]^T$.

Example 69. Find the characteristic function of $y = x_1 + x_2 + x_3$, where all x_i are independent. Specifically, x_1 follows a uniform distribution from -1 to 1, x_2 is a standard zero-mean normal distribution (with variance $\sigma^2 = 2$), and x_3 is an exponential distribution with parameter $\lambda = 1$.

Solution. As derived above

$$\varphi_{X_1}(t) = \frac{e^{jt} - e^{-jt}}{2jt} = \frac{\sin(t)}{t} \tag{2.166}$$

$$\varphi_{X_2}(t) = e^{-t^2} \tag{2.167}$$

$$\varphi_{X_3}(t) = \frac{1}{1-t}. \tag{2.168}$$

Therefore, the characteristic function of the summation is

$$\varphi_Y(t) = \frac{\sin(t)e^{-t^2}}{t(1-t)}. \tag{2.169}$$

If we need to find the probability density function of y, we should perform the inverse Fourier transform as:

$$f_Y(y) = \frac{1}{2\pi}\int_{-\infty}^{\infty} \frac{\sin(t)e^{-t^2}}{t(1-t)} e^{-jyt}\,dt. \tag{2.170}$$

Example 70. What is the characteristic function of $y = \alpha x + \beta$, where x is a random variable with an arbitrary distribution $f_X(x)$?

Solution. Applying the general algorithm of the characteristic function as:

$$\varphi_Y(t) = E[e^{jyt}] = \int_{-\infty}^{\infty} e^{jyt} f_Y(y)dy = \int_{-\infty}^{\infty} e^{jyt} f_X\left(\frac{y-\beta}{\alpha}\right)dy. \tag{2.171}$$

With a change of variables as $x = \frac{y-\beta}{\alpha}$ and $dx = \frac{dy}{\alpha}$, we obtain

$$\varphi_Y(t) = e^{j\beta t}\int_{-\infty}^{\infty} e^{j\alpha xt} f_X(x)dx. \tag{2.172}$$

When the characteristic function of x is given as $\varphi_X(t)$ therefore, the characteristic function of y is given by

$$\varphi_Y(t) = e^{j\beta t}\varphi_X(\alpha t). \tag{2.173}$$

Example 71. Using the results of the previous example, what will be the distribution of y, when x is normal distribution with mean μ_X and variance σ^2.

Solution. The characteristic function can be derived as:

$$\varphi_Y(t) = e^{j\beta t}\varphi_X(\alpha t)$$
$$\because \varphi_X(t) = e^{(j\mu t - \sigma^2 t^2/2)}$$

$$\therefore \varphi_Y(t) = e^{j\beta t} e^{(j\mu_x \alpha t - \sigma^2\alpha^2 t^2/2)}$$
$$= e^{(j(\mu_x \alpha + \beta)t - \sigma^2\alpha^2 t^2/2)}.$$

It is obvious from the mathematical form of the characteristic function that the distribution of y is also normal but with the mean = $\alpha\mu_x + \beta$ and variance = $(\alpha\sigma)^2$.

The following example is prominent as it includes several useful concepts for the coming chapters.

Example 72. Assume three correlated random variables with multivariate normal distribution, the mean $\mu_X = [0, 1, -2]^T$ and the covariance matrix

$$\mathbf{R}_{XX} = \begin{bmatrix} 2 & 0.5 & -0.2 \\ 0.5 & 4 & 1 \\ -0.2 & 1 & 1 \end{bmatrix}.$$

Find:
- The formula of the joint probability density function, $f_X(x_1, x_2, x_3)$. Show in detail how to substitute the numbers.
- The characteristic function $\varphi_X(t)$.
- For the following system of linear equations, $y_1 = 2x_1 - 7x_3 + 1$ and $y_2 = x_1 - x_2 + 2x_3 - 4$, find the characteristic function $\varphi_Y(t)$.
- Use the previous characteristic function to determine the joint probability density function $f_Y(y_1, y_2)$. Are y_1 and y_2 correlated?
- Write a simulation code to validate the above analytical approach.

Solution.
- To substitute the values of the formula of the multivariate normal distribution as given in eq. (2.111), we should find the determinant and the inverse of the covariance matrix as follows:

$$\det(\mathbf{R}_{XX}) = \det\left(\begin{bmatrix} 2 & 0.5 & -0.2 \\ 0.5 & 4 & 1 \\ -0.2 & 1 & 1 \end{bmatrix}\right) = 5.390$$

$$\mathbf{R}_{XX}^{-1} = \begin{bmatrix} 2 & 0.5 & -0.2 \\ 0.5 & 4 & 1 \\ -0.2 & 1 & 1 \end{bmatrix}^{-1} = \begin{bmatrix} 0.557 & -0.130 & 0.241 \\ -0.130 & 0.364 & -0.390 \\ 0.241 & -0.390 & 1.438 \end{bmatrix}.$$

Therefore, the joint distribution is given by:

$$f_{\mathbf{X}}(x_1, x_2, x_3) = \frac{1}{(2\pi)^{1.5} 5.39^3}$$

$$\times \exp\left(-\frac{1}{2}[x_1\ x_2 - 1\ x_3 + 2] \begin{bmatrix} 0.557 & -0.130 & 0.241 \\ -0.130 & 0.364 & -0.390 \\ 0.241 & -0.390 & 1.438 \end{bmatrix} \begin{bmatrix} x_1 \\ x_2 - 1 \\ x_3 + 2 \end{bmatrix} \right).$$

- The characteristic function formula of multivariate normal distribution is given in eq. (2.165) as

$$\varphi_X(\mathbf{t}) = \exp\left(j[t_1\ t_2\ t_3] \begin{bmatrix} 0 \\ 1 \\ -2 \end{bmatrix} - \frac{1}{2}[t_1\ t_2\ t_3] \begin{bmatrix} 2 & 0.5 & -0.2 \\ 0.5 & 4 & 1 \\ -0.2 & 1 & 1 \end{bmatrix} \begin{bmatrix} t_1 \\ t_2 \\ t_3 \end{bmatrix} \right)$$
$$= \exp(jt_2 - 2jt_3 - t_1^2 - 0.5t_1t_2 + 0.2t_1t_3 - 2t_2^2 - t_2t_3 - 0.5t_3^2).$$

- We may express the linear equations in more compact form as $\mathbf{Y} = \mathbf{AX} + \mathbf{B}$, where $\mathbf{Y} = \left[\begin{smallmatrix} y_1 \\ y_2 \end{smallmatrix}\right]$, the matrix $\mathbf{A} = \left[\begin{smallmatrix} 2 & 0 & -7 \\ 1 & -1 & 2 \end{smallmatrix}\right]$, the vector $\mathbf{X} = [x_1\ x_2\ x_3]^T$, and the vector $\mathbf{B} = [1\ -4]^T$.
  The characteristic function can be computed as:

$$\varphi_Y(\mathbf{v}) = E[e^{j\mathbf{v}^T\mathbf{Y}}] = E[e^{j\mathbf{v}^T\mathbf{AX}}e^{j\mathbf{v}^T\mathbf{B}}]$$
$$= e^{j\mathbf{v}^T\mathbf{B}}E[e^{j\mathbf{v}^T\mathbf{AX}}] = e^{j\mathbf{v}^T\mathbf{B}}\varphi_X(\mathbf{A}^T\mathbf{v}).$$

  Please note that $\mathbf{v}$ here has a dimension of 2. The detailed expression of this formula is

$$\mathbf{t} = \begin{bmatrix} t_1 \\ t_2 \\ t_3 \end{bmatrix} = \mathbf{A}^T\mathbf{v} = \begin{bmatrix} 2 & 1 \\ 0 & -1 \\ -7 & 2 \end{bmatrix} \begin{bmatrix} v_1 \\ v_2 \end{bmatrix} = \begin{bmatrix} 2v_1 + v_2 \\ -v_2 \\ -7v_1 + 2v_2 \end{bmatrix}.$$

  Substituting t values in the above characteristic function of $\mathbf{X}$ and add $j(v_1 - 4v_2)$ we obtain the characteristic function of $\mathbf{Y}$.
- The general shape of the characteristic function of $\mathbf{Y}$ is

$$\varphi_Y(\mathbf{v}) = e^{j\mathbf{v}^T\mathbf{B}}\varphi_X(\mathbf{A}^T\mathbf{v}) = e^{j\mathbf{v}^T\mathbf{B}}e^{(j\mathbf{v}^T\mathbf{A}\mu_X - \mathbf{v}^T\mathbf{A}\mathbf{R}_{XX}\mathbf{A}^T\mathbf{v}/2)}$$
$$= e^{(j\mathbf{v}^T(\mathbf{B}+\mathbf{A}\mu_X) - \mathbf{v}^T\mathbf{A}\mathbf{R}_{XX}\mathbf{A}^T\mathbf{v}/2)}.$$

  Hence the distribution of $\mathbf{Y}$ is also multivariate normal distribution with mean

$$\mu_Y = \mathbf{B} + \mathbf{A}\mu_X$$
$$= \begin{bmatrix} 1 \\ -4 \end{bmatrix} + \begin{bmatrix} 2 & 0 & -7 \\ 1 & -1 & 2 \end{bmatrix} \begin{bmatrix} 0 \\ 1 \\ -2 \end{bmatrix} = \begin{bmatrix} 15 \\ -9 \end{bmatrix}.$$

  The covariance matrix is deduced as

$$\mathbf{R}_{YY} = \mathbf{A}\mathbf{R}_{XX}\mathbf{A}^T$$

$$= \begin{bmatrix} 2 & 0 & -7 \\ 1 & -1 & 2 \end{bmatrix} \begin{bmatrix} 2 & 0.5 & -0.2 \\ 0.5 & 4 & 1 \\ -0.2 & 1 & 1 \end{bmatrix} \begin{bmatrix} 2 & 1 \\ 0 & -1 \\ -7 & 2 \end{bmatrix}$$

$$= \begin{bmatrix} 62.6 & -3.4 \\ -3.4 & 4.2 \end{bmatrix}.$$

Since the covariance matrix is not diagonal, therefore, y_1 and y_2 must be correlated.

– Next we show the simulation codes for this problem in both Octave and Scilab, respectively.

```
%Octave Code
clear all
N=100000; %number of generated RVs
Cov=[2 0.5 -0.2; 0.5 4 1; -0.2 1 1]; %Covariance
    Matrix
Mean=[0;1;-2]; %mean values
X=mvnrnd(Mean,Cov,N);
A=[2 0 -7;1 -1 2];
B=[1;-4];
Y=A*X'+B;
CovY=cov(Y')
MeanY=mean(Y')
```

```
// Scilab Code
clear
N=100000; //number of generated RVs
Cov=[2 0.5 -0.2; 0.5 4 1; -0.2 1 1]; // Covariance
    Matrix
Mean=[0;1;-2]; // Mean vector
X=grand(N,'mn',Mean,Cov);
A=[2 0 -7;1 -1 2];
B=[1;-4]
Y=A*X+B*ones(1,N);
CovY=cov(Y')
MeanY=mean(Y,2)
```

The results based on the simulation were MeanY=[15.00, –8.99] and the covariance of Y was

$$CovY = \begin{bmatrix} 62.97 & -3.39 \\ -3.39 & 4.20 \end{bmatrix}.$$

It is clear that the simulation results are fully compatible with analytical results.

Based on the previous example, we may formulate the following Theorems without further proofs.

Theorem 1. *For any linear system of equations* $\mathbf{Y} = \mathbf{AX}+\mathbf{B}$, *where* $\mathbf{X} \subset \mathbb{R}^{n\times1}$ *is a multivariate normal distribution with vector mean* $\mu_{\mathbf{X}} \subset \mathbb{R}^{n\times1}$ *and covariance matrix* $\mathbf{R}_{XX} \subset \mathbb{R}^{n\times n}$, $\mathbf{A} \subset \mathbb{R}^{m\times n}$ *is a deterministic matrix,* $\mathbf{B} \subset \mathbb{R}^{m\times1}$ *is a deterministic vector, therefore* $\mathbf{Y} \subset \mathbb{R}^{m\times1}$ *will be also a multivariate normal distribution with mean*

$$\mu_{\mathbf{Y}} = \mathbf{B} + \mathbf{A} \times \mu_{\mathbf{X}} \tag{2.174}$$

and Covariance

$$\mathbf{R}_{YY} = \mathbf{A}\mathbf{R}_{XX}\mathbf{A}^T. \tag{2.175}$$

It is also possible to extend the previous theorem to include the case where **B** is a multivariate normal distribution, as explained by the next theorem.

Theorem 2. *For any linear system of equations represented by* $\mathbf{Z} = \mathbf{AX} + \mathbf{Y}$, *where* $\mathbf{X} \subset \mathbb{R}^{n\times1}$ *is a multivariate normal distribution with vector mean* $\mu_{\mathbf{X}} \subset \mathbb{R}^{n\times1}$ *and covariance matrix* $\mathbf{R}_{XX} \subset \mathbb{R}^{n\times n}$, $\mathbf{A} \subset \mathbb{R}^{m\times n}$ *is the deterministic matrix,* $\mathbf{Y} \subset \mathbb{R}^{m\times1}$ *is a multivariate normal distributed vector with vector mean* $\mu_{\mathbf{Y}} \subset \mathbb{R}^{m\times1}$ *and covariance matrix* $\mathbf{R}_{YY} \subset \mathbb{R}^{m\times m}$, $\boldsymbol{X}$ *and* $\boldsymbol{Y}$ *are independent of each other, so* $\mathbf{Z} \subset \mathbb{R}^{m\times1}$ *will also be a multivariate normal distribution with mean*

$$\mu_{\mathbf{Z}} = \mu_{\mathbf{Y}} + \mathbf{A} \times \mu_{\mathbf{X}} \tag{2.176}$$

and Covariance

$$\mathbf{R}_{ZZ} = \mathbf{A}\mathbf{R}_{XX}\mathbf{A}^T + \mathbf{R}_{YY}. \tag{2.177}$$

The Theorem 2 could be easily proved using the characteristic function of two summed multivariate random variables, such as: $\varphi_Z(\mathbf{t}) = \varphi_X(\mathbf{A}^T\mathbf{t})\varphi_Y(\mathbf{t})$.

The above two theorems will be very useful when we talk about Kalman filters later.

2.8 Limits and bounds

Sequences of random variables can converge to a limit, but it is often not possible to provide a strict deterministic proof for this limit. For example, if we have a deterministic sequence defined as $x_n = 0.5 + \frac{1}{n^2}$, it is straightforward to prove that $\lim_{n\to\infty} y_n = 0.5$ because $\frac{1}{n^2}$ will definitely converge to zero as $n \to \infty$. However, consider the case where we throw a fair coin n times and let $N_H(n)$ be the number of heads that appear. We can mathematically formulate $N_H(n)$ as

$$N_H(n) = \sum_{k=1}^{n} \mathbf{1}_A(k),$$

where $\mathbf{1}_A(k) = 1$ when the k^{k^th} outcome of the experiment A is heads, and 0 otherwise, i. e., tails; in this case, $A \in \{\text{Head}, \text{Tail}\}$. We know that for a fair coin, the limit of $\frac{N_H(n)}{n}$ should be 0.5 for very large n. However, we cannot provide a rigorous proof that the limit will is exactly 0.5. If you simulate this experiment on a computer, you will approach $0.5 \pm \Delta$. However, the number obtained will never remain fixed at exact 0.5, regardless of how large n is. Therefore, for random sequences, we use mainly use three different limit definitions to study the convergence behavior, they are:

- Convergence in probability:
  We say that a random sequence x_n converges to x in probability if

$$\lim_{n\to\infty} P(|x_n - x| > \epsilon) = 0 \quad \text{for any real number } \epsilon > 0. \tag{2.178}$$

- Almost certain convergence:
  We say that the random sequence x_n almost certainly converges to x if

$$P\Big(\lim_{n\to\infty} x_n = x\Big) = 1. \tag{2.179}$$

- Convergence in the mean square:
  If the second moments of x and x_n are exist and finite, i. e. $E[x^2] < \infty$ and $E[x_n^2] < \infty\ \forall n$, we say that the random sequence x_n is mean-square convergent to x if

$$\lim_{n\to\infty} E\big[(x_n - x)^2\big] = 0. \tag{2.180}$$

Determining the statistical properties and behavior of random variables requires accurate knowledge of, for example, the probability density function. However, in many situations, such knowledge does not exist, or only rough estimates are available, or accurate analysis is very complicated. Nevertheless, we can build some bounds about the results. For example, to say the random variable x is bounded above (or below) by a certain number. A few useful bounds are given next.

2.8.1 The Chebyshev bound

As we have seen before, we can determine the probability of certain events happening if we know the probability density (or mass) function of the random variables. However, if we do not know about the probability density function and only the mean and variance are available; hence, Chebyshev inequality could be a useful alternative.

Let X be a random variable with an unknown probability density function $f_X(x)$. However, we know (perhaps from long-term observations) its mean and variance as μ_X and σ_X^2 respectively. Hence

$$\sigma_X^2 = \int_{-\infty}^{\infty} (x - \mu_X)^2 f_X(x)dx \geq \delta^2 \int_{|x-\mu_X|\geq\delta} f_X(x)dx$$
$$= \delta^2 P(|x - \mu_X| \geq \delta).$$

Therefore, we may express the Chebyshev bound as:

$$P(|x - \mu_X| \geq \delta) \leq \frac{\sigma_X^2}{\delta^2}. \tag{2.181}$$

This inequality is valid for any probability density function with a finite mean and variance. The Chebyshev inequality gives the upper bound on the probability that the distance between the random variable and its mean is greater than a specific value. It has many applications; moreover, it is useful for several proofs.

Example 73. A positive random number x with zero mean and unit variance and an unknown distribution. Find
- The upper bound of the probability that $P(x \geq 3)$.
- The upper bound of the probability that $P(x \leq 1)$.

Solution. Since $\mu_X = 0$ and $\sigma^2 = 1$, therefore:
- $P(x \geq 3) \leq \frac{1}{9}$.
- $P(x \leq 1) = 1 - P(x \geq 1) \geq 1 - \frac{1}{1} = 0$, in this example, the Chebyshev inequality is informationless, because every probability must be at least 0.

The Chebyshev inequality can be a rather loose bound. For example, when $\sigma^2 \geq \delta^2$, the bound is greater than 1.

Example 74. Assume x_k is a random variable sequence obtained from any distribution with a finite mean and variance. Using the Chebyshev inequality, prove that

$$\lim_{N\to\infty} \frac{1}{N} \sum_{k=1}^{N} x_k \to \mu_X = E[x].$$

Solution. Let's define the random variable y as

$$y = \frac{1}{N} \sum_{k=1}^{N} x_k.$$

Applying Chebyshev bound as

$$P(|y - \mu_X| \geq \delta) = P\left(\left|\frac{1}{N} \sum_{k=1}^{N} x_k - \mu_X\right| \geq \delta\right) \leq \frac{\sigma_Y^2}{\delta^2}.$$

The variance of y could be computed as follows:

$$\sigma_Y^2 = E\left[\left(\frac{\sum_{k=1}^{N} x_k}{N} - \mu_X\right)^2\right] = E\left[\left(\frac{\sum_{k=1}^{N} x_k}{N} - \frac{N\mu_X}{N}\right)^2\right]$$
$$= \frac{1}{N^2}E\left[\left(\sum_{k=1}^{N} x_k - N\mu_X\right)^2\right] = \frac{N\sigma_X^2}{N^2} = \frac{\sigma_X^2}{N}.$$

By substituting in the Chebyshev bound we obtain

$$P\left(\left|\frac{1}{N}\sum_{k=1}^{N} x_k - \mu_X\right| \geq \delta\right) \leq \frac{\sigma_X^2}{N\delta^2}.$$

It is clear that as $N \to \infty$ the upper bound

$$\frac{\sigma_X^2}{N\delta^2} \to 0 \quad \forall \delta > 0.$$

Therefore,

$$\lim_{N\to\infty} P\left(\left|\frac{1}{N}\sum_{k=1}^{N} x_k - \mu_X\right| \geq \delta\right) = 0.$$

It means that $y = \frac{1}{N}\sum_{k=1}^{N} x_k$ will converge in probability (see eq. (2.178)) to the exact μ_X for very large N under the condition that x has finite mean and variance.

Example 75. In eq. (2.10), we showed that by repeating an experiment with uncertain outcomes many times and counting the relative number of times a certain outcome (event) occurs, the probability of that event approaches $N \to \infty$. This is the main concept behind Monte Carlo simulation. Prove the concept of eq. (2.10) using Chebyshev inequality.

Solution. Here, our target is to prove that the following limit is true in some sense

$$P(E) = \lim_{N\to\infty} \frac{N_E}{N}. \tag{2.182}$$

Let's define a random variable X_k such that $X_k = 1$ if the event E occurs at iteration k and 0 otherwise. Next we define another random variable y such as:

$$y = \frac{\sum_{k=1}^{N} X_k}{N}.$$

The expected value of X_k is

$$E[X_k] = 1 \times P(E) + 0 \times [1 - P(E)] = P(E)$$
$$\Rightarrow \mu_y = E[y] = \frac{NP(E)}{N} = P(E).$$

Furthermore,

$$E[X_k^2] = 1^2 \times P(E) + 0^2 \times [1 - P(E)] = P(E).$$

But, $\sigma_y^2 = E[y^2] - \mu_y^2$. We may find $E[y^2]$ as follows,

$$\begin{aligned}
E[y^2] &= \frac{1}{N^2} E\left[\sum_{k=1}^{N}\sum_{i=1}^{N} X_i X_k\right] \\
&= \frac{1}{N^2} E\left[\sum_{i=1}^{N} X_i^2 + \sum_{k=1}^{N}\sum_{i \neq k}^{N} X_i X_k\right] \\
&= \frac{1}{N^2}\left[NP(E) + N(N-1)P(E)^2\right] \\
\Rightarrow \sigma_y^2 &= \frac{P(E)}{N}.
\end{aligned}$$

Applying Chebyshev bound, we obtain:

$$P\left(\left|\frac{\sum_{k=1}^{N} X_k}{N} - P(E)\right| \geq \delta\right) \leq \frac{\sigma_y^2}{\delta^2} = \frac{P(E)}{\delta^2 N}. \tag{2.183}$$

It is clear that as $N \to \infty$ the relative repetition of the experiment will converge to the exact probability with probability 1, for any nonzero δ.

Example 76. Consider an experiment where the precise probability of a specific event, denoted as E, remains unknown, but it is constrained to lie between 0.2 and 0.8. How many trials, represented by N, are necessary to ensure a result within an accuracy of 0.01 with a probability of at least 0.95?

Solution. We can utilize the formula obtained in the previous example as

$$P\left(\left|\frac{\sum_{k=1}^{N} X_k}{N} - P(E)\right| \leq \delta\right) \geq \left(1 - \frac{P(E)}{\delta^2 N}\right). \tag{2.184}$$

Here, $\delta = 0.01$, and $(1 - \frac{P(E)}{\delta^2 N}) = 0.95$. Therefore,

$$N = \frac{P(E)}{0.05\delta^2}. \tag{2.185}$$

For the worst case, the number of N should be computed for $P(E) = 0.8$, which gives: $N = \frac{0.8}{0.05 \times 0.01^2} = 160,000$. This number is valid regardless of the distribution of the original random variable. However, in practice, we may need much less than this number. Furthermore, observe that this number is highly dependent on the required accuracy. For example, reducing the accuracy to be 0.05, will reduce N to 6400 trails.

2.8.2 The Chernoff bound

The Chebyshev bound has a notable drawback: it tends to be a loose upper bound. In contrast, the Chernoff bound offers a tighter estimation, albeit it requires slightly more complex computations. It relies on the fundamental premise that the unit step is consistently less than or equal to the exponential function, i. e., for $u(x) = 1$ for $x \geq 0$ and 0 elsewhere, it is always true that

$$u(x) \leq e^{\lambda x}, \quad \text{for any } x, \text{ and } \forall \lambda \geq 0. \tag{2.186}$$

Therefore,

$$\begin{aligned} P(X \geq 0) &= \int_0^{\infty} f_X(x)dx \\ &= \int_{-\infty}^{\infty} u(x)f_X(x)dx \leq \int_{-\infty}^{\infty} e^{\lambda x} f_X(x)dx = E[e^{\lambda x}]. \end{aligned}$$

An interesting point here is that $E[e^{\lambda x}]$ could be obtained from the moment generating function. Furthermore, to make the bound very tight, we may derive the optimal $\hat{\lambda}$ as:

$$\hat{\lambda} = \arg\min_{\lambda > 0} E[e^{\lambda x}].$$

Therefore, the following bound applies to any distribution:

$$P(X \geq 0) \leq E[e^{\hat{\lambda} x}]. \tag{2.187}$$

Example 77. 1000 random discrete time samples with numbers −10, −1, 6 and percentages 0.7, 0.2, 0.3, respectively. However, the original distribution is unknown. Find the probability of having positive samples.

Solution. We may find the expected value as

$$E[e^{\hat{\lambda} x}] = e^{-10\hat{\lambda}} \times 0.3 + e^{-\hat{\lambda}} \times 0.2 + e^{6\hat{\lambda}} \times 0.3.$$

The probability inequality in (2.187) is valid for any $\hat{\lambda}$, so to get a tight estimate of the boundary, we should find $\hat{\lambda}$ that minimize the expectation. Figure 2.29 shows the expectation versus λ. From the figure, we can see that $P(X \geq 0) \leq 0.572$.

2.8.3 The Markov inequality

The Markov inequality is useful for nonnegative random variables, i. e., $x \geq 0$. In this case

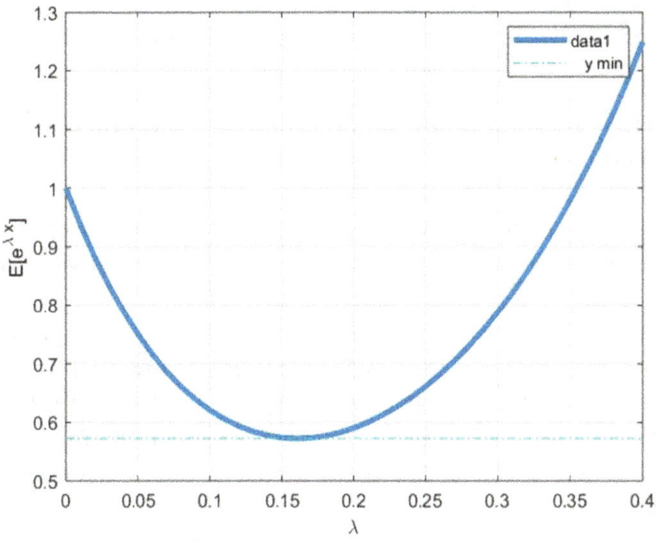

Figure 2.29: Chernoff example.

$$E[X] = \int_0^\infty x f_X(x)dx$$
$$= \int_0^a x f_X(x)dx + \int_a^\infty x f_X(x)dx.$$

Therefore,

$$E[X] \geq \int_a^\infty x f_X(x)dx \geq a \int_a^\infty f_X(x)dx = aP(x \geq a).$$

Hence, the upper bound of the probability for nonnegative random variables x is:

$$P(x \geq a) \leq \frac{E[x]}{a} = \frac{\mu_X}{a}. \tag{2.188}$$

Where $a > 0$.

Example 78. Consider a positive random number sequence with unknown distribution as: 1, 2.5, 0.7, 3, 1, x. Find the upper bound of the probability that $P(x \geq 4)$?

Solution. Hence, we have no more information than the random variables are always positive, we may estimate the mean from the observation (sample mean) as:

$$\mu_X = \frac{1 + 2.5 + 0.7 + 3 + 1}{5} = 1.64. \tag{2.189}$$

Using the Markov inequality $P(x \geq 4) \leq \frac{1.64}{4} = 0.41$.

2.9 Stochastic processes

The stochastic, or random, process encompasses the unpredictable outcomes of an experiment tied to an independent variable such as time, distance, or frequency. For instance, consider the coin-tossing experiment yielding either 'heads' or 'tails.' To translate these events into a random variable, let's assign $x = 1$ for 'heads' and $x = -1$ for 'tails.' However, this variable holds meaning solely within the context of the coin toss; outside of that context, such as when the coins are in your pocket or being used for transactions, it loses significance.

However, when the experiment is repeated at regular intervals, creating a sequence over time (let's say every second), it transforms a random or stochastic process. Imagine a random square wave signal generator producing ± 1 volts at a frequency of 1 MHz, much like a coin flip determining the voltage output every microsecond. This results in a signal that fluctuates over time. Understanding and analyzing such random processes requires robust methods.

Questions arise regarding the spectral properties of these signals, their Fourier transforms, and the system outputs when exposed to random inputs. Dynamic systems, described by mathematical differential equations, prompt inquiries about differentiation or integration of random signals. Exploring the distinction between averaging in the time domain versus the probabilistic domain becomes crucial.

This section delves into these inquiries and more, offering insight without unnecessarily complicating the subject matter.

Let's consider a simple scenario: someone throws a die every second, where the resulting random variable represents the number of dots on the die. This can also be envisioned as a random signal generator producing signals $x(t) \in \{1, 2, 3, 4, 5, 6\}$ volts at a frequency of 1 Hz. In Figure 2.30, we observe a single instance of the output of such a signal. However, there are numerous potential realizations; imagine thousands of individuals throwing similar dice, each with different speeds or probability distributions. The collective set of all these time-based realizations is called an *ensemble*. Figure 2.31 displays four distinct time-based realizations of this random process involving die tossing.

We define random processes, denoted as $X(A, t)$, as functions with two variables: an event A and time t (or other independent variables). At a specific time t_k, the random process $X(A, t_k)$ simplifies to a random variable $X(t_k)$, contingent upon the event. Its nature can be described by a probability density function (PDF) if it is continuous, or by a probability mass function (PMF) if it is discrete (such as in the example of throwing a die). Conversely, for a particular event A_j, we observe a singular time-based function realization as $X(A_j, t) = X_j(t)$.

Example 79. For the case of a four-dice random process shown in Figure 2.31, construct the sample space of the random variable $X(t_k)$.

Answer. In this case, at any given point in time, the problem is reduced to a four-dice throwing experiment. Therefore, the sample space is

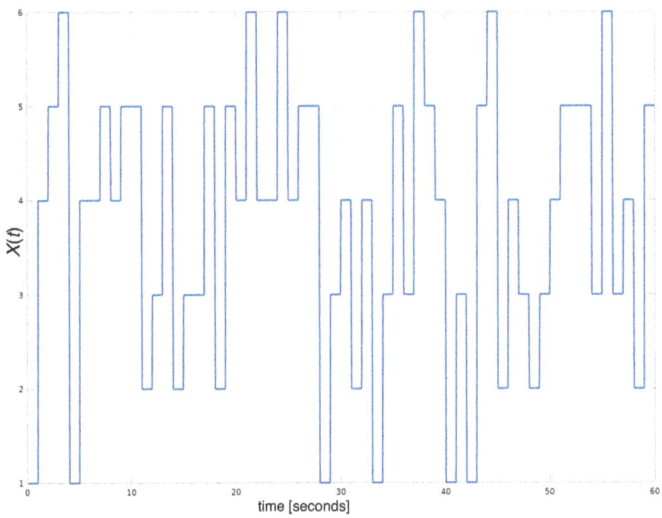

Figure 2.30: Die tossing random process.

$$\Omega = \{(1,1,1,1), (1,1,1,2), \ldots, (6,6,6,6)\}. \tag{2.190}$$

In the preceding example, our sample space is finite, given that it involves a discrete and finite number of dice. However, the concept extends to the probability density function in the realm of continuous random variables and infinite time realizations. Random processes manifest themselves in various scenarios, and we will explore some of them based on the straightforward experiment of sequentially rolling four dice.

- Each time realization $X_j(t)$ may exhibit partial or complete correlation with other realizations, such as $X_k(t)$ where $k \neq j$.
- The statistical parameters—the probability of events, mean, variance, and even the probability density (mass) function associated with each realization $X_j(t)$ can either remain constant over time or fluctuate. For instance, consider a scenario with a single roll of the dice: if the probability distribution of outcomes changes over time, or the frequency of occurrence changes, this results in a time-varying realization.
- Each realization could have its own properties. For example, in Figure 2.31, it is clear that $X(A_4, t)$ has lower frequency than the others.

After incorporating time as an independent variable, it becomes crucial to include it in computing all statistical parameters, including mean and variance. To differentiate random process signals from deterministic or ergodic signals (to be explained soon), it's customary to include an event symbol such as $x(A, t)$ or $x(\omega, t)$. Nevertheless, throughout this book, we'll predominantly utilize the notation $x(t)$.

The formulation of the cumulative distribution function (CDF) now appears as follows:

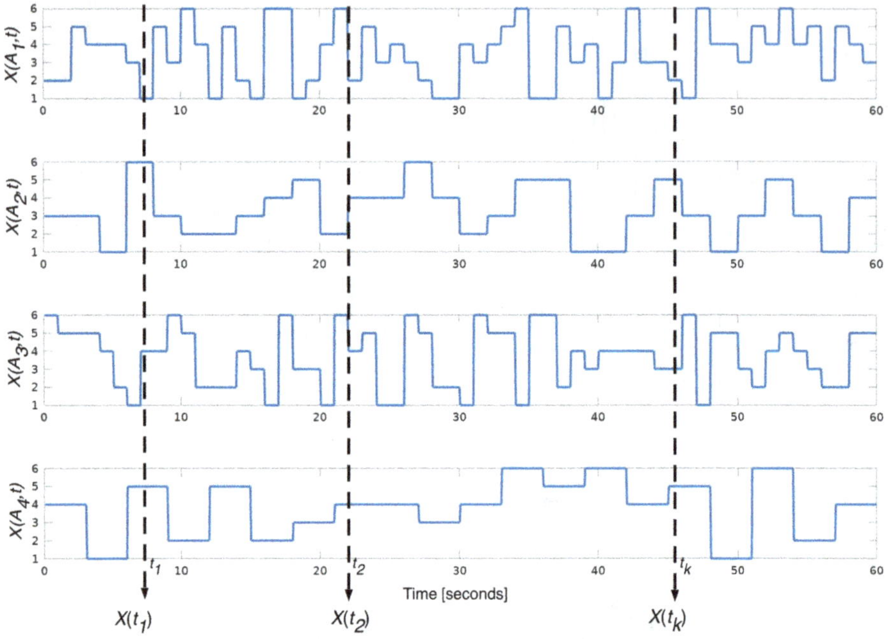

Figure 2.31: 4 Dice tossing random process.

$$F_X(x_0; t_k) = P(x(t_k) \leq x_0). \tag{2.191}$$

Thus, it's crucial to consider the time parameter, which reflects the moment when the experiment is conducted. Keep in mind that our analysis consistently revolves around two dimensions: the event dimension and the time dimension.

In general, the probability density function becomes time-dependent as follows:

$$f_X(x, t_k) = \left.\frac{dF_X(x_0; t_k)}{dx_0}\right|_{x=x_0}. \tag{2.192}$$

Hence, the calculation for the mean of the random process $X(t_k)$ is expressed as:

$$E[X(t_k)] = \int_{-\infty}^{\infty} x f_{X_t}(x, t_k) dx = \mu_X(t_k). \tag{2.193}$$

The higher moment computations should also take the time variable into consideration. For instance $E[X(t_1)X(t_2)]$ could be formulated as:

$$E[X(t_1)X(t_2)] = \int_{-\infty}^{\infty}\int_{-\infty}^{\infty} x_{t_1} x_{t_2} f_{X_t X_t}(x_{t_1}, t_1; x_{t_2}, t_2) dx_{t_1} dx_{t_2}. \tag{2.194}$$

To comprehend the meaning of $f_{X_t X_t}(x_{t1}, t_1; x_{t2}, t_2)$, let's establish the second-order cumulative distribution function for a time-dependent random variable at two distinct points in time. In simpler terms, for a random process $x(t)$, this represents the probability that $x(t)$ is less than or equal to x_1 at time t_1 *and* less than or equal to x_2 at time t_2, formulated as:

$$F(x_1, t_1; x_2, t_2) = P(x(t_1) \leq x_1, x(t_2) \leq x_2). \tag{2.195}$$

Therefore, the joint time-varying probability density function is:

$$f(x_{t_1}, t_1; x_{t_2}, t_2) = \frac{\partial^2 F(x_1, t_1; x_2, t_2)}{\partial x_1 \partial x_2}\bigg|_{x_{t_1} = x_1; x_{t_2} = x_2}. \tag{2.196}$$

The integration of the time variable into the analysis of random processes is essential. The variance, traditionally defined as $E[x^2] - \mu_X^2$, is not complete when considering a random process $x(t)$. This is due to the dynamic nature of $x(t)$, which can exhibit time-dependent properties. Hence, instead of variance, we introduce the autocovariance for random processes as follows:

$$\begin{aligned} \mathrm{cov}_{XX}(t_1, t_2) &= E[(X(t_1) - \mu_X(t_1))(X(t_2) - \mu_X(t_2))] \\ &= E[X(t_1)X(t_2)] - \mu_X(t_1)\mu_X(t_2) \\ &= R_{XX}(t_1, t_2) - \mu_X(t_1)\mu_X(t_2). \end{aligned} \tag{2.197}$$

Where $R_{XX}(t_1, t_2) = E[X(t_1)X(t_2)]$ is called the autocorrelation function. Therefore, the formulation of variance aligns as a special case when examining specific instances of certain random processes, effectively reducing it to a random variable. Equation (2.197) illustrates that when the mean is zero for at least one time instance, the autocovariance function coincides with the autocorrelation. In this particular scenario:

$$\mathrm{cov}_{XX}(t_k, t_k) = R_{XX}(t_1, t_2).$$

Example 80. Consider a time-varying random voltage generator that produces voltage levels of ±1 volts. Each second, it generates these levels based on a probability distribution given by

$$P(x_k = +1) = \frac{1+k}{1+2k},$$

where k represents the time iteration ($k = 0, 1, 2, \dots$). The aim is to analyze the statistical characteristics of this generator both analytically and through simulation.

Answer. It is evident that this stochastic process exhibits time variation. Since the generator has only two possible outcomes, hence,

$$P(x_k = -1) = 1 - \frac{1+k}{1+2k} = \frac{k}{1+2k}.$$

Now, when we compute the mean, we should also consider the time instance as:

$$\mu_X(k) = +1 \times \frac{1+k}{1+2k} - 1 \times \frac{k}{1+2k} = \frac{1}{1+2k}.$$

This implies that if many generators sharing these characteristics are used, and their outcomes are measured at time iteration k, the average at that specific time will converge to $\frac{1}{1+2k}$ as the number of generators increases. The autocorrelation of this generator is $R_{XX}(k,m) = E[x_k x_m]$. For $m = k$,

$$R_{XX}(k,k) = E[x_k^2] = +1 \times \frac{1+k}{1+2k} + 1 \times \frac{k}{1+2k} = 1.$$

For $m \neq k$, we have,

$$R_{XX}(k,m) = E[x_k]E[x_m] = \frac{1}{(1+2k)(1+2m)}.$$

Finally, the autocovariance function for $k = m$ is

$$\text{cov}_{XX}(k,m) = 1 - \frac{1}{(1+2k)^2}$$

and for $k \neq m$, the autocovariance $\text{cov}_{XX}(k,m) = 0$.

Now, let's develop a simulation code to model this scenario and subsequently compare the results obtained with the aforementioned analytical findings.

```
clear all
N=10000; %Number of random events
T=59; %maximum time iterations
for k=0:T
p=(k+1)/(1+2*k);
x(k+1,:)=sign(p-rand(1,N));
end
k=0:T;
Mean=mean(x,2);% Mean based on simulation
Mu=1./(1+2*k);%Mean based on Analytical analysis
Cov=cov(x'); % Autocovariance based on simulation
Cov_Analytical=diag((1-1./(1+2*k).^2));
plot(k,Mean,"Linewidth",4,k,Mu,"Linewidth",4); xlim([0
   T])
figure
subplot(211),
mesh(Cov','FaceLighting','gouraud','LineWidth',2);
```

```
    zlim([-0.05 1])
    subplot(212), mesh(Cov_Analytical,'FaceLighting',
        'gouraud', 'LineWidth',2);
```

The comparison between the results of the simulation code and the derived analytical outcomes is illustrated for the time-varying mean and the autocovariance in Figures 2.32 and 2.33, respectively.

The cross-correlation, which quantifies the similarity between two distinct random processes with a time delay between them, is defined as:

$$R_{XY}(t_1,t_2) = E[X(t_1)Y(t_2)] = \int_{-\infty}^{\infty}\int_{-\infty}^{\infty} x_{t_1} y_{t_2} f_{XY}(x,t_1;y,t_2)\,dx_{t_1}\,dy_{t_2}. \tag{2.198}$$

Random processes X and Y are orthogonal processes when $R_{XY}(t_1,t_2) = 0$. The cross-covariance between two random processes is defined in the same way:

$$\mathrm{Cov}_{XY}(t_1,t_2) = R_{XY}(t_1,t_2) - \mu_X(t_1)\mu_Y(t_2). \tag{2.199}$$

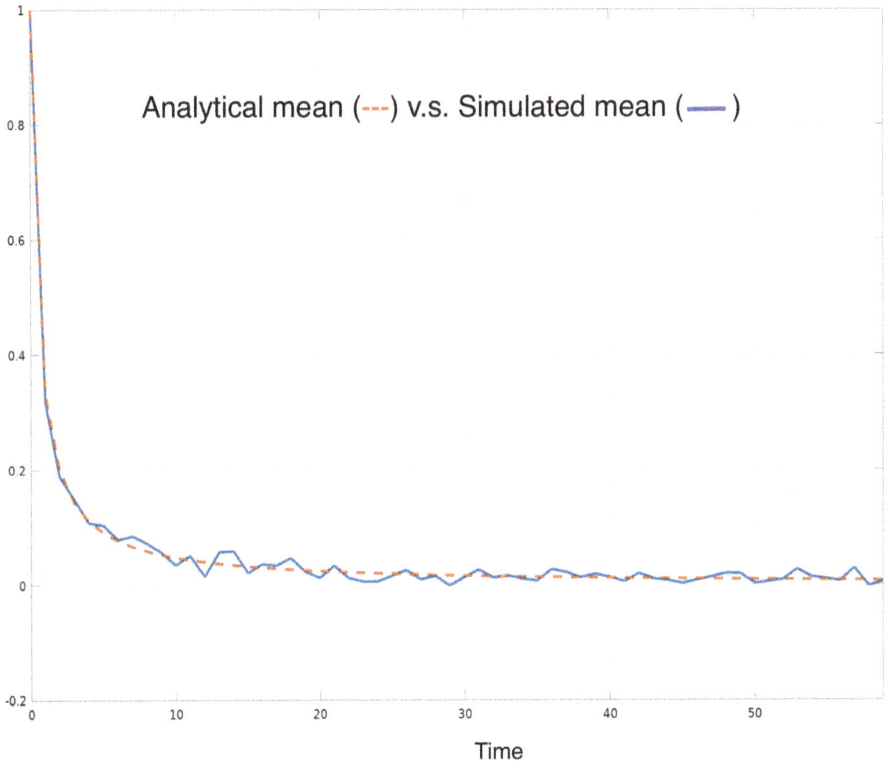

Figure 2.32: Time varying mean.

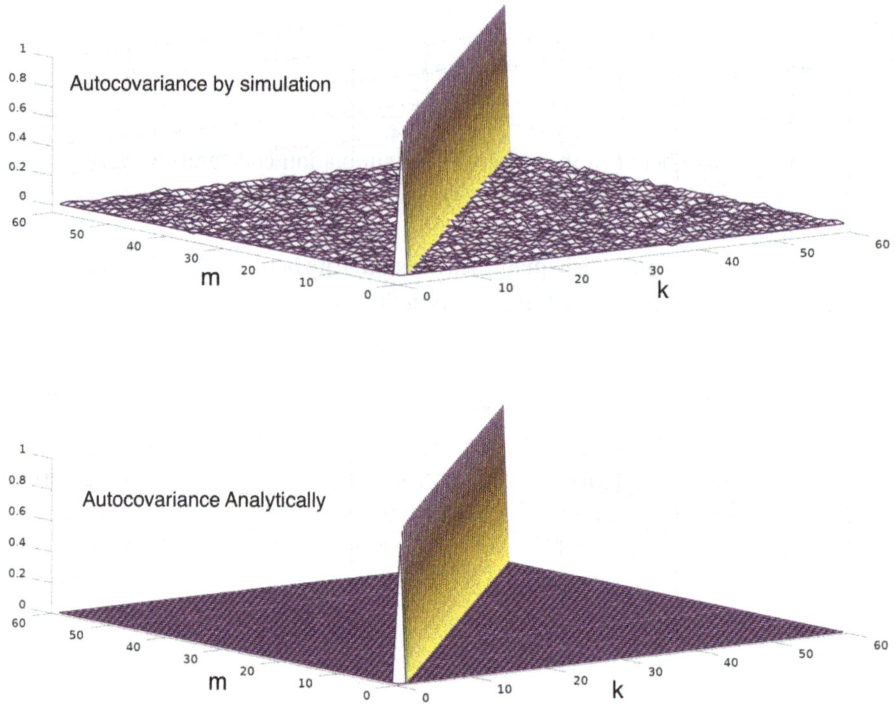

Figure 2.33: Autocovariance by simulation and analysis.

Random processes X and Y are independent processes when

$$\mathrm{Cov}_{XY}(t_1, t_2) = 0. \tag{2.200}$$

2.9.1 The stationary process

The nature of general random processes is inherently time-dependent, which often poses challenges in their analysis, unless we have access to the time-dependent probability density or mass function, as demonstrated in the last example. However, finding this function is typically a complex task. Despite this, certain random processes exhibit time-invariant properties, where their probability density or mass function remains constant with time, expressed as $f_X(x,t) = f_X(x)$. This characteristic signifies that all statistical parameters and moments remain unchanged over time, a property known as *stationarity in the strict sense*. Yet, establishing strict stationarity in a natural random process can be a challenging endeavor. In many practical applications, particularly in fields like electrical engineering, the primary interest lies in the first and second moments. For instance, focusing on the average voltage (or current) and correlation properties enables comprehensive signal and system analysis. Therefore, demonstrating a random process's stationarity in the first and second moments can be sufficient, termed as *wide-sense stationarity* (WSS). The key properties of wide-sense stationary (WSS) processes include:

$$E[X(t)] = \mu_X, \quad \forall t \tag{2.201}$$

$$\begin{aligned} E[X(t_1)X(t_2)] &= E[X(t_0)X(t_0 - (t_2 - t_1))] \\ &= R_{XX}(t_2 - t_1). \end{aligned} \tag{2.202}$$

Consequently, within a wide sense stationary (WSS) framework, the autocorrelation remains independent of time and solely relies on the time difference $t_2 - t_1$. Fortunately, many real-world random processes can be effectively be approximated as WSS, at least within a defined time frame. The autocorrelation function plays a pivotal role in handling random processes. The autocorrelation function of a stationary process

$$R_{XX}(\tau) = E[X(t)X(t + \tau)].$$

Autocorrelation serves as a metric for quantifying the randomness inherent in a random process. A highly correlated random process indicates a lower rate of random changes, resulting in a broader duration for the autocorrelation function. On the other hand, a shorter duration of the autocorrelation indicates higher random fluctuation in the signal. The extreme case is an uncorrelated random process, where the autocorrelation function collapses to a Dirac delta. To illustrate the relationship between a random process and its autocorrelation function, we have generated two examples. The first exhibits high correlation, visualized by its autocorrelation $R_{XX}(\tau)$ in Figure 2.34. The second is nearly uncorrelated, depicted in Figure 2.35, with its autocorrelation function resembling a Dirac delta $\delta(\tau)$.

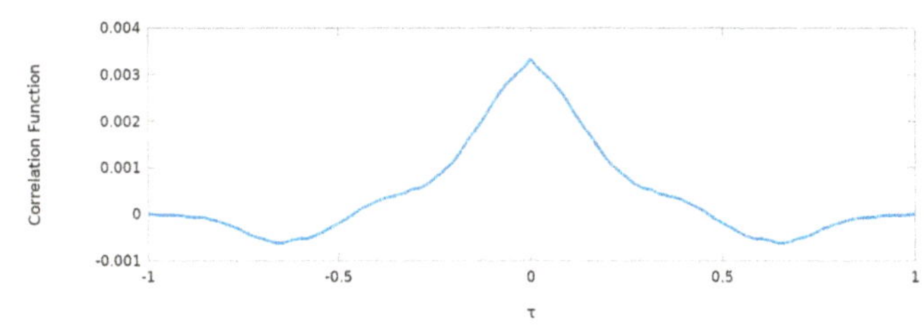

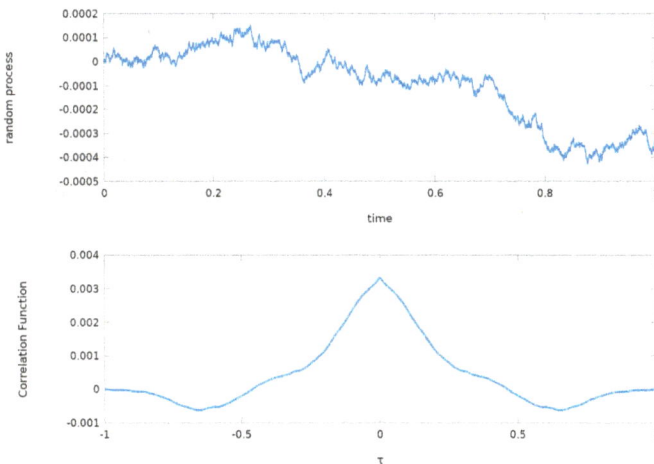

Figure 2.34: Correlated random process.

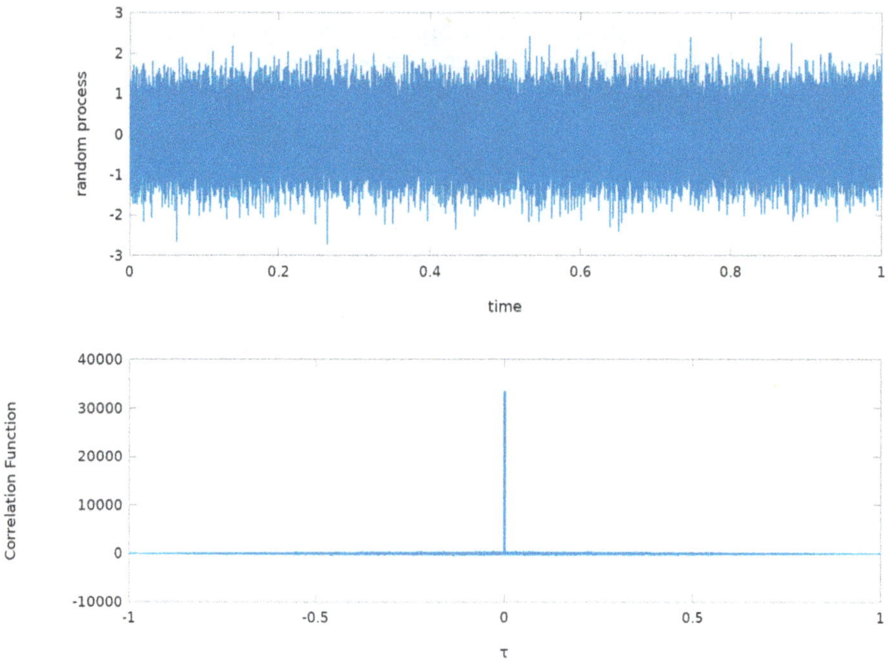

Figure 2.35: Uncorrelated random process.

The figures highlight that predicting a highly correlated random process is achievable with higher accuracy. However, for an uncorrelated random process, even knowing its present state and history doesn't allow accurate future predictions. In the case of an uncorrelated random process following a normal distribution, the best estimate is always its mean, irrespective of its current value or history. Notably, the highest correlation of a random signal occurs at the same moment, i. e., at $R_{XX}(0) = E[X(t)^2]$, i. e., the total average power. Observe that the autocorrelation function is an even function as $R_{XX}(\tau) = R_{XX}(-\tau)$.

The Fourier transform of the autocorrelation function is a window into the spectrum of a random process. When a process exhibits stronger correlations, its spectrum tends to be narrower due to fewer rapid changes. Conversely, an uncorrelated random process generates an infinite spectrum, earning the term *White Noise*—akin to white light that includes all colors. This transformation of the autocorrelation function is known as the power (or energy) spectral density, a concept we'll explore later.

2.9.2 The ergodic process

The term *ergodic* describes a random process where a single, sufficiently long observation holds enough information to reveal the statistical characteristics of that process.

Specifically, a process is considered ergodic in the mean if:

$$\mu_X = E[x] = \int_{-\infty}^{\infty} x f_X(x)dx = \lim_{T\to\infty} \frac{1}{T} \int_{-T/2}^{T/2} X(t)dt. \tag{2.203}$$

Moreover, the randomness of the process is deemed ergodic in its autocorrelation when:

$$R_{XX}(\tau) = E[x(t)x(t-\tau)] = \lim_{T\to\infty} \frac{1}{T} \int_{-T/2}^{T/2} X(t)X(t-\tau)dt. \tag{2.204}$$

Ergodicity belongs to the class of stationary processes. This implies that every ergodic process must be stationary, but the opposite isn't always the case. Embracing the ergodic assumption notably streamlines the analysis of real systems. It allows us to determine the statistical properties of a random process, with a certain level of accuracy, without requiring knowledge of the underlying probability density or mass function governing the random signal. For instance, consider a single time-realized random sequence such as $0, -1, 2, 1, 12, -2$. Consequently, we can easily compute that the mean of this sequence is 2. As the sequence lengthens, we can reasonably infer that the computed mean (i. e., the averaging over time) will approach the actual statistical mean (if it exists). While the assumption of ergodicity simplifies the analysis, it can lead to significant modeling errors if the studied system's physical behavior doesn't align with ergodicity.

Example 81. Consider a random process given by

$$x(t) = a_t \cos(2\pi f t),$$

where a_t is a stationary random signal with exponential distribution with $\lambda = 2$. Find:
- The average of the process $E[x(t)]$.
- The autocorrelation function $R_{XX}(t, \tau)$.
- The autocovariance function.
- What do you say about the type of the random process $x(t)$.

Solution. Since the cosine wave is deterministic, it is not included in the expectation function, i. e.,

$$E[x(t)] = E[a_t \cos(2\pi f t)] = E[a_t] \cos(2\pi f t) = \frac{1}{2} \cos(2\pi f t).$$

It is clear that the process is not stationary. The mean is function in time. The autocorrelation function is computed as:

$$\begin{aligned} R_{XX}(t,\tau) &= E[x(t)x(\tau)] = E[a_t^2 \cos(2\pi f t) \cos(2\pi f \tau)] \\ &= E[a_t^2] \cos(2\pi f t) \cos(2\pi f \tau) = \frac{1}{2} \cos(2\pi f t) \cos(2\pi f \tau). \end{aligned}$$

The autocovariance function is derived as

$$\begin{aligned}\mathrm{Cov}_{XX}(t,\tau) &= R_{XX}(t,\tau) - \mu_X(t)\mu_X(\tau)\\ &= \frac{1}{2}\cos(2\pi ft)\cos(2\pi f\tau) - \frac{1}{4}\cos(2\pi ft)\cos(2\pi f\tau)\\ &= \frac{1}{4}\cos(2\pi ft)\cos(2\pi f\tau).\end{aligned}$$

2.9.3 Energy and power spectral densities

The frequency domain is a reservoir of critical and important information regarding a signal's spectrum. It unveils essential details such as the signal's useful bandwidth. A fundamental signal in this realm is the complex exponential $e^{j2\pi f_0 t}$, which is characterized by a sole frequency component at $f = f_0$ and a bandwidth approaching zero, denoted as $\delta(f - f_0)$. Consequently, any signal $x(t)$ can have its frequency domain revealed by correlating it with this complex exponential signal and averaging over all its frequency components. This widely recognized process is known as the Fourier transform.

$$\hat{X}(f) = \int_{-\infty}^{\infty} x(t)e^{-j2\pi ft}\,dt, \tag{2.205}$$

where $\hat{X}(f)$ represents the frequency domain of $x(t)$. When $x(t)$ is a random process, its Fourier transform naturally assumes a random process as well. Since $\hat{X}(f)$ is just a transformation of $x(t)$ without altering its scale (because $|e^{2j\pi ft}| = 1$), therefore, the energy must be the same in both time and frequency domains, i. e.,

$$\int_{-\infty}^{\infty} x(t)^2 dt = \int_{-\infty}^{\infty} |\hat{X}(f)|^2 df. \tag{2.206}$$

We compute the absolute value of the frequency domain signal, typically a complex function, as is standard practice. It's important to note that, in reality, we don't encounter complex signals in the time domain; instead, we are dealing with delays. Temporal delays manifest themselves as phase shifts in the frequency domain, and these shifts are represented mathematically using complex numbers. Assuming that the signal $x(t)$ has finite energy (e. g., exists for a finite duration) and follows ergodic principles, we can define the energy spectral density as follows:

$$\Psi(f) = |\hat{X}(f)|^2. \tag{2.207}$$

However, if the signal has infinite energy, then the average power (for an ergodic signal) is

$$\lim_{T\to\infty}\frac{1}{T}\int_{-T/2}^{T/2}x(t)^2dt=\lim_{T\to\infty}\int_{-T/2}^{T/2}\left(\frac{x(t)}{\sqrt{T}}\right)^2dt. \tag{2.208}$$

Since we are computing for certain time duration, we can compute the Fourier transform as:

$$\hat{X}_T(f)=\int_{-T/2}^{T/2}\frac{x(t)}{\sqrt{T}}e^{-j2\pi ft}dt. \tag{2.209}$$

The power spectral density could be defined as:

$$\Psi(f)=\lim_{T\to\infty}|\hat{X}_T(f)|^2. \tag{2.210}$$

Random signals are usually power signals (i. e., they have infinite energy for an infinite time duration), so we will always use $\Psi(f)$ for the power spectral density.

In order to derive the power spectral density, we need to determine its relationship to the autocorrelation function. Let's find the Fourier transform of the autocorrelation function in equation (2.204):

$$\begin{aligned}
£(R_{XX}(\tau)) &= \lim_{T\to\infty}\int_{-T/2}^{T/2}R_{XX}(\tau)e^{-j2\pi f\tau}d\tau \\
&= \lim_{T\to\infty}\frac{1}{T}\int_{-T/2}^{T/2}X(t)\left[\int_{-T/2}^{T/2}X(t-\tau)e^{-j2\pi f\tau}d\tau\right]dt \\
&= \lim_{T\to\infty}\int_{-T/2}^{T/2}\frac{X(t)}{\sqrt{T}}\left[\int_{-T/2}^{T/2}\frac{X(t-\tau)}{\sqrt{T}}e^{-j2\pi f\tau}d\tau\right]dt \\
&= \lim_{T\to\infty}\int_{-T/2}^{T/2}\frac{X(t)}{\sqrt{T}}\Phi(t)dt.
\end{aligned}$$

Let $\lambda = t-\tau$, therefore,

$$\begin{aligned}
\Phi(t(\tau)) &= \lim_{T\to\infty}\int_{T/2}^{T/2}\frac{X(t-\tau)}{\sqrt{T}}e^{-j2\pi f\tau}d\tau \\
&= e^{-j2\pi ft}\lim_{T\to\infty}\int_{-T/2}^{T/2}\frac{X(\lambda)}{\sqrt{T}}e^{j2\pi f\lambda}d\lambda \\
&= e^{-j2\pi ft}\hat{X}_T^*(f),
\end{aligned}$$

where $\hat{X}_T^*(f)$ is the conjugate of $\hat{X}_T(f)$. Hence,

$$£(R_{XX}(\tau)) = \hat{X}_T^*(f) \lim_{T\to\infty} \int_{-T/2}^{T/2} \frac{X(t)}{\sqrt{T}} e^{-j2\pi ft} dt = \left|\hat{X}_T(f)\right|^2. \tag{2.211}$$

This outcome highlights that the Fourier transform of the autocorrelation function yields the power spectral density. Interestingly, this remains true even for nonergodic and nonstationary random signals. However, in these scenarios, the average should be computed over the probability density function rather than over a singular realization. For a general nonstationary random process, the power spectral density is calculated as follows:

$$\Psi(t,f) = £(R_{XX}(t, t+\tau)). \tag{2.212}$$

The computation can be rather complicated as:

$$\Psi(t,f) = \int_{-\infty}^{\infty}\int_{-\infty}^{\infty}\int_{-\infty}^{\infty} x_t x_{t+\tau} f_{XX}(x, t; x, t+\tau) e^{-2j\pi f\tau} dx_t dx_{t+\tau} d\tau.$$

Therefore, in the case of nonstationary random processes, the power spectral density will generally exhibit time-varying characteristics.

Example 82. Assuming a zero-mean random process $x(t)$ with an autocorrelation function $R_{XX}(\tau) = 2r^{-|\tau|}$, solve the following:
- Is the process $x(t)$ stationary?
- What is the power spectral density of $x(t)$?
- What is the average power of $x(t)$?
- If we aim to design an ideal low-pass filter that allows 90 % of the average power of this random process to pass through, what bandwidth should this filter have?

Solution.
- The process is stationary because the autocorrelation function depends only on the time difference of τ. It does not depend on the time of the observation.
- The power spectral density is the Fourier transform of the autocorrelation function as shown in eq. (2.211):

$$\begin{aligned}\Psi(f) &= \int_{-\infty}^{\infty} 2e^{-|\tau|} e^{-j2\pi f\tau} d\tau \\ &= 2\int_{-\infty}^{0} e^{(1-j2\pi f)\tau} d\tau + 2\int_{0}^{\infty} e^{-(1+j2\pi f)\tau} d\tau\end{aligned}$$

performing the integration we obtain the final result as:

$$\Psi(f) = \frac{2}{1-j2\pi f} + \frac{2}{1+j2\pi f} = \frac{4}{1+(2\pi f)^2}.$$

- The average power

$$E[x(t)^2] = R_{XX}(0) = 2e^0 = 2.$$

- The average power of stationary random process could be computed as:

$$E[x(t)^2] = R_{XX}(0) = \int_{-\infty}^{\infty} \Psi(f)df.$$

In this example, we need to determine the bandwidth B_w that include 90 % of the average power in this random process, i. e., $0.9 \times 2 = 1.8$ Watt. It can be computed by performing the following integration:

$$\int_{-B_w}^{B_w} \Psi(f)df = \int_{-B_w}^{B_w} \frac{4}{1+(2\pi f)^2} df = 1.8. \tag{2.213}$$

This integration can be solved easily by substituting $2\pi f = \tan(\theta)$. Hence, $df = \frac{1}{2\pi}\sec(\theta)^2 d\theta$. Therefore, the integration becomes

$$\frac{2}{\pi} \int_{-\tan^{-1}(2\pi B_w)}^{\tan^{-1}(2\pi B_w)} \frac{1}{1+\tan(\theta)^2} \sec(\theta)^2 d\theta.$$

But, from the basic trigonometric we know $1+\tan(\theta)^2 = \sec(\theta)^2$, hence,

$$\int_{-B_w}^{B_w} \Psi(f)df = \frac{4}{\pi}\tan^{-1}(2\pi B_w) = 1.8.$$

Therefore, the bandwidth is calculated as

$$B_w = \frac{1}{2\pi}\tan\left(\frac{1.8 \times \pi}{4}\right) = 1.0048\,\text{Hz}. \tag{2.214}$$

This example demonstrates that while the random process $x(t)$ theoretically possesses infinite bandwidth, 90 % of its average power is concentrated within a mere 2 Hz of that bandwidth. This finding holds significant implications in wireless communications, where the efficient utilization of bandwidth stands as a crucial requirement.

Example 83. Discrete random process x_k that has the autocorrelation function $R(k) = a^{|k|}$, where $|a| < 1$.

Find the average power and the power spectral density.

Solution. The average power is $R(0) = 1$. The power spectral density is the DTFT of the autocorrelation function as

$$\Psi(f) = \sum_{m=-\infty}^{\infty} a^{|m|} e^{-j2\pi fm}$$
$$= \sum_{m=-\infty}^{-1} a^{-m} e^{-j2\pi fm} + \sum_{m=0}^{\infty} a^{m} e^{-j2\pi fm}.$$

With some math manipulation, the above equation can be rewritten as

$$\Psi(f) = \frac{ae^{j2\pi f}}{1 - ae^{j2\pi f}} + \frac{1}{1 - ae^{-j2\pi f}} = \frac{1 - a^2}{1 + a^2 - 2a\cos(2\pi f)}.$$

In various simulation programs, the ability to create random processes with specific distributions and tailored correlation traits is crucial. Achieving this is relatively straightforward in Octave and Scilab. One approach is to generate an uncorrelated white noise signal and pass it through a linear filter. This filter's impulse response or transfer function can be chosen or designed to fulfill the desired specifications.

Example 84. Consider a wide sense stationary discrete random process x_k characterized by a zero mean and an autocorrelation function expressed as $R_{XX}(k) = 0.5^{|k|}$. Determine the mean and autocorrelation function for the following process:

$$y_k = x_k + 0.5x_{k-1} + 1.$$

Is y_k stationary or not?

Solution. The mean could be found as

$$E[y_k] = E[x_k] + 0.5[x_{k-1}] + 1 = 0 + 0 + 1 = 1.$$

The autocorrelation function of the process y_k is derived as:

$$R_{YY}(k, m) = E[y_k y_m] = E[(x_k + 0.5x_{k-1} + 1)(x_m + 0.5x_{m-1} + 1)]$$
$$= E[x_k x_m] + 0.5E[x_k x_{m-1}] + 0.5E[x_{k-1} x_m] + 0.25E[x_{k-1} x_{m-1}] + 1$$
$$R_{YY}(k - m) = R_{YY}(n) = 1.25 \times 0.5^{|n|} + 0.5 \times 0.5^{|n+1|} + 0.5 \times 0.5^{|n-1|} + 1.$$

2.9.4 White noise

White noise is a distinct type of random process characterized by an exceptionally wide and flat power spectral density. In theory, it is considered to have infinite frequency width, encompassing the entire frequency band. However, there are practical limitations. In electrical engineering, thermal noise serves as a primary source of white noise, which manifests itself as minute random voltage fluctuations across conductors. This voltage results from the erratic movement of electrons within the conductor induced by temperature, such as room temperature. This voltage behaves as a zero-mean random process and follows a normal distribution. Another source of white noise, albeit with a lower cutoff frequency, is known as shot noise. Shot noise becomes apparent when an electric current traverses a semiconductor junction. It arises due to the discrete charges (electrons) carrying the current. Unlike thermal noise, shot noise necessitates current flow to be observed. However, in normal conductors, shot noise is not observed because electron-phonon scattering smooths out the current fluctuations arising from the discrete nature of electrons, leaving only thermal noise. Beyond its physical origins, white noise holds essential applications in modeling and analyzing stochastic systems. Its broad noise spectrum with uniform density qualifies it as a reference random signal. Signals without a broad uniform spectrum are termed "colored noise." Any colored noise can be generated as the output of a filter, with white noise as the input. The properties of the resulting colored noise hinge on the characteristics of the filter. The key attribute of white noise is its flat power spectral density. As established in previous analyses, the inverse Fourier transform of the power spectral density yields the autocorrelation function. For white noise, with a fixed value of $\Psi_{XX}(f) = 1$, the inverse Fourier transform produces the autocorrelation function $R_{XX}(\tau) = \delta(\tau)$, where $\delta(\tau)$ is the Dirac delta function, elucidated further in the following chapter (see Figure 3.1). The Dirac delta is characterized by the following two equations:

$$\delta(\tau) = \begin{cases} \infty & \text{at } \tau = 0, \\ 0 & \text{at } \tau \neq 0. \end{cases} \tag{2.215}$$

$$\text{Furthermore,} \int_{-\infty}^{\infty} \delta(\tau) d\tau = 1. \tag{2.216}$$

This highlights a crucial aspect of white noise: it's an uncorrelated process. Consequently, regardless of how far back you track its history, it won't help you predict its future values. Therefore, the most reliable estimate for an uncorrelated process is its mean, regardless of the historical data available. This might prompt inquiries about its smoothness. The smoothness and continuity of signals are directly related to the feasibility of differentiability and integrability. To mathematically model white noise, it's been articulated as the generalized mean-square derivative of Brownian motion as explained soon.

2.10 Dynamic stochastic process

This section introduces fundamental concepts for establishing a mathematical foundation for random processes. The previous section covered properties such as stationarity, ergodicity, autocorrelation, and spectral density. However, when the goal is to model uncertainty based on theoretical concepts or observations, these discussions might fall short.

Random processes can exist in continuous or discrete forms. In continuous cases, understanding the governing mathematical model for the dynamics of a random process is essential. For instance, dropping a tiny particle into liquid within a container creates a 3D random process as the particle traverses from top to bottom. Even under identical conditions, each repetition yields a different trajectory. Simply studying the statistical properties of such random dynamics might prove insufficient in various applications. Therefore, constructing the mathematics that underpin these random trajectories becomes crucial. Yet, this requires the definition of fundamental principles to achieve this objective.

In general, working with discrete random processes tends to be more manageable than working with continuous cases. However, even for these discrete processes, establishing structured frameworks to articulate their dynamics remains crucial. Consider, for instance, the number of newly diagnosed diabetics in a hospital each day—a prime example of a discrete random process. Another instance involves the categorization of weather conditions at specific times, such as beautiful, normal, or bad, representing three defined statuses daily.

In these discrete scenarios, the simplest approach is often to assume that each random outcome is independent of its history. This assumption streamlines the handling of the random process significantly. Consequently, describing the entire probability density function for a sequence of n samples becomes a straightforward multiplication of the probability density functions for each individual sample. In the case of independent and identical distribution, we will have:

$$f_{\mathbf{X}}(x_1, \ldots, x_n) = [f_X(x)]^n. \tag{2.217}$$

This simplification may be appropriate for certain scenarios, but could introduce significant bias in others. For instance, assuming that the daily weather status is independent from preceding days oversimplifies the model. Such an approach might render the model inadequate for predicting future weather conditions. Conversely, factoring in an extensive history (like considering a whole month in the weather case) could unnecessarily complicate the modeling process. The Markov process offers a valuable compromise between the assumption of independence and full historical dependence in dynamic systems. It serves as a widely adopted method for modeling both continuous and discrete systems.

2.10.1 The Markov process

The Markov process serves as a middle ground between oversimplified independence assumptions and the complexity of full historical dependence. When dealing with discrete observation times, such as $t_0 \leq t_1 \leq \cdots \leq t_{n-1} \leq t_n$. Therefore, we call the random process $x(t)$ a Markov process if it is always valid to say:

$$P(x(t_n) \leq a \mid x(t_0), \ldots, x(t_{n-1})) = P(x(t_n) \leq a \mid x(t_{n-1})). \tag{2.218}$$

This implies that all statistics of the random process (referred to as state) x at time instance t_n depend solely on the immediately preceding time instance t_{n-1}, independent of any past states before t_{n-1}. This concept simplifies the analysis, yet provides sufficient power to effectively model and describe various real-world uncertain problems.

To maintain generality, we'll denote x_i instead of $x(t_i)$ within this section. In general, x_i can be either a countable or an uncountable real number. Time can be defined at specific known instances (e. g., with defined sampling periods) and can exist as a continuous entity.

In the case of discrete time and a countable number of states, the definition $x_k = i$ signifies that at time instance k, the system exists in state i. Consequently, there are a countable number of states, ensuring that the system must occupy be in one these states at each sampling time. If we assume that the system transitions between states solely at these sampling time instances, we can then establish the transition probability P_{ij}, as follows,

$$P_{ij} = P(x_{k+1} = j \mid x_{k+1} = i). \tag{2.219}$$

This model is commonly referred to as a Markov chain, as depicted in Figure 2.36. At each time instance k, the system is compelled to either remain in its current state or transition to one of the other allowed states within the system, i. e.,

$$\begin{aligned} &P_{ij} \geq 0 \quad \forall i,j \\ &\sum_{j=0}^{\infty} P_{ij} = 1, \quad \forall i = 0, 1, \ldots \end{aligned} \tag{2.220}$$

Based on that, we can define the probability transition matrix as

$$\mathbf{P} = \begin{bmatrix} P_{00} & P_{01} & P_{02} & \cdots \\ P_{10} & P_{11} & P_{12} & \cdots \\ \vdots & \vdots & \vdots & \cdots \\ P_{i0} & P_{i1} & P_{i2} & \cdots \\ \vdots & \vdots & \vdots & \cdots \end{bmatrix}. \tag{2.221}$$

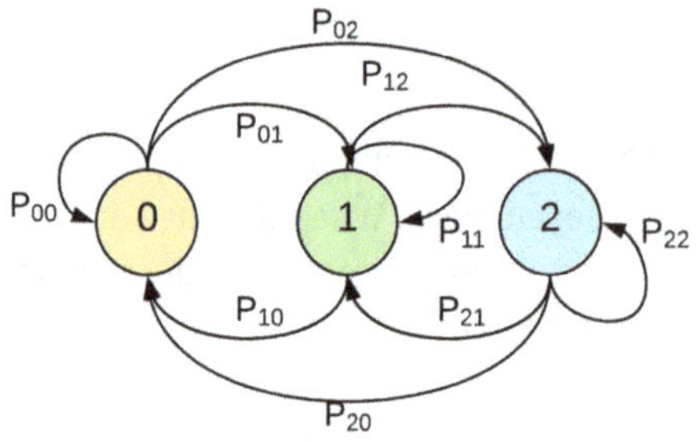

Figure 2.36: Markov chain with three states.

The probability transition matrix simplifies the computation of probabilities for multiple repetitions of the Markov process. Assume that $\hat{P}_{ij}^n$ as the probability to be in state j starting from state i after n transitions, i. e.,

$$\hat{P}_{ij}^n = P(x_{k+n} = j \mid x_k = i). \tag{2.222}$$

To calculate the probability of being in state j after n transitions, determining all possible intermediate transitions is necessary. The Chapman–Kolmogorov equations offer a method to compute these transition probabilities, as follows:

$$\hat{P}_{ij}^{n+m} = \sum_{k=0}^{\infty} \hat{P}_{ik}^n \hat{P}_{kj}^m, \quad \forall n,\ m \geq 0. \tag{2.223}$$

To explain how it works, let's consider the simplest two-state problem with a probability matrix

$$\mathbf{P} = \begin{bmatrix} P_{00} & P_{01} \\ P_{10} & P_{11} \end{bmatrix}.$$

If we start from state 0, what is the probability of being in the same state after three transitions? In this simple two-state example, it is a easy to compute manually. There are three possible scenarios to start from state 0 and being in the same state after three transitions as:

- The first possible scenario is to stay in the same state three times. This will have the probability of P_{00}^3.
- The second scenario is to stay in the same state 0 in the first transition, then move to state 1, and finally return to state 0. The probability of having such a scenario is $P_{00}P_{01}P_{10}$.
- The third possible scenario is to transit from state 0 to 1 and to stay one transition in 1 and then return to 0. The probability of having this scenario is $P_{01}P_{11}P_{10}$.
- The last possible scenario is to transit from state 0 to 1 then from 1 to 0 and finally to self-transit to 0. The probability of this scenario is $P_{01}P_{10}P_{00}$.

Based on the above scenarios, we may conclude that the probability of being in state 0 after three transitions, given that we started from state 0 is: $P_{00}^3 + 2P_{00}P_{01}P_{10} + P_{01}P_{11}P_{10}$. If we take the power 3 of the matrix $\mathbf{P}$ as:

$$\mathbf{P}^3 = \begin{bmatrix} x_{00} & x_{01} \\ x_{10} & x_{11} \end{bmatrix} \tag{2.224}$$

$$\begin{aligned}
x_{00} &= P_{00}^3 + 2P_{00}P_{01}P_{10} + P_{01}P_{11}P_{10} \\
x_{01} &= P_{01}P_{11}^2 + P_{01}^2P_{10} + P_{00}P_{11}P_{01} + P_{00}^2P_{01} \\
x_{10} &= P_{00}^2P_{10} + P_{10}^2P_{01} + P_{11}^2P_{10} + P_{00}P_{11}P_{10} \\
x_{11} &= P_{11}^3 + 2P_{10}P_{11}P_{01} + P_{00}P_{01}P_{10}.
\end{aligned}$$

It's evident that the first element of the matrix matches our calculated probability. Generally, it can be verified that the probability of transitioning over n steps in a discrete Markov process can be computed by raising the probability matrix to the nth power. This principle will be exemplified later through numerical illustrations. When all states in a Markov chain can communicate with each other, the limit $\lim_{n\to\infty} \hat{P}_{ij}^n = \pi_j$ exists and is independent of i. Here, π_j signifies the probability of being in state j after numerous iterations of the Markov process. One possible mathematical approach to determine the steady-state probabilities is:

$$\begin{bmatrix} \pi_0 \\ \pi_1 \\ \vdots \end{bmatrix} = \begin{bmatrix} P_{00} & P_{01} & \cdots \\ P_{10} & P_{11} & \cdots \\ \vdots & \vdots & \vdots \end{bmatrix}^T \begin{bmatrix} \pi_0 \\ \pi_1 \\ \vdots \end{bmatrix} \tag{2.225}$$

$$\sum_{i=0} \pi_i = 1.$$

One effective method to derive the steady-state probabilities involves utilizing the pseudo-inverse, such as:

$$\begin{bmatrix} \pi_0 \\ \pi_1 \\ \vdots \end{bmatrix} = \begin{bmatrix} \mathbf{P}^T - \mathbf{I} \\ \mathbf{1}^T \end{bmatrix}^+ \begin{bmatrix} 0 \\ \vdots \\ 0 \\ 1 \end{bmatrix}, \tag{2.226}$$

where $[\bullet]^+$ is the pseudo-inverse, and $\mathbf{1} = [1\ldots1]^T$. The pseudo-inverse of matrix $\mathbf{A}$ could be found in Octave and Scilab using pinv($\mathbf{A}$).

We define a state j as accessible from state i if, for some nonnegative integer n greater than or equal to 0, there exists a nonzero probability p_{ij} of transitioning from state i to state j in n steps is possible, denoted as $i \to j$. When two states, i and j, are accessible to each other, they are said to communicate. If all states within the Markov chain communicate with one another, we classify the chain as irreducible.

Example 85. Let's begin with a straightforward example demonstrating how to model specific scenarios using Markov chains. Consider an urn containing two balls, which can be either white or red. Each draw involves selecting one ball; if it's red, it gets replaced with a white ball with a probability of 0.7, and if it's white, it gets replaced with a red ball with a probability of 0.5. Draw the state diagram, determine the probability of having two red balls in the urn after 4 iterations, assuming the initial state was $x_0 = 1$. Lastly, calculate the probability of having balls of different colors after numerous iterations.

To model the problem using a Markov framework, the initial step involves defining the Markov 'state.' An intuitive approach is to designate the state based on the number of white balls in the urn. Consequently, we'll have three states: 0, 1, or 2 white balls. The state diagram will closely resemble Figure 2.36. However, for clarity, we'll reproduce the diagram with the associated probability values in Figure 2.37.

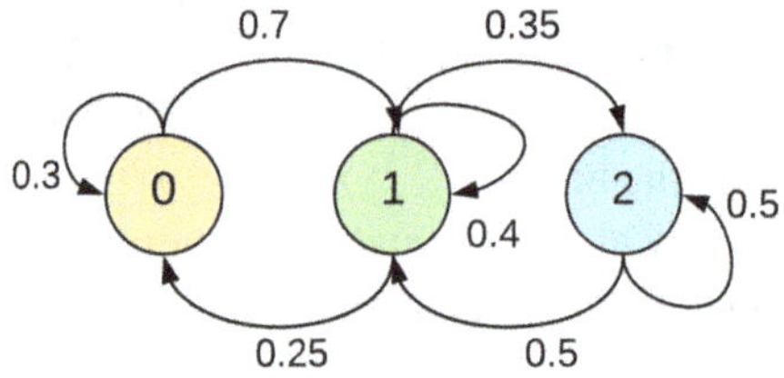

Figure 2.37: Markov chain for Example 85.

The probability values are derived from the example. For instance, when in state 1 (representing a red ball and a white ball), the probability of transitioning to state 0 involves the event of drawing a white ball *AND* replacing it with a red ball. This probability is computed as $\frac{1}{2} \times \frac{1}{2} = 0.25$. Conversely, the likelihood of transitioning to state 2 (where both balls are white) in the next step is determined by the event of drawing a red ball *AND* replacing it with a white ball, resulting in a calculation of $\frac{1}{2} \times 0.7 = 0.35$. Furthermore, the probability of remaining in state 1 can easily be computed as $1 - 0.25 - 0.35 = 0.4$. Alternatively, it can also be calculated as the event of drawing a red ball AND replacing it with a red ball OR drawing a white ball AND replacing it with a white ball, resulting in $\frac{1}{2} \times 0.3 + \frac{1}{2} \times \frac{1}{2} = 0.4$. This, of course, yields the same result as expected. The probability matrix for this problem is as follows:

$$\mathbf{P} = \begin{bmatrix} 0.3 & 0.7 & 0 \\ 0.25 & 0.4 & 0.35 \\ 0 & 0.5 & 0.5 \end{bmatrix}. \tag{2.227}$$

After 4 steps, the probability matrix becomes:

$$\mathbf{P}^4 = \begin{bmatrix} 0.1866 & 0.4900 & 0.3234 \\ 0.1750 & 0.4876 & 0.3374 \\ 0.1650 & 0.4820 & 0.3530 \end{bmatrix}. \tag{2.228}$$

It is clear that the probability of being in state 0 (two reds) after four steps starting from state 1 is $\hat{P}^4_{10} = 0.1750$. Finally, the steady-state probabilities of each state are: $\pi_0 = 0.1736$, $\pi_1 = 0.4861$, and $\pi_2 = 0.3403$. The values can be found simply by taking a large power for the probability matrix $\mathbf{P}$ or by solving the equations in (2.225).

In numerous practical modeling scenarios aimed at resolving system uncertainties, Markov chains are often employed in a reverse manner. This approach involves deriving transition probabilities for the model based on the historical data available in a database.

Example 86. Consider a living cell that experiences four distinct health states, as outlined below:
- State A: The cell is in a healthy condition.
- State B: The cell exhibits a specific defect 1.
- State C: The cell manifests a distinct defect 2.
- State D: The cell has ceased to function and is considered dead.

The referenced figure (Figure 2.38) illustrates the transitions between these states, showcasing probabilities based on daily averages. For example, a healthy cell typically maintains its state with a 99 % probability day to day, i. e., $P_{AA} = 0.99$. However, there's a 0.9 % probability of developing defect 1, i. e., $P_{AB} = 0.009$, and an extremely minimal 0.1 % chance of sudden cell failure, i. e., $P_{AD} = 0.001$. In state B, which represents the presence of defect 1, there's a 60 % likelihood of the cell undergoing self-repair and reverting to the healthy state A, i. e., $P_{BA} = 0.6$. Further transition probabilities are specified as $P_{BB} = 0.2$, $P_{BC} = 0.15$, and $P_{BD} = 0.05$. State C represents the presence of defect 2, which could be developed from defect 1. From defect 2, the cell has a probability to cease $P_{CD} = 0.4$. Other transition probabilities from state C are $P_{CC} = 0.3$ and $P_{CA} = 0.3$. It's crucial to note that state D indicates an absorbed state, signifying cell termination, i. e., $P_{DD} = 1$. Construct the transition probability matrix of this process. Starting from the

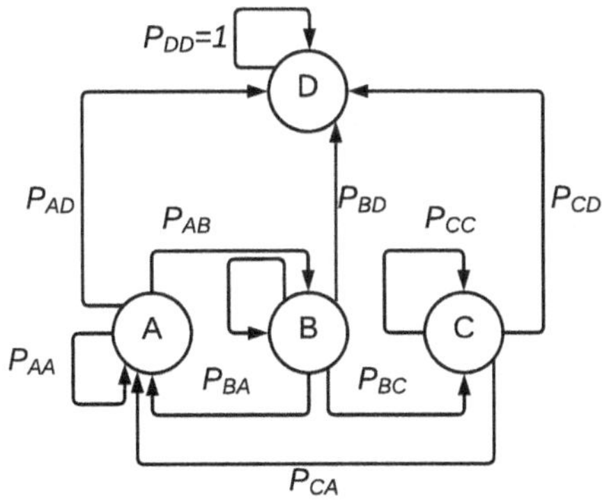

Figure 2.38: Living cell transition model in Example 86.

healthy State A, what is the likelihood of cell death within 2 days? Again starting from the state A, what is the probability of reaching state C after 5 days? Validate the obtained results through simulations without using ready-made Markov functions.

Solution. The transition matrix can be constructed as:

$$\mathbf{P} = \begin{bmatrix} 0.990 & 0.009 & 0 & 0.001 \\ 0.600 & 0.200 & 0.150 & 0.050 \\ 0.300 & 0 & 0.300 & 0.400 \\ 0 & 0 & 0 & 1 \end{bmatrix}. \quad (2.229)$$

Starting from the healthy state A, the likelihood of cell death within 2 days can happen in these scenarios: 1. The cell moves from healthy to death on the first day and stays there, this has probability 0.001. 2. On the first day the healthy state moves to itself and then on the second day moves to death, this scenario has probability of $0.99 \times 0.001 = 0.00099$. 3. On the first day it moves from healthy state to defect 1 state and then on the second day from state B to death, this will happen with probability of $0.009 \times 0.05 = 0.00045$. There is no other scenario that the healthy cell will move to the death state within 2 days other than these three scenarios. Therefore, the probability will be the summation all these cases as $0.001 + 0.00099 + 0.00045 = 0.00244$. The same result is obtained by taking the square of the matrix **P** as:

$$\mathbf{P}^2 = \begin{bmatrix} 0.9855 & 0.0107 & 0.0013 & \mathbf{0.0024} \\ 0.7590 & 0.0454 & 0.0750 & 0.1206 \\ 0.3870 & 0.0027 & 0.0900 & 0.5203 \\ 0 & 0 & 0 & 1 \end{bmatrix}. \quad (2.230)$$

In the same manner we can evaluate the transition matrix after 5 days by taking the power 5 of the matrix **P** as:

$$\mathbf{P}^5 = \begin{bmatrix} 0.977 & 0.011 & \mathbf{0.002} & 0.009 \\ 0.813 & 0.009 & 0.005 & 0.173 \\ 0.419 & 0.005 & 0.003 & 0.573 \\ 0 & 0 & 0 & 1 \end{bmatrix}. \quad (2.231)$$

Therefore starting from the healthy state, the probability of being in the defect 2 state is $P^5_{AC} = 0.002$. Here, we present a concise Octave code for simulating the cell-health scenario. Validating the results through simulation is achievable. Moreover, by making minor adjustments to the code, you can determine the average lifespan of the cell before its demise.

```
# Cell-Health Markov Example
clear all
P=[0.99 0.009 0 0.001;0.6 0.2 0.15 0.05; 0.3 0 0.3
    0.4; 0 0 0 1];
```

```
N=1000; %repeating
%States A=1, B=2, C=3, D=4
T=0;
s=0;
StartState=3;
EndState=4;
NumberofDays=5;
NumberofDays=NumberofDays+1;
while (T<=N)
S(1)=StartState; %starting from State A
t=1;
while (t<NumberofDays)
p=rand(1); %Uniform random number 0 to 1
if S(t) ==1,
if p<= P(1,1),
S(t+1)=1;
elseif p>=P(1,1) && p<=P(1,1)+P(1,2),
S(t+1)=2;
else
S(t+1)=4;
end
elseif S(t) ==2,
if p<= P(2,1),
S(t+1)=1;
elseif p>=P(2,1) && p<=P(2,1)+P(2,2),
S(t+1)=2;
elseif  p>=P(2,1)+P(2,2) && p<=P(2,1)+P(2,2)+P(2,3),
S(t+1)=3;
else
S(t+1)=4;
end
elseif S(t) ==3,
if p<= P(3,1),
S(t+1)=1;
elseif p>=P(3,1) && p<=P(3,1)+P(3,3),
S(t+1)=3;
else
S(t+1)=4;
end
elseif S(t) ==4,
S(t+1)=4;
end
```

```
    t=t+1;
    end
    if S(NumberofDays)==EndState,
    s=s+1;
    end
    T=T+1;
    end
    %Probability of being in EndState starting
    %from StartState after NumberofDays is:
    Pr=s/N
```

Example 87. Let's categorize wireless channels into two states: 'Excellent' and 'Bad.' However, these channels tend to persist in a given state for a duration influenced by the channel's coherence time. For instance, assume that the transmitted packets last a third of the channel's coherence time. We aim to model the uncertainty surrounding the next channel state that the subsequent packet might encounter. The subsequent state relies on the preceding three states, enabling us to characterize it as a Markov process with eight states as shown in Figure 2.39. For instance, if the current state is denoted as BGG, it signifies that within the last three packets, the first packet encountered a Bad channel, the second encountered a Good channel, and the third (current) packet faced a Good channel. What would be the anticipated next state in this scenario? There are two possibilities: if the next packet encounters a Good channel, the new state will be (GGG); alternatively, if it encounters a Bad channel, the new state will be (GGB). It's important

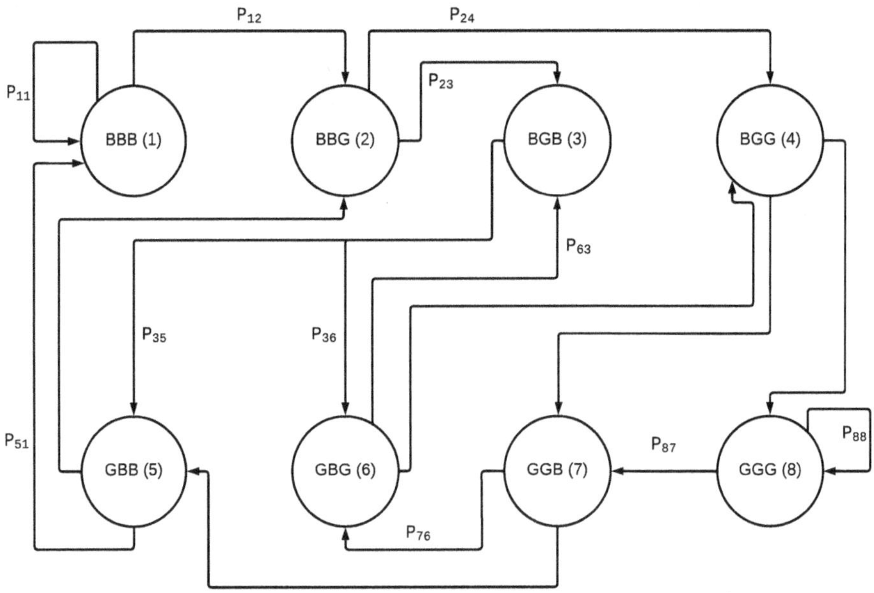

Figure 2.39: Markov model of correlated channels (Example 87).

to note that transitioning to other states is not feasible in this context. Let's number the states as State 1 for BBB, State 2 for BBG, State 3 for BGB, State 4 for BGG, State 5 for GBB, State 6 for GBG, State 7 for GGB, and State 8 for GGG. Show the state diagram and construct the matrix transition probability. Starting from the State 1 (BBB), what is the probability that the channel moves to the best state (GGG) after three cycles? Find the steady state probabilities for the given transition probability values.

Solution. The probability transition matrix is

$$\mathbf{P} = \begin{bmatrix} P_{11} & P_{12} & \dots & P_{18} \\ P_{21} & P_{22} & \dots & P_{28} \\ \dots & \dots & \dots & \dots \\ P_{81} & P_{82} & \dots & P_{88} \end{bmatrix}. \tag{2.232}$$

Let's assume that $P_{11} = 0.3$, $P_{23} = 0.1$, $P_{35} = 0.8$, $P_{48} = 0.7$, $P_{51} = 0.75$, $P_{63} = 0.2$, $P_{75} = 0.5$, $P_{88} = 0.9$. Therefore, the probability transition matrix becomes:

$$\mathbf{P} = \begin{bmatrix} 0.3 & 0.7 & 0 & 0 & 0 & 0 & 0 & 0 \\ 0 & 0 & 0.1 & 0.9 & 0 & 0 & 0 & 0 \\ 0 & 0 & 0 & 0 & 0.8 & 0.2 & 0 & 0 \\ 0 & 0 & 0 & 0 & 0 & 0 & 0.3 & 0.7 \\ 0.75 & 0.25 & 0 & 0 & 0 & 0 & 0 & 0 \\ 0 & 0 & 0.2 & 0.8 & 0 & 0 & 0 & 0 \\ 0 & 0 & 0 & 0 & 0.5 & 0.5 & 0 & 0 \\ 0 & 0 & 0 & 0 & 0 & 0 & 0.1 & 0.9 \end{bmatrix}. \tag{2.233}$$

Starting from state 1, the probability of being in state 8 in three transitions is computed as $P_{18}^3 = 0.4410$. Finally the steady state probabilities of all channel states could be computed using (2.226). The results are given as:

$$\pi = [0.058, 0.054, 0.015, 0.085, 0.054, 0.046, 0.086, 0.600]^T. \tag{2.234}$$

This result shows that on the long run the channel will be in the 8th state about 60 % of time.

2.10.2 The Brownian motion

The field of stochastic differential equations finds widespread applications across various scientific domains, primarily in the modeling of random dynamical systems. Each observation derived from a stochastic differential equation exhibits a unique random variation. However, to model such dynamic systems effectively, it's crucial to establish robust mathematical foundations for continuous random processes. Stochastic differential equations can be thought of as deterministic differential equations perturbed by an uncorrelated random process, often referred to as white noise. The requirement for uncorrelated noise arises from the fact that any correlation observed in the outcomes should be inherent in the properties of the differential equation itself. One might question the necessity of modeling equations whose outputs cannot be precisely determined. Despite the inherent unpredictability of outputs in stochastic dynamic systems, it remains possible to determine their statistical properties. Consequently, it becomes feasible to specify the probability of an output trajectory will fall within a certain space over a given duration. The forthcoming chapter will delve into stochastic differential equations for modeling purposes. However, this section introduces foundational concepts related to the modeling of uncorrelated noise, rooted in a simple notion known as a random walk. In this context, if we consider a one-dimensional random movement or increment of a particle at instant i as $x_i = +\Delta x$ with a probability p_+ and a decrement $x_i = -\Delta x$ with a probability $p_- = 1 - p_+$, the particle's position after n successive random movements is described as $X_n = \sum_{i=1}^{n} x_i$. Assuming the particle starts from the origin ($x_0 = 0$), the model for the random walk is as follows:

$$X_n = X_{n-1} + x_n. \tag{2.235}$$

In fact, the variable x_n remains independent of the time instant n, its increments occur independently, with a probability of increase denoted by p_+ and a corresponding probability of decrease, given by $p_- = 1 - p_+$. The graphical representation in Figure 2.40 illustrates a single realization of the random walk under the condition $p_+ = 0.5$, with each step taking ± 1 and an increment time of 0.05 seconds. Figure 2.41 depicts the same random walk, but with a significantly reduced increment time of 0.0005 seconds.

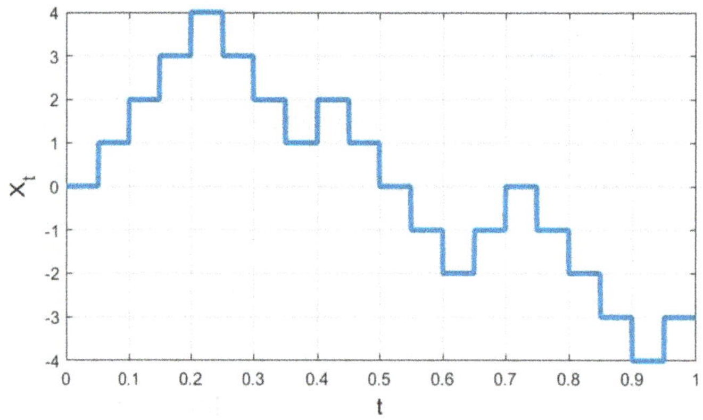

Figure 2.40: Discrete time random walk.

As the increment time approaches zero, we observe the emergence of a continuous random walk, a phenomenon known as *Brownian motion*, or alternatively, the Wiener

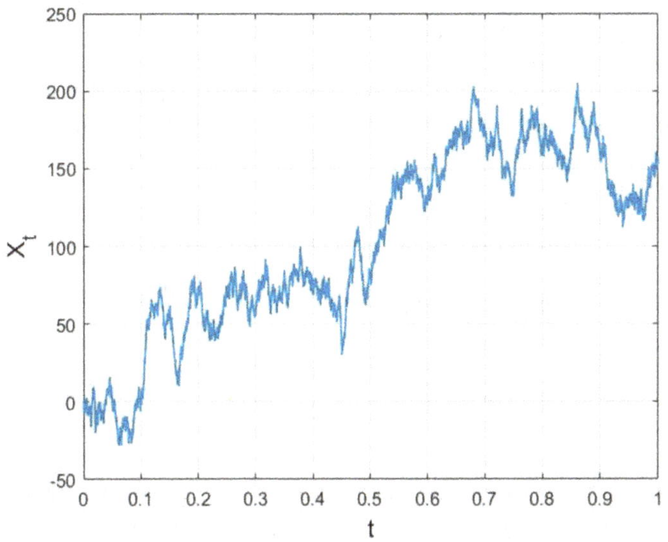

Figure 2.41: Continuous time random walk, *Brownian motion.*

or Wiener-Levy process. Assuming X_0 as the initial position of the random walk, equation (2.235) can be reformulated as:

$$X_n = X_0 + \sum_{i=1}^{n} x_i. \tag{2.236}$$

Observe that $E[x_i] = \Delta x \times p_+ - \Delta x \times (1 - p_+) = 2\Delta x p_+ - \Delta x$. Therefore, for $p_+ = \frac{1}{2}$ then the average is $E[x_i] = 0$. Furthermore, $E[x_i^2] = \Delta x^2 p_+ + \Delta x^2(1 - p_+) = \Delta x^2$ for any $0 \le p_+ \le 1$.

Example 88. Assume a random walk process with a periodic time increments each Δt to $\pm\Delta x$ with probabilities $p_+ = p_- = \frac{1}{2}$, and starting with $X_0 = 0$. After time $T = n\Delta t$, find the mean $E[X_n]$ and the variance $E[X_n^2] - (E[X_n])^2$ of this random process.

The mean could be easily computed using (2.236) with $X_0 = 0$ such as: $E[X_n] = \sum_{i=1}^{n} E[x_i] = 0$, hence the mean of such a random process is always zero regardless of the observation time T. To compute the variance after time T, we should compute $E[X_n^2]$. Hence,

$$\begin{aligned} E[X_n^2] &= E\left[\left(\sum_{i=1}^{n} x_i\right)^2\right] \\ &= E\left[\sum_{i=1}^{n}\sum_{j=1}^{n} x_i x_j\right] = \sum_{i=1}^{n}\sum_{j=1}^{n} E[x_i x_j]. \end{aligned}$$

We can separate two events when $i = j$ and $i \neq j$ as:

$$E[X_n^2] = \sum_{i=1}^{n} E[x_i^2] + \sum_{i=1}^{n} \sum_{j=1, j \neq i}^{n} E[x_i x_j].$$

Since x_i and x_j are independent for $i \neq j$, hence, $E[x_i x_j] = E[x_i]E[x_j] = 0$. Therefore,

$$E[X_n^2] = \sum_{i=1}^{n} E[x_i^2] = n\Delta x^2.$$

But since $T = n\Delta t$, therefore, we can represent the variance in terms of observation time as:

$$\sigma_T^2 = E[X_n^2] = \frac{T\Delta x^2}{\Delta t} = \sigma^2 T, \tag{2.237}$$

where $\sigma^2 = \frac{\Delta x^2}{\Delta t}$ could be considered as the unit time variance, and this result is important in the analysis and modeling of dynamic random systems. Since n is directly related to observation time T, hence, we can use X_T instead.

In the previous example, we have derived the first and second moments of the random walk process. However, in order to know all statistical properties, we should know the probability density function. As has been stated previously, the *central limit theorem* states that: the resulting distribution of the summation of many independent and identically distributed random variables will converge to a normal distribution. Hence, for large n $X_T = \sum_{i=1}^{n} x_i$ will converge to a normal distribution. However, if you like to see a mathematical proof for that, see the next example.

Example 89. Prove that the limit for a random walk for a large number of steps converges to a normal distribution.

Solution. We've previously explored how the combined probability density function resulting from the summation of multiple random variables can be derived through successive convolution integrals of their individual probability density functions. Additionally, we highlighted that this complex process can be simplified by leveraging the multiplication of their characteristic functions. This part was discussed in Section 2.7. Hence, we can say that the characteristic function of X_T is:

$$\varphi_{X_T}(v) = \prod_{i=1}^{n} \Phi_{x_i}(v).$$

The characteristic function of x_i is

$$E[e^{jvx_i}] = (p_+ e^{jv\Delta x} + (1 - p_+)e^{-jv\Delta x}).$$

For $p_+ = \frac{1}{2}$,

$$\varphi_{X_i} = \left(\frac{e^{jv\Delta x} + e^{-jv\Delta x}}{2}\right) = \cos(v\Delta x).$$

Now we can derive the characteristic function of the process X_T as

$$\varphi_{X_T}(v) = \cos(v\Delta x)^n. \quad (2.238)$$

For very small time steps, i. e., $\Delta t \to 0$, we can approximate the characteristic function as

$$\varphi_{X_T}(v) \approx \left(1 - \frac{v^2\Delta x^2}{2}\right)^n.$$

But since $n = \frac{T}{\Delta t}$, and $\sigma^2 = \frac{\Delta x^2}{\Delta t}$, therefore,

$$\varphi_{X_T}(v) \approx \left(1 - \frac{v^2\sigma^2\Delta t}{2}\right)^{T/\Delta t}.$$

Hence, from the limit theorem, it is straightforward to prove that

$$\lim_{\Delta t \to 0}\left(1 - \frac{v^2\sigma_T^2\Delta t}{2}\right)^{T/\Delta t} = e^{-v^2\sigma^2 T/2}. \quad (2.239)$$

The resulting characteristic function is identical to the characteristic functions of the normal distribution (see (2.164)) with zero mean and variance $\sigma_T^2 = \sigma^2 T$.

Since the Brownian motion converges to a zero-mean normal distribution, its statistical properties become readily computable. The probability density function describing Brownian motion is expressed as follows:

$$f_{X_T}(x, T) = \frac{1}{\sqrt{2\pi\sigma^2 T}} e^{-\frac{x^2}{2\sigma^2 T}}. \quad (2.240)$$

This probability density function differs from our prior discussions due to the inclusion of time, signifying an increase in the variance of the random process over time. Despite this variation, the formula proves immensely valuable in estimating and modeling random dynamical systems.

Example 90. A continuous random process that can be modeled as Brownian motion started at $X_0 = 0$ with $\sigma^2 = 0.1\,m^2/s$. After 1 minute, what is the probability that the object location is between $1 \le X_T \le 2$?

Since the process is a normal distribution, then the probability is:

$$P(1 \le x \le 2) = \int_1^2 \frac{1}{\sqrt{12\pi}} e^{-\frac{x^2}{12}}\, dx. \quad (2.241)$$

It can be evaluated using an error function or Q-function as shown in Example 32, and the result will be $P(1 \le x \le 2) = 0.134$ It is possible to emulate this Brownian motion with Octave as shown next. Figure 2.42 shows three different realizations of the Brownian motion. The following code shows how to generate such random motion easily.

```
% Octave Code
T=60; % Observation time
n=10000; % number of increments
dT=T/n; % time steps
x=sign(randn(n,1)); %independent incemments
sigma2=0.1; % Sigma^2
dx=sqrt(sigma2*dT); %step size
k=1:n;
X(1)=0; % initial X
for i=2:n,
X(i)=X(i-1)+dx*x(i); %Motion
end
t=0:dT:T-dT;
plot(t,X) %plot
```

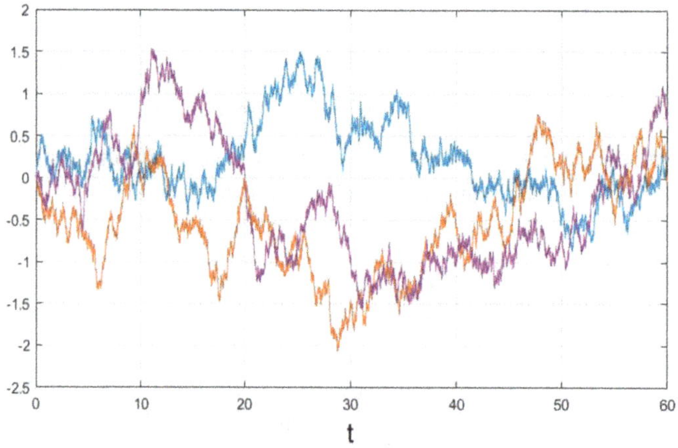

Figure 2.42: Three realizations of Example 90.

Since Brownian motion is a random process, we should find its autocorrelation function in order to evaluate its time/frequency properties. Based on our initial assumptions, the Brownian motion has independent increments. Hence, the evaluation of the autocorrelation function $R_{XX}(t,\tau) = E[X_t X_\tau]$ could be performed as follows:

In the case of $t \ge \tau$, we can express $X_t = X_\tau + X_{(t-\tau)}$, where $X_{(t-\tau)}$ refers to the increments from time τ until time t. Therefore,

$$E[X_tX_\tau] = E[(X_\tau + X_{(t-\tau)})X_\tau] = E[X_\tau X_\tau + X_\tau X_{(t-\tau)}]$$
$$= E[X_\tau X_\tau] + E[X_\tau X_{(t-\tau)}].$$

But since $X_{(t-\tau)}$ is future independent of X_τ, therefore,

$$E[X_\tau X_{(t-\tau)}] = E[X_\tau]E[X_{(t-\tau)}] = 0.$$

Then for $t \geq \tau$,

$$E[X_tX_\tau] = E[X_\tau X_\tau] = \sigma^2\tau.$$

Exactly the same procedure could be repeated for the case of $\tau > t$, and we obtain the result of

$$E[X_tX_\tau] = E[X_tX_t] = \sigma^2 t.$$

Therefore the general formula for the autocorrelation of the Brownian motion is:

$$E[X_tX_\tau] = \sigma^2 \min(t, \tau). \tag{2.242}$$

The evidence indicates that Brownian motion does not exhibit stationarity in the traditional sense. Nevertheless, a crucial observation emerges: for any time instants t, τ, and λ, the difference between X_t and X_τ shares an identical distribution with $X_{t+\lambda} - X_{\tau+\lambda}$. This revelation endows the process with stationary increment properties. Additionally, by considering $\min(t, \tau)$ as the overlapping time and acknowledging that the process's mean is consistently zero, we establish a connection between the covariance of the Brownian process and the duration of the overlapping time period.

The importance of Brownian motion lies in its ability to model white noise—an essential concept for expressing uncertainties in various modeling scenarios. An intriguing question emerges: can Brownian motion be differentiated? What conditions must a random process meet to be considered differentiable or integrable? The answer holds crucial implications for elucidating uncertainties in dynamic systems, which will be explored in the following chapter.

In conventional calculus, the notion of smoothness defines the differentiability of functions. However, random processes lack explicit functions to assess smoothness. Consequently, we rely on assessing smoothness through convergence tests, detailed in Section 2.8. Specifically, convergence in the mean square suggests convergence in probability—a relationship easily proven using Chebyshev's inequality. Therefore, subsequent convergence tests will be grounded in the mean square criterion.

Brownian motion, as a second-order process denoted by $E[X_T^2] = \sigma^2 T < \infty\ \forall T < \infty$, is continuous in the mean square sense if the following limit function converges to zero in the mean square sense:

$$\lim_{h\to 0} E[(X_{t+h} - X_t)^2] = 0. \tag{2.243}$$

However, by expanding the above expectation we obtain

$$\begin{aligned}&E[X_{t+h}X_{t+h} - X_{t+h}X_t - X_tX_{t+h} + X_tX_t]\\ &\quad = R_{XX}(t+h, t+h) - R_{XX}(t+h, t) - R_{XX}(t, t+h) + R_{XX}(t, t).\end{aligned}$$

Therefore, the main condition for a random process to be continuous at time t is that its autocorrelation function must be continuous at time t, i. e.,

$$\lim_{h\to 0, \delta\to 0} [R_{XX}(t+h, t+\delta) - R_{XX}(t, t)] = 0. \tag{2.244}$$

Note that the absence of the expectation operator in the preceding equation is due to the deterministic nature of the autocorrelation function. Extending these findings, we can assess the differentiability of a random process at a given time t by examining the existence of the following limit in the mean square sense:

$$\lim_{h\to 0} E\left[\left(\frac{X_{t+h} - X_t}{h}\right)^2\right]. \tag{2.245}$$

This result means that a random process X_t is differentiable at time t if the following limit exists:

$$\lim_{h\to 0, \delta\to 0} \frac{R_{XX}(t+h, t+\delta) - R_{XX}(t, t)}{h\delta} = \frac{\partial^2 R_{XX}(t, \tau)}{\partial t \partial \tau}. \tag{2.246}$$

Therefore, a random process X_t is differentiable at time t if and only if its autocorrelation function is differentiable at both time dimensions.

Now, what are the conditions required for the existence of the integrability of a random process X_t? There are different mathematical formulations for integration. Let's start with the Riemann integral, which is equivalent to the summation of infinitely small sectors of the function. In general for a deterministic function $f(t)$ to be integrated over a period of time from t_1 to t_2 is given by:

$$\int_x^y f(t)dt = \lim_{a\to 0} \sum_{k=0}^{n-1} f(t_k)\Delta t_k. \tag{2.247}$$

Where $\Delta t_k = t_{k+1} - t_k$, $a = \max(\Delta t_k)$, and $x = t_0 \leq t_1 \cdots \leq t_{n-1} \leq t_n = y$. Based on this general Riemann integral form, we can say that a second-order random process is integrable if the following limit on the mean square sense exists:

$$\lim_{a\to 0} E\left[\left(\sum_{k=0}^{n-1} X_{t_k}\Delta t_k - \int_x^y X_t dt\right)^2\right] = 0. \tag{2.248}$$

By expanding the limit in the above equation and since:

$$E\left[\left(\sum_{k=0}^{n-1} X_{t_k}\Delta t_k\right)^2\right] = E\left[\sum_{k=0}^{n-1}\sum_{m=0}^{n-1} X_{t_k}\Delta t_k X_{t_m}\Delta t_m\right]$$

that leads to:

$$E\left[\left(\sum_{k=0}^{n-1} X_{t_k}\Delta t_k\right)^2\right] = \sum_{k=0}^{n-1}\sum_{m=0}^{n-1} R_{XX}(t_k, t_m)\Delta t_k \Delta t_m.$$

Therefore, the main condition necessary for the integration of a second-order random process X_t to exist within the range x to y is that its autocorrelation function is Riemann integrable as:

$$\int_x^y\int_x^y R_{XX}(t,\tau)dt d\tau. \tag{2.249}$$

Consequently, the key conditions for a second-order random process X_t to exhibit differentiability or integrability hinge upon the differentiability and integrability of its autocorrelation function, respectively. This leads us to assess the differentiability and integrability criteria for Brownian motion.

The differentiation of the autocorrelation function (2.242) is

$$\frac{\partial^2}{\partial t \partial \tau}[\sigma^2 \min(t,\tau)] = \sigma^2\delta(t-\tau). \tag{2.250}$$

Absolutely, here's an improved version:

Given that the Dirac delta function doesn't strictly conform to conventional function criteria, it follows that the characterization of Brownian motion as entirely smooth and differentiable might not hold. This inference aligns with the intrinsic stochastic independence ingrained within Brownian motion. Moreover, when deriving the autocorrelation function of Brownian motion's derivative, $\frac{dX_t}{dt}$, it corresponds to the autocorrelation function of the derivative of the Brownian process.

In the context of white noise, recognized for its uncorrelated nature, its autocorrelation function manifests itself as a Dirac delta function. This holds significance in expressing $w(t) = \frac{dX_t}{dt}$, with $w(t)$ representing white noise. However, it's crucial to acknowledge a mathematical nuance: while using this representation, it's mathematically more accurate to articulate it as

$$dX_t = w_t dt. \tag{2.251}$$

Here, dX_t denotes the increment of the Brownian motion, while w_t represents the white noise at time t. Nevertheless, from an engineering standpoint, the choice between these forms isn't typically a significant concern as long as the outcomes or results remain consistent.

Example 91. Prove that the derivative of the autocorrelation of a second-order and differentiable random process X_t is identical to the autocorrelation of the derivative of X_t.

Solution. Let $X_t' = \frac{dX_t}{dt}$, hence

$$\begin{aligned} R_{X'X'}(t,\tau) &= E[X_t' X_\tau'] = E\left[\frac{dX_t}{dt}\frac{dX_\tau}{d\tau}\right] \\ &= \frac{\partial^2}{\partial t \partial \tau} E[X_t X_\tau] = \frac{\partial^2}{\partial t \partial \tau} R_{XX}(t,\tau). \end{aligned}$$

For the integrability test, since the autocorrelation of the Brownian motion in integrable, then we can say that the process is also integrable without specific conditions.

3 Dynamic systems modeling

The purpose of this chapter is not to provide a complete and detailed explanation of dynamical systems. Rather, the goal is to introduce the topic very briefly and place it in the context of the book, which is about the importance of modeling in reducing uncertainty when studying different systems and for various applications. There are many excellent and comprehensive books that deal with the subject of dynamical systems, for example [2].

After completing this chapter and working through the provided examples, you will:

- Understand the necessity and challenges of modeling systems with both deterministic and uncertain components.
- Grasp the fundamentals of linear time-invariant (LTI) systems, including their properties, impulse response, and transfer function.
- Utilize the state-space representation to model and analyze the dynamics of various systems, from linear to nonlinear and time-invariant to time-varying.
- Analyze the impact of random input signals on the output of LTI systems and derive statistical properties of the output.
- Employ stochastic differential equations (SDEs) to model and understand the statistical behavior of dynamic systems with inherent uncertainties.

As mentioned earlier, uncertainty is a result of insufficient knowledge, often stemming from an inaccurate model or incomplete information about the system's inputs or parameters. Prior to the formulation of Newton's laws of motion (classical mechanics) describing the movement of objects on the ground, through the air, or even planetary motion (albeit less accurately) was challenging. Even in fully deterministic systems, the absence of a precise mathematical model introduces uncertainty, making prediction and description difficult.

Since most systems have both deterministic and uncertain components, it's crucial to model the deterministic aspect accurately and specify the uncertain part. Understanding the relationship between these deterministic and uncertain elements is equally important. While many physical systems can be modeled deterministically with sufficient information, there remains an inherent uncertain component, representing modeling errors or unresolved uncertainties. Every model operates within a specific range of validity and relies on a set of assumptions.

In some cases, relying solely on the deterministic part of the model is feasible, with the uncertain component manifesting itself as minor fluctuations around the deterministic trajectory. However, this isn't universally applicable. For instance, accurate deterministic models for stock market fluctuations are nonexistent. Disregarding inherent uncertainties can lead to erroneous conclusions.

Additionally, it's important to understand how systems influence random processes. For instance, when a system receives a random process as input, how does it affect the system's output? Conversely, if we observe a random signal as the system's output,

https://doi.org/10.1515/9783111585055-003

what can we infer about the characteristics of the random input signals? In practice, all observations result from the outputs of various systems, whether linear or nonlinear.

In Section 2.6, we delved into the influence of linear or nonlinear functions on random variables. However, when addressing random processes, we confront variables that evolve with an independent variable, such as time. Despite this extension, the principles we discussed earlier remain applicable when dealing with memoryless functions, where the current output solely relies on the present inputs. This holds true for various scenarios, including linear relationships like

$$y(t) = ax(t) + b$$

and nonlinear relationships such as

$$y(t) = a\log(x(t) + b).$$

Here, we can leverage the techniques explored in Section 2.6 to analyze the characteristics of $y(t)$. However, this approach becomes insufficient when the system's output is influenced by both current and historical inputs and outputs—characteristic of dynamic systems. For instance, electrical circuits with storage elements such as capacitors or inductors exhibit a dependence on both current and past inputs due to the stored electrical energy. This concept extends to numerous real-world systems, including stock markets, mechanical systems, biological processes, and even social relationships. Dynamic systems, by their nature require a more sophisticated modeling approach, often involving differential or integral equations. These equations can be either linear or nonlinear, reflecting the corresponding nature of the dynamic systems they represent. Thus, the study of dynamic systems encompasses a broader and more intricate framework, allowing us to capture the nuanced interplay between current and historical factors in the evolution of these systems.

Many real systems are complex, making it challenging to develop an analytical model with the necessary precision. While this chapter doesn't aim to exhaustively explore dynamic systems, it offers a swift and concise overview of key concepts and fundamental aspects.

It's worth noting that our focus, in particular, will be on linear dynamic systems, and we will delve into these with a level of detail. The objective here is not to guide readers through the process of solving such models comprehensively. Instead, we aim to present fundamental principles of modeling methods that effectively address uncertainties inherent in dynamic systems.

While a thorough discussion of solving differential equations is readily available in numerous textbooks, we may touch on some relevant methods in this chapter to offer a glimpse into the broader landscape of solving dynamic system models. Our emphasis, however, remains on laying the groundwork for understanding the modeling techniques crucial for handling the intricacies of dynamic systems, rather than providing an exhaustive guide to their solution.

3.1 Linear time-invariant systems

Linear systems can be categorized as either linear time-invariant (LTI) or linear time-variant systems. Extensive research over many decades has thoroughly examined linear time-invariant systems, resulting in a comprehensive set of theorems that fully elucidate their behavior. While this section does not aim to encompass all facets of LTI systems, its purpose is to explore essential foundations that will facilitate comprehension of the subsequent chapters in this book.

Linear systems are characterized by their adherence to the *superposition* property. Suppose $y_1(t)$ represents the output response when the input was $x_1(t)$, and $y_2(t)$ is the output when the input was $x_2(t)$. In the context of linear systems, when the input is expressed as:

$$ax_1(t) + bx_2(t) \tag{3.1}$$

the output for linear systems follows:

$$ay_1(t) + by_2(t).$$

Here, a and b denote constants. Moreover, a linear system is termed time-invariant when the system's response remains unaffected by the time at which the input is applied. For instance, if the system yields an output of $y(t)$ with an input of $x(t)$, then for time-invariant systems, the output would be $y(t - \tau)$ when subjected to the input $x(t - \tau)$, where τ represents a time shift.

In system analysis, linear systems are easier to analyze than nonlinear ones, mainly because they obey the superposition principle. This property, rooted in the concept of composing signals as a sum (potentially infinite) of fundamental "stem" signals, simplifies the determination of system responses. Specifically, understanding the response of a fundamental stem signal enables us to compute the output for any other signal. It's important to note that our focus here is on smooth deterministic signals.

Now, the crucial question arises: is there a universal root signal that can effectively represent arbitrary signals $x(t)$? The answer is affirmative, and this signal is commonly known as the Dirac delta, denoted as $\delta(t)$. An intuitive explanation follows. Envision a rectangular pulse signal that is active (ON) exclusively within a time period from $-\frac{\tau}{2}$ to $\frac{\tau}{2}$, with an amplitude given by:

$$x(t) = \frac{1}{\sqrt{\tau}}.$$

The normalized power and energy content of this pulse signal are, respectively:

$$x(t)^2 = \begin{cases} \frac{1}{\tau} & \text{at } |t| \le \frac{\tau}{2}, \\ 0 & \text{elsewhere}, \end{cases}$$

$$E = \int_{-\infty}^{\infty} x(t)^2 dt = 1 \quad \forall \tau.$$

Now, consider the behavior as τ approaches zero ($\tau \to 0$). The signal exhibits an extremely large but undefined magnitude, with a time-width close to zero at $t = 0$ and zero values elsewhere. However, its area (energy) remains constant at 1, as captured by the integral:

$$\int_{-\infty}^{\infty} \delta(t)dt = \int_{0^-}^{0^+} \delta(t)dt = 1. \tag{3.2}$$

Figure 3.1 visually illustrates these concepts and includes a depiction of a Dirac delta signal shifted to the right by λ. While an ideal Dirac delta signal is theoretical, it serves as a valuable model for spiky signals, such as those resembling electrical discharges or lightning in practical scenarios.

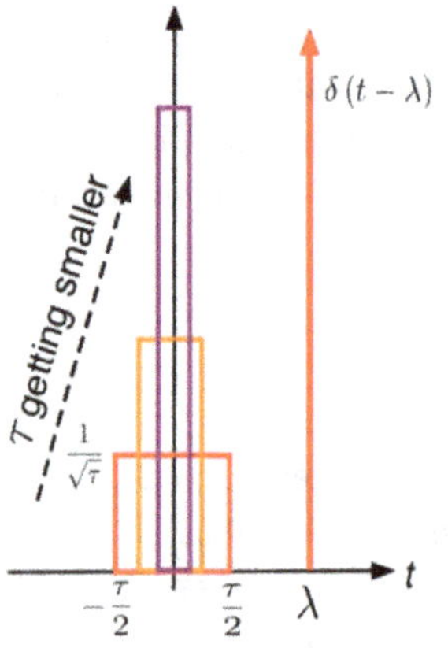

Figure 3.1: Dirac Delta.

Since Dirac delta (impulse) is zero everywhere except at its moment of existence, the multiplication of any signal $x(t)$ by a shifted delta $\delta(t - \lambda)$, as shown in Figure 3.2, leads to

$$x(t) \times \delta(t - \lambda) = x(\lambda) \times \delta(t - \lambda).$$

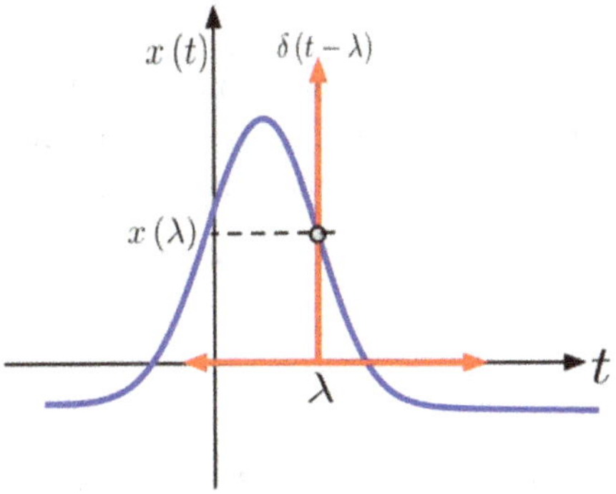

Figure 3.2: Dirac Delta Multiplication.

However, the area under the impulse is one, therefore,

$$\int_{-\infty}^{\infty} x(t)\delta(t-\lambda)dt = x(\lambda). \tag{3.3}$$

This result is very important since it indicates that it would be possible to express any continuous signal in terms of a Dirac delta such as:

$$x(t) = \int_{-\infty}^{\infty} x(\lambda)\delta(t-\lambda)d\lambda. \tag{3.4}$$

It is worth remembering that the Dirac delta has the property of being an *even* function (although the Dirac delta is not a function in the strict sense), i. e., $\delta(t-\lambda) = \delta(\lambda - t)$.

Building on the previous introduction, acquiring knowledge about the response of a Linear Time-Invariant (LTI) system when subjected to a Dirac delta input becomes pivotal. Leveraging the superposition property, this understanding empowers us to deduce the system's response to any other arbitrary signal. The response of a linear system when the input signal is a Dirac delta is called the *impulse response*. The impulse response of the LTI is usually indicated as $h(t)$, i. e., when the input of the LTI system is $\delta(t)$ the output of the system will be $h(t)$. And since it is time-invariant, it is also true to say that the output will be $h(t-\lambda)$ when the input is $\delta(t-\lambda)$. Therefore, the LTI system response of any signal $x(t)$ will be the infinitesimal summation (integration) of all weighted responses of single Dirac deltas as:

$$\begin{aligned} y(t) &= \int_{-\infty}^{\infty} x(\lambda)h(t-\lambda)d\lambda \\ &= \int_{-\infty}^{\infty} x(t-\lambda)h(\lambda)d\lambda. \end{aligned} \tag{3.5}$$

This integral equation stands out as one of the pivotal equations in linear system theory, commonly known as the *Convolution Integral.* Another compelling rationale for choosing the Dirac delta as a reference signal lies in its unique properties. The Fourier transform of the Dirac delta unveils the entire frequency spectrum of the impulse, making it a particularly insightful selection such as:

$$\int_{-\infty}^{\infty} \delta(t)e^{-j2\pi ft}dt = 1 \quad \forall f. \tag{3.6}$$

The outcome reveals that the frequency domain of $\delta(t)$ encompasses all frequencies within the useful spectrum, each with equal amplitude. Hence, the impulse response of a linear system reflects the system's response across all spectrum frequencies. The Fourier transformation of the LTI system's impulse response is commonly referred to as the *Transfer function*, $H(f)$. The inverse Fourier transform of the transfer function gives the impulse response. It is prevalent in practice to build systems' internal models through input excitation and to observe the output. This is called *system identification.* There are various approaches to tackle the system identification task, each with its own set of challenges. While one might consider employing a Dirac delta impulse and recording the resulting output, this approach, while theoretically sound, presents practical difficulties. Generating a perfect Dirac delta signal is inherently challenging in real-world scenarios.

As a viable alternative, a method involving a signal with a broad and uniformly distributed bandwidth can be employed, at least on an average scale. This signal should mimic the characteristics of a Dirac delta in the frequency domain. An example of such a signal is white noise, as discussed in Section 2.9.4. White noise exhibits a power spectral density that remains flat across a wide range of useful frequencies. Consequently, a practical and effective approach to system identification involves stimulating the unknown system with a signal similar to white noise (called PN sequence) and observing the ensuing system response. In practical scenarios, it's important to note that not all systems can be effectively stimulated using our designed input signal. Consequently, alternative methods for system identification become necessary.

For the case of *time varying* linear systems, both the impulse response and the transfer function exhibit time dependence, denoted as $h(t,\tau)$ and $H(f,\tau)$, respectively. This implies that the system's impulse response and transfer function are contingent on the operational time, τ. So far, we've examined the response of the Linear Time-Invariant (LTI) system in the time domain, as described in (3.5). Given this temporal perspective, the question arises: what is the corresponding relationship in the frequency domain? Determining this relationship is straightforward by analyzing the output in the frequency domain, employing tools such as the Fourier transform:

$$Y(f) = \int_{-\infty}^{\infty} y(t)e^{-j2\pi ft}dt$$

$$\because y(t) = \int_{-\infty}^{\infty} x(t-\lambda)h(\lambda)d\lambda$$

$$\therefore Y(f) = \int_{-\infty}^{\infty}\left[\int_{-\infty}^{\infty} x(t-\lambda)e^{-j2\pi ft}dt\right]h(\lambda)d\lambda.$$

But since:

$$\int_{-\infty}^{\infty} x(t-\lambda)e^{-j2\pi ft}dt = X(f)e^{-j2\pi f\lambda}.$$

Therefore,

$$Y(f) = X(f)\int_{-\infty}^{\infty} h(\lambda)e^{-j2\pi f\lambda}d\lambda = X(f)H(f). \tag{3.7}$$

The final relation in (3.7) holds paramount importance within Linear Time-Invariant (LTI) systems. This relation establishes the input-output connection in the frequency domain: it states that the system's output in the frequency domain results from the input in the frequency domain multiplied by the system's transfer function. In linear system analysis, it's a common practice to address problems in the frequency domain due to the simplicity of multiplication compared to solving convolution integrals, especially in cases involving multiple cascaded linear systems. Up to this point, we've encountered two representations detailing the connection between the input and output of LTI systems. Both rely on the convolution integral and its transformation to the frequency domain through multiplication by the transfer function. However, there is a third way that arises directly from mathematical models based on physical behavior: the use of differential equations. This approach provides a different perspective on understanding the relationship between inputs and outputs in LTI systems. The relation between the input $x(t)$ and the output $y(t)$ for a general LTI system can be represented as (usually $n > m$):

$$y^{(n)} + a_{n-1}y^{(n-1)} + \cdots + a_0 y = b_m x^{(m)} + b_{m-1}x^{(m-1)} + \cdots + b_1 x' + b_0 x. \tag{3.8}$$

In more compact form it becomes:

$$\sum_{i=0}^{n} a_i \frac{d^i y(t)}{dt^i} = \sum_{j=0}^{m} b_j \frac{d^j x(t)}{dt^j}, \quad a_n = 1. \tag{3.9}$$

Where:

$$y^{(n)} = \frac{d^n y(t)}{dt^n}$$
$$x^{(m)} = \frac{d^m x(t)}{dt^m}$$

and a_i and b_j are all constants $\forall i = 0, 1, \ldots n-1$ and $j = 0, 1, \ldots m$, respectively. The largest index n of a non-zero coefficient a_i determines what is called the *order* of the differential equation.

A pertinent question arises: What is the connection between the differential equation and the system's impulse response or transfer function? Given that all these formulations describe the same system, a direct and clear relation must exist. To unveil this connection, let's initiate the process by exploring the Fourier transform of a derivative. If the Fourier transform of $y(t)$ is expressed as $Y(f)$, and the inverse Fourier transform is given by:

$$y(t) = \int_{-\infty}^{\infty} Y(f) e^{j2\pi ft}\, df. \tag{3.10}$$

Therefore,

$$\frac{dy(t)}{dt} = \int_{-\infty}^{\infty} \frac{d}{dt} Y(f) e^{j2\pi ft}\, df = \int_{-\infty}^{\infty} [j2\pi f Y(f)] e^{j2\pi ft}\, df. \tag{3.11}$$

This result implies that the Fourier transform of $\frac{dy(t)}{dt}$ is simply:

$$\frac{dy(t)}{dt} \Longleftrightarrow j2\pi f Y(f).$$

Generally speaking, the Fourier transform of the nth derivatives is

$$y^{(n)}(t) \Longleftrightarrow (j2\pi f)^n Y(f).$$

In the field of linear systems theory, the preferred method for solving systems of differential equations is often the *Laplace* transform. This transformation employs the symbol $s = \sigma + j2\pi f$, where s is composed of a real number σ. The Laplace transform holds a broader scope compared to the Fourier transform because it considers the initial conditions of the system. The Laplace transform is typically defined for functions defined on the interval from 0 to infinity. This domain restriction implies that the Laplace transform is not directly applicable to functions defined for negative time values within its standard definition. In fact, the Fourier transform can be seen as a specific instance of the Laplace transform, which occurs when $\sigma = 0$ and time is positive. Leveraging the Laplace transform with all initial conditions set to zero:

$$y^{(n)}(t) \Longleftrightarrow s^n Y(s).$$

Transforming the LTI system differential equation in eq. (3.8) into frequency domain yields (assuming that both $y(t)$ and $x(t)$ have valid Laplace transform):

$$\frac{Y(s)}{X(s)} = H(s) = \frac{b_m s^m + b_{m-1}s^{m-1} + \ldots + b_1 s + b_0}{s^n + a_{n-1}s^{n-1} + \ldots + a_1 s + a_0}. \tag{3.12}$$

The same equation could also be represented in the frequency domain, f, as:

$$H(f) = \frac{b_m(j2\pi f)^m + b_{m-1}(j2\pi f)^{m-1} + \ldots + b_1(j2\pi f) + b_0}{(j2\pi f)^n + a_{n-1}(j2\pi f)^{n-1} + \ldots + a_1(j2\pi f) + a_0}. \tag{3.13}$$

Therefore, this last equation completes the relation. The inverse Laplace transform of the eq. (3.12) or the inverse Fourier transform of the eq. (3.13) gives the impulse response of the LTI systems.

Example 92. Figure 3.3 depicts a basic electrical circuit, serving not only as a literal representation of its physical components but also as a versatile model. Beyond its electrical nature, this circuit can encapsulate broader systems, such as mechanical structures, business frameworks, or even social dynamics. Let's establish initial conditions: the capacitor's initial voltage, $V_c(0)$, is set at 1 volt, while the initial inductor current, $I_L(0)$, starts at 0. Assume that the input source, $v_{\text{in}}(t)$, becomes active at $t = 0$, having been open-circuited before this point. To elucidate the dynamics governing the current $i(t)$, we'll derive the circuit's differential equation. This equation will unveil the interplay between its components. It's essential to note that we are considering ideal components in this analysis. Upon obtaining the differential equation, we'll proceed to derive both the transfer function and the impulse response. These representations will illuminate how the system responds to inputs and evolves over time, offering valuable insights into its behavior.

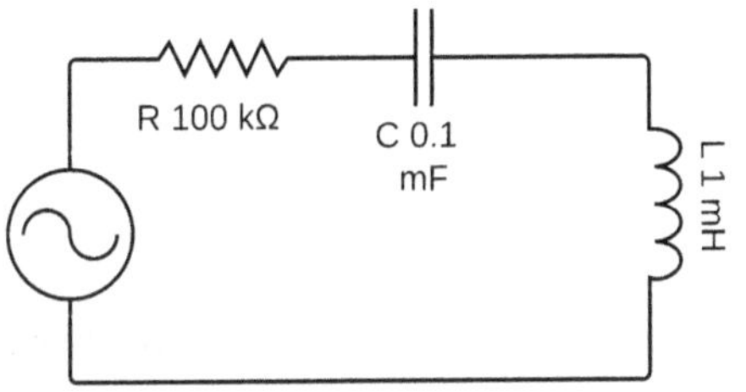

Figure 3.3: Circuit Example.

From basic electrical circuit analysis and using KVL we obtain:

$$v_{\text{in}}(t) = Ri(t) + \frac{1}{C}\int_0^t i(\tau)d\tau + L\frac{di(t)}{dt}. \tag{3.14}$$

Differentiating the equation again to remove the integration term, we have:

$$L\frac{d^2 i(t)}{dt^2} + R\frac{di(t)}{dt} + \frac{1}{C}i(t) = \frac{dv_{\text{in}}(t)}{dt}. \tag{3.15}$$

Now by dividing all terms by L, we obtain the standard 2nd order differential equation that describes the dynamics of this circuit as:

$$\frac{d^2i(t)}{dt^2} + \frac{R}{L}\frac{di(t)}{dt} + \frac{1}{LC}i(t) = \frac{1}{L}\frac{dv_{\text{in}}(t)}{dt}. \tag{3.16}$$

From fundamental linear differential theory, we can address the aforementioned differential equation by focusing on the input signal $v_{\text{in}}(t)$. This solution unfolds in two sequential steps, beginning with the homogeneous solution, which elucidates the transient properties of the circuit when $V_{\text{in}}(t) = 0$ and derives the circuit initial conditions. Subsequently, the steady-state solution, known as the particular solution, is determined. The total solution emerges as the sum of both the homogeneous and particular solutions. It's noteworthy that the homogeneous solution inherently manifests itself as a function of the eigenfunction of the linear differential equation, denoted as e^{at}, where a may assume real or complex values. This underlines the pervasive applicability of the eigenfunction in characterizing the system's behavior. To prove this property and to find the value of a that solves the homogeneous solution, simply substitute $i(t) = e^{at}$ in the above differential equation (3.16) with $v_{\text{in}}(t) = 0$. As we know from high school math that

$$\frac{d^n}{dt^n}e^{at} = a^n e^{at}. \tag{3.17}$$

Hence the value of a that solves the second-order homogeneous differential equation is

$$a^2 + \frac{R}{L}a + \frac{1}{LC} = 0. \tag{3.18}$$

Solving the last equation to find a, we reach to

$$a = -\frac{R}{2L} \pm \frac{R}{2L}\sqrt{1 - \frac{4L}{R^2C}}. \tag{3.19}$$

The homogeneous response of the circuit depends on the initial condition values and the circuit components. Analyzing (3.19), it is clear that for:

$$\frac{4L}{R^2C} > 1$$

we will have a decayed oscillatory response with a decay rate of

$$e^{-\frac{R}{2L}t}$$

and a frequency of

$$\sqrt{\omega_n^2 - \frac{R^2}{4L^2}}\ \text{rad/s},$$

where $\omega_n = \frac{1}{\sqrt{LC}}$ is called the natural resonance frequency of the circuit. In this example,

$$\frac{4L}{R^2C} = \frac{4 \times 1 \times 10^{-3}}{(100 \times 10^3)^2 \times 0.1 \times 10^{-3}} = 4 \times 10^{-9} < 1.$$

This implies that the homogeneous solution of this circuit tends to decay without oscillation, commonly termed as the over-damped response. Conversely, when the decay is accompanied by oscillation, it's referred to as an underdamped response. While it's relatively straightforward to determine the general particular solution for step or sinusoidal inputs in the above differential equation, different input scenarios necessitate individualized solutions. Given the circuit's Linearity Time-Invariance (LTI), when the input $v_{\text{in}}(t)$ is sinusoidal, the output mirrors this sinusoidal pattern with an identical frequency. However, the circuit's influence is predominantly observed in the output's amplitude and phase. For non-sinusoidal periodic signals, Fourier series expansion can express them as an infinite sum of sinusoidal signals, enabling the evaluation of the differential equation's response. Additionally, non-periodic signals can be analyzed in the frequency domain. Taking the Laplace transform of (3.16), assuming $v_{\text{in}}(0) = 0$, provides a way to assess the output characteristics.

$$s^2 I(s) - si(0) - \left.\frac{di(t)}{dt}\right|_{t=0} + \frac{R}{L} sI(s) - \frac{R}{L} i(0) + \frac{1}{LC} I(s) = \frac{1}{L} sV_{\text{in}}(s). \tag{3.20}$$

We know that $i(0) = 0$, and since $V_c(0) = 1$ and $V_L(0) = L\frac{di(t)}{dt}|_{t=0}$, therefore, $\frac{di(t)}{dt}|_{t=0} = 10^3$. Hence, we have

$$s^2 I(s) + \frac{R}{L} sI(s) + \frac{1}{LC} I(s) = \frac{1}{L} sV_{\text{in}}(s) + 10^3 \tag{3.21}$$

$$\therefore I(s) = \frac{1/LsV_{\text{in}}(s)}{s^2 + R/Ls + 1/LC} + \frac{10^3}{s^2 + R/Ls + 1/LC}. \tag{3.22}$$

The initial term represents the output current in the complex frequency domain as a direct response to the input source. Meanwhile, the second term signifies the output resulting from the initial condition of the circuit, which gradually decreases and eventually dissipates over time. If the focus is on the output of interest driven by the input, specifically the loop current $i(t)$, then the transfer function governing the relationship between the input and this particular output is expressed as:

$$H(s) = \frac{I(s)}{V_{\text{in}}(s)} = \frac{1/Ls}{s^2 + R/Ls + 1/(LC)}. \tag{3.23}$$

Substituting the component values we get:

$$H(s) = \frac{10^3 s}{s^2 + 10^8 s + 10^7}. \tag{3.24}$$

Both Octave and Scilab offer robust tools similar to Matlab for handling linear systems effectively. In Octave, for instance, you can effortlessly define the transfer function mentioned above using a simple command like $H = tf([1e3\ 0], [1\ 1e8\ 1e7])$. Subsequently, examining the impulse response becomes a matter of using the command impulse(H), while the step response is assessed through step(H). For analyzing the system's response to any input, say x, over time (defined by the time vector, t), you can use lsim(H, x, t).

Scilab, on the other hand, presents a distinct yet equally robust and user-friendly approach to simulating linear systems, as illustrated in the following example.

Example 93. Consider a physical system characterized by the following LTI system

$$2y''(t) + y'(t) + 4y(t) = x'(t) + 0.25x(t).$$

Assume that the system has zero initial conditions. Derive the transfer function and then by using inverse Laplace derive the impulse response. Using Scilab, display the impulse and step responses of this system. Finally, for an input given by

$$x(t) = te^{-t}\cos(20\pi t), \quad t \geq 0$$

simulate the system output.

Solution. We can determine the transfer function by utilizing eq. (3.12) as:

$$H(s) = \frac{s + 0.25}{2s^2 + s + 4}. \tag{3.25}$$

The impulse response simply results from the inverse Laplace transform of the transfer function $H(s)$. By referencing Laplace transform tables, we have identified that

$$\frac{s - a}{(s - a)^2 + b^2} \rightleftharpoons e^{at}\cos(bt). \tag{3.26}$$

Therefore, the determination of a as -0.25 and b as $\frac{\sqrt{31}}{2\sqrt{2}}$ is straightforward. Consequently, the impulse response is expressed as follows:

$$h(t) = \frac{1}{2}e^{-0.25t}\cos\left(\frac{\sqrt{31}}{2\sqrt{2}}t\right), \quad t \geq 0. \tag{3.27}$$

The transfer function has been implemented in Scilab using the command syslin(). In the following code you'll find the code for simulating the impulse response and the corresponding output for the specified input.

```
// Scilab Code
clear
s=poly(0,'s');
```

```
H=syslin("c",(s+0.25)/(2*s^2+s+4)); // c mean continuous
    time
t=0:.01:25;
// The plot of the impulse response
plot(t, csim('impulse',t,H));
xlabel("Time [s]");
ylabel("Impulse Response");
title("Impulse Response");
xgrid(1, 1, 10);
figure // to start new figure
// Input signal
u=t.*exp(-0.5*t).*cos(10*t);
plot(t, csim(u,t,H));
xlabel("Time [s]");
ylabel("System Output");
xgrid(1, 1, 10);
```

Figures 3.4 and 3.5 illustrate the simulated impulse response and output response respectively.

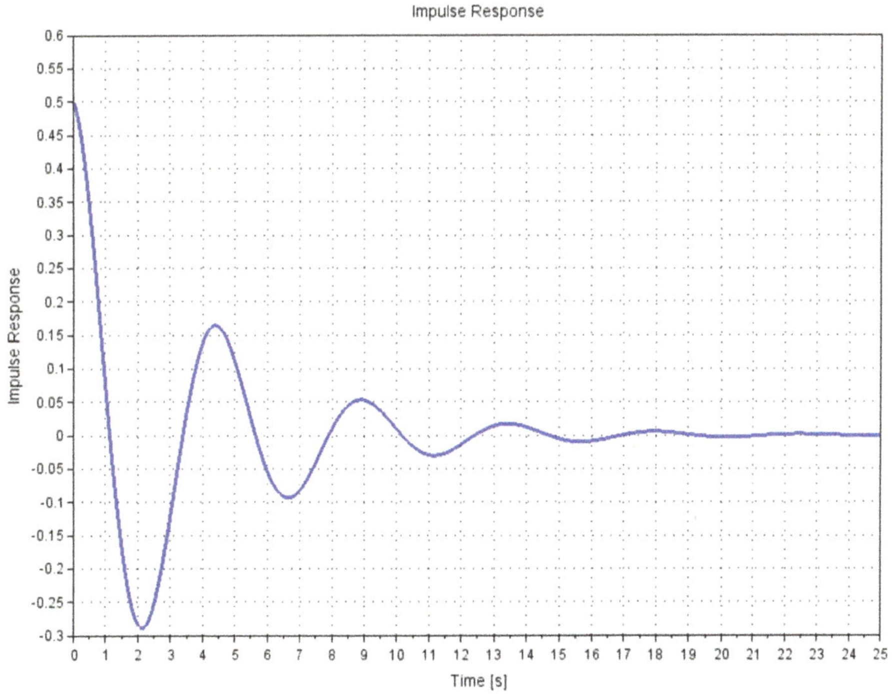

Figure 3.4: Impulse Response of Example 93.

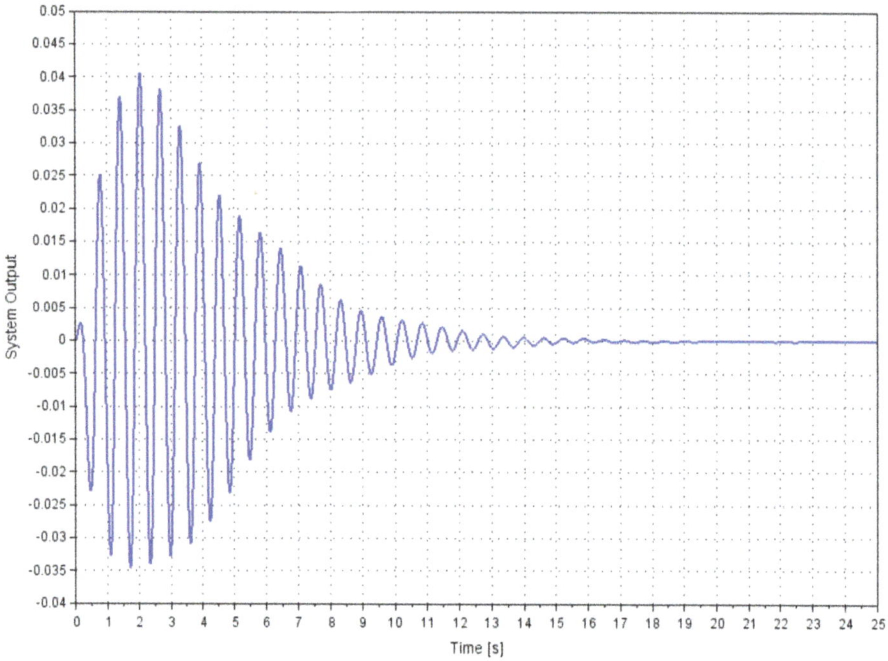

Figure 3.5: System Output of Example 93.

Next, we present a modeling example illustrating a straightforward nonlinear dynamic system.

Example 94. In this example, we provide a straightforward model of a physical system. The system comprises a tank with a uniform cross-section, which is fed at the top by a flow that is subjected to variations. Liquid is withdrawn from the bottom through a controllable valve, as illustrated in the Figure 3.6.

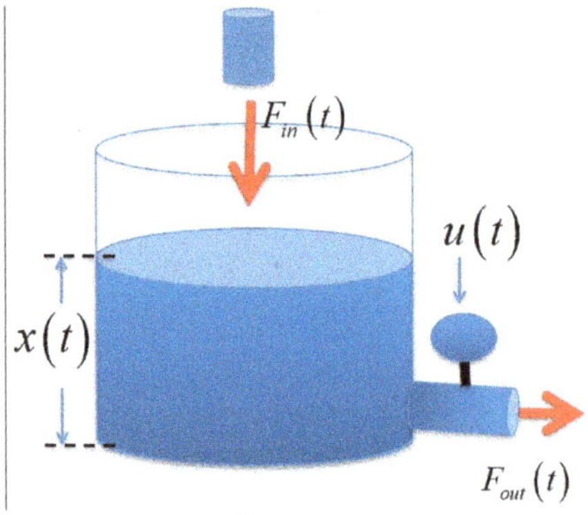

Figure 3.6: Modeling Example.

Assume that the cross-sectional area of the tank is A_0, which is constant. Hence, with ignoring the tiny liquid inside the orifice tube, the total volume of the liquid inside the tank at time t is $V(t) = A_0 x(t)$, where $x(t)$ is the height of the liquid in the tank at time t. It is well-known that the liquid flows through the orifice with the speed of free fall as $\sqrt{2gx(t)}$, where $g = 9.8\,\text{m/s}^2$. This model assumes uniform free fall. However, in reality, there should be terms depending on the viscosity of the liquid, the density, and the tank area. Since liquid viscosity depends on the temperature, hence, it should be included in a more accurate model. However, the aim of this example is only to show how to formulate a simple analytical dynamic model. The variations of the liquid volume at time t depend on the current level $x(t)$ and the input flow as well as the output flow, such as:

$$\Delta V(t) = F_{\text{in}}(t) - F_{\text{out}}(t). \tag{3.28}$$

It is clear that output volume is

$$F_{\text{out}}(t) = A_u \sqrt{2gx(t)} = s_u, \quad \sqrt{x(t)} \tag{3.29}$$

where A_u is the cross-sectional area of the orifice with the valve, and $s_u = A_u \sqrt{2g}$. Therefore, the level model becomes

$$\begin{aligned} A_0 dx(t) &= [F_{\text{in}}(t) - s_u \sqrt{x(t)}]dt \\ \therefore A_0 \frac{dx(t)}{dt} &+ s_u \sqrt{x(t)} = F_{\text{in}}(t). \end{aligned} \tag{3.30}$$

Therefore, equation (3.30) presents a robust analytical model for predicting the liquid level at any given time t, considering the varying inflow and outflow dynamics. However, as previously noted, this model remains an approximation, lacking the incorporation of numerous other influencing factors. Despite these simplifications, the dynamic model outlined is inherently nonlinear. While solving this nonlinear differential equation through numerical methods poses no issue, applying the concepts of impulse response and transfer functions directly becomes challenging. However, it can be linearized around an operating point x_0. Moreover, if the inflow and/or outflow exhibit stochastic fluctuations, they ought to be modeled as stochastic nonlinear differential equations. In such instances, the objective shifts from determining the precise level at time t to analyzing statistical properties such as averages, covariances, and more. This particular aspect will be addressed in a later section of this chapter.

3.2 State space representation

The state-space representation stands out as a fundamental technique for constructing dynamic models, applicable to linear, nonlinear, LTI (Linear Time-Invariant), or time-varying systems. Its versatility allows for straightforward extension to accommodate modeling for multiple inputs and multiple outputs. Additionally, it serves as the primary model for estimating system dynamics, particularly when utilizing Kalman filters, as elaborated in Chapter 5. The essence of the state-space representation lies in defining the internal states of a system, characterizing the system's internal behavior that shapes its observed dynamics. Typically, the output is correlated with at least some of the system states. It's important to note that the state-space representation is not unique; in other words, numerous distinct (and potentially infinite) state-space representations may characterize the same system. Consequently, these representations yield identical responses for identical inputs and operating conditions. The system state, represented as a vector of internal variables, relies on its present values to determine future states. This encapsulates the comprehensive influence of past states on the subsequent behavior of the system.

As previously discussed, nearly any continuous dynamic system can be characterized by either a single or a system of differential equations. These equations can be linear or nonlinear, time-invariant or time-varying. Considering the general form of a continuous LTI system as described in (3.8), an nth order differential equation can be transformed into n equations of first-order differential equations. One way to make such transformation is based on the following steps. First, check equation (3.8), and let's define

$$\dot{z}_1 = \frac{dz_1}{dt} = b_0 x - a_0 y = \sum_{i=1}^{n} a_i \frac{d^i y(t)}{dt^i} - \sum_{j=1}^{m} b_j \frac{d^j x(t)}{dt^j}.$$

Integrating both sides leads to

$$z_1 = \sum_{i=1}^{n} a_i \frac{d^{i-1} y(t)}{dt^{i-1}} - \sum_{j=1}^{m} b_j \frac{d^{j-1} x(t)}{dt^{j-1}}.$$

Next let's define

$$\dot{z}_2 = z_1 - a_1 y + b_1 x = \sum_{i=2}^{n} a_i \frac{d^{i-1} y(t)}{dt^{i-1}} - \sum_{j=2}^{m} b_j \frac{d^{j-1} x(t)}{dt^{j-1}}.$$

Again, integrating both sides leads to

$$z_2 = \sum_{i=2}^{n} a_i \frac{d^{i-2} y(t)}{dt^{i-2}} - \sum_{j=2}^{m} b_j \frac{d^{j-2} x(t)}{dt^{j-2}}.$$

Then we define

$$\dot{z}_3 = z_2 - a_2 y + b_2 x = \sum_{i=3}^{n} a_i \frac{d^{i-2} y(t)}{dt^{i-2}} - \sum_{j=3}^{m} b_j \frac{d^{j-2} x(t)}{dt^{j-2}}.$$

We continue this process until we reach to

$$\dot{z_n} = z_{n-1} - a_{n-1}y + b_{n-1}x = \dot{y}$$
$$\Rightarrow y = z_n.$$

Once all first order equations have been calculated, we put the state-space form as:

$$\dot{\mathbf{z}} = \mathbf{A}\mathbf{z} + \mathbf{B}\mathbf{x}, \tag{3.31}$$

where $\mathbf{z} = [z_1, z_2, \ldots, z_n]^T$ is called the system state, $\mathbf{A}$ is an $n \times n$ matrix called *System Matrix*, $\mathbf{B}$ is $n \times m$ input matrix, and $\mathbf{x}$ is the input vector. The behavior and the dynamics of the system is characterized by the matrix $\mathbf{A}$. The system output or observation can be referred to the system state. If the output of the system is $y(t)$ in (3.8), then $y(t) = z_n$, or $y = \mathbf{C}\mathbf{z}$, with $\mathbf{C} = [0, 0, \ldots, 0, 1]$ as $1 \times n$ vector. However, in general we may express the output vector as $\mathbf{y} = \mathbf{C}\mathbf{z} + \mathbf{D}\mathbf{x}$. In this general representation, the output vector $\mathbf{y}$ is $k \times 1$, $\mathbf{C}$ is called the output matrix with dimension $k \times n$, and $\mathbf{D}$ is called feedforward matrix with dimension $k \times m$. For linear but time varying systems, equation parameters ($a's$ and $b's$) in (3.8) can be time varying. Hence, the state-space representation in this case is given as

$$\dot{\mathbf{z}} = \mathbf{A}(t)\mathbf{z} + \mathbf{B}(t)\mathbf{x} \tag{3.32}$$
$$\mathbf{y} = \mathbf{C}(t)\mathbf{z} + \mathbf{D}(t)\mathbf{x}. \tag{3.33}$$

Of course, $\mathbf{z}$, $\mathbf{x}$, and $\mathbf{y}$ are all function in time as well. The following examples illustrate the transformation process:

Example 95. Find the state-space representation of a dynamic systems represented by a LTI differential equation as

$$\frac{d^2y}{dt^2} + 2\frac{dy}{dt} - 8y + 7\frac{dx}{dt} - x = 0.$$

Solution. Initially, it is evident that the given differential equation is of second order. Consequently, we will reformulate it down into two first-order differential equations. The procedural steps outlined earlier can be seamlessly applied: commence by defining

$$\dot{z_1} = x + 8y = \frac{d^2y}{dt^2} + 2\frac{dy}{dt} + 7\frac{dx}{dt}$$

Integrate both sides

$$\rightsquigarrow z_1 = \frac{dy}{dt} + 2y + 7x.$$

Again define

$$\dot{z_2} = z_1 - 2y - 7x = \frac{dy}{dt}$$
$$\therefore y = z_2.$$

In matrix form, the resulting state space form is:

$$\dot{\mathbf{z}} = \begin{bmatrix} \dot{z}_1 \\ \dot{z}_2 \end{bmatrix} = \begin{bmatrix} 0 & 8 \\ 1 & -2 \end{bmatrix} \begin{bmatrix} z_1 \\ z_2 \end{bmatrix} + \begin{bmatrix} 1 \\ -7 \end{bmatrix} x \tag{3.34}$$
$$y = [0 \;\; 1]\mathbf{z}.$$

This is the state-space form of the same system represented by the above differential equation.

As previously noted, the state-space representation is not unique. Let's define new states for the same system as $\mathbf{z} = \mathbf{Sg}$, where matrix $\mathbf{S}$ is any $n \times n$ nonsingular matrix and $\mathbf{g}$ is the new state vector. Substitute in the original state space equation as:

$$\mathbf{S}\dot{\mathbf{g}} = \mathbf{ASg} + \mathbf{Bx}.$$

Therefore, the new state-space representation becomes:

$$\dot{\mathbf{g}} = \mathbf{S}^{-1}\mathbf{ASg} + \mathbf{S}^{-1}\mathbf{Bx}.$$

The output equation becomes:

$$\mathbf{y} = \mathbf{CSg} + \mathbf{Dx}.$$

In the Example 95 and for $\mathbf{S} = \left[\begin{smallmatrix} 1 & -1 \\ 2 & 0 \end{smallmatrix}\right]$, the new state-space representation for the same system becomes:

$$\dot{\mathbf{g}} = \begin{bmatrix} \dot{g}_1 \\ \dot{g}_2 \end{bmatrix} = \begin{bmatrix} -1.5 & -0.5 \\ -17.5 & -0.5 \end{bmatrix} \begin{bmatrix} g_1 \\ g_2 \end{bmatrix} + \begin{bmatrix} -3.5 \\ -4.5 \end{bmatrix} x$$
$$y = [2 \;\; 0]\mathbf{g}.$$

Despite introducing these new states, the equations continue to represent the same system. This characteristic offers significant flexibility in the representation, allowing for useful applications such as system matrix diagonalization.

Example 96. Find the state-space representation of the RLC circuit shown in Figure 3.3.

Solution. It's worth noting that using Octave or Scilab makes it straightforward to find the state-space matrices in continuous or discrete time with a single command, such as: *tf2ss*(). Nevertheless, this guide explains the manual steps to accomplish this

$$\dot{z}_1 = -\frac{1}{LC} i(t) = \frac{d^2 i(t)}{dt^2} + \frac{R}{L}\frac{di(t)}{dt} - \frac{1}{L}\frac{dv_{\text{in}}(t)}{dt}$$
$$\therefore z_1 = \frac{di(t)}{dt} + \frac{Ri(t)}{L} - \frac{v_{\text{in}}(t)}{L}$$
$$\dot{z}_2 = z_1 - \frac{Ri(t)}{L} - \frac{v_{\text{in}}(t)}{L} = \frac{di(t)}{dt}$$

$$\therefore z_2 = i(t).$$

Finally, we can put these two first order differential equation in matrix form as:

$$\dot{\mathbf{z}} = \begin{bmatrix} \dot{z}_1 \\ \dot{z}_2 \end{bmatrix} = \begin{bmatrix} 0 & -\frac{1}{LC} \\ 1 & -\frac{R}{L} \end{bmatrix} \begin{bmatrix} z_1 \\ z_2 \end{bmatrix} + \begin{bmatrix} 0 \\ -\frac{1}{L} \end{bmatrix} v_{\text{in}(t)}$$

$$y_o = [0\,1] \begin{bmatrix} z_1 \\ z_2 \end{bmatrix}.$$

Example 97. Derive the state-space representation of a dynamic system described by a linear time-varying differential equation with two inputs given as

$$\frac{d^3y}{dt^3} + 2\sin(5t)\frac{dy}{dt} - 7tx_1 - x_2 = 0.$$

The output of the system is given by:

$$y_o = y + \frac{dy}{dt} - x_2.$$

Solution. First, it is clear that the order of this differential equation is 3.

It is quite straightforward to transform this problem into state space by the following steps:

$$\begin{aligned} z_1 &= y \\ z_2 &= \dot{z}_1 \\ z_3 &= \dot{z}_2 \\ \therefore \dot{z}_3 &= \frac{d^3y}{dt^3} = -2\sin(5t)z_2 + 7tx_1 + x_2. \end{aligned}$$

The state-space form of this problem is:

$$\dot{\mathbf{z}} = \begin{bmatrix} \dot{z}_1 \\ \dot{z}_2 \\ \dot{z}_3 \end{bmatrix} = \begin{bmatrix} 0 & 1 & 0 \\ 0 & 0 & 1 \\ 0 & -2\sin(5t) & 0 \end{bmatrix} \begin{bmatrix} z_1 \\ z_2 \\ z_3 \end{bmatrix} + \begin{bmatrix} 0 & 0 \\ 0 & 0 \\ 7t & 1 \end{bmatrix} \begin{bmatrix} x_1 \\ x_2 \end{bmatrix}$$

$$y_o = [1\,1\,0] \begin{bmatrix} z_1 \\ z_2 \\ z_3 \end{bmatrix} + [0\ -1] \begin{bmatrix} x_1 \\ x_2 \end{bmatrix}.$$

The state space representation of a Linear Time-Invariant (LTI) system can be readily transformed into a transfer function or impulse response. In the presence of nonzero initial conditions, the Laplace transform of the state equations can be determined as follows:

$$\dot{\mathbf{z}}(t) \rightleftharpoons s\mathbf{Z}(s) - \mathbf{Z}(0).$$

Therefore, the Laplace transformation of any state-space representation becomes:

$$s\mathbf{Z}(s) - \mathbf{Z}(0) = \mathbf{A}\mathbf{Z}(s) + \mathbf{B}\mathbf{x}(s).$$

Hence, it can be rearranged as:

$$\mathbf{Z}(s) = (s\mathbf{I} - \mathbf{A})^{-1}\mathbf{B}\mathbf{X}(s) + (s\mathbf{I} - \mathbf{A})^{-1}\mathbf{Z}(0). \tag{3.35}$$

Now, we substitute $\mathbf{Z}(s)$ in the output equations as

$$\mathbf{Y}(s) = \left[\mathbf{C}(s\mathbf{I} - \mathbf{A})^{-1}\mathbf{B} + \mathbf{D}\right]\mathbf{X}(s) + \mathbf{C}(s\mathbf{I} - \mathbf{A})^{-1}\mathbf{Z}(0).$$

The first term represents the output in the frequency domain as a response to the input, and the second term represents the output due to the initial states of the system. Hence, for single input single output (SISO) LTI systems with zero initial condition, the transfer function in terms of state space matrices is given by:

$$\frac{Y(s)}{X(s)} = \mathbf{C}(s\mathbf{I} - \mathbf{A})^{-1}\mathbf{B} + \mathbf{D}. \tag{3.36}$$

Once the system has been modeled in the state space representation, it can be solved using various methods. For a linear system, the resulting state vector:

$$\mathbf{z}(t) = \mathbf{z}_{zi}(t) + \mathbf{z}_{zs}(t),$$

where $\mathbf{z}_{zi}(t)$ is the system response for zero input, i. e., when $\mathbf{x}(t) = 0$, and $\mathbf{z}_{zs}(t)$ is the system response when the initial conditions are zero, i. e., when $\mathbf{Z}(0) = \mathbf{0}$. Determining the dynamics of the system output involves solving for the system states $\mathbf{z}(t)$. Referring to equation (3.35) the Laplace domain states are directly influenced by the term $(s\mathbf{I}-\mathbf{A})^{-1}$ which is multiplied in the first term with the input and in the second term with the initial conditions. To obtain the state space representation in the time domain, the inverse Laplace transform is applied. It's essential to note that multiplication in the frequency domain corresponds to convolution in the time domain. For instance, for

$$a(t) \rightleftharpoons A(s) \tag{3.37}$$

$$b(t) \rightleftharpoons B(s) \tag{3.38}$$

$$A(s) \times B(s) \rightleftharpoons a(t) \otimes b(t). \tag{3.39}$$

Where $\rightleftharpoons$ and $\otimes$ denote the Laplace transform and the convolution operation respectively. It is rather straightforward (using for example, Taylor series expansion) to prove that

$$(s\mathbf{I}-\mathbf{A})^{-1} \rightleftharpoons e^{\mathbf{A}t}. \tag{3.40}$$

Therefore, the states vector at any time t can be solved as:

$$\mathbf{z}(t) = e^{\mathbf{A}(t-t_0)}\mathbf{z}(t_0) + \int_{t_0}^{t} e^{\mathbf{A}(t-\lambda)}\mathbf{B}\mathbf{x}(\lambda)d\lambda. \tag{3.41}$$

The convolution integral is the second term. Finally, we can substitute this equation into the output equation to determine the output as:

$$\mathbf{y}(t) = \mathbf{C}e^{\mathbf{A}(t-t_0)}\mathbf{z}(t_0) + \int_{t_0}^{t} \mathbf{C}e^{\mathbf{A}(t-\lambda)}\mathbf{B}\mathbf{x}(\lambda)d\lambda + \mathbf{D}\mathbf{x}(t). \tag{3.42}$$

This model is primarily applicable to Linear Time-Invariant (LTI) systems. Its accuracy is contingent upon the system's adherence to the LTI assumption. In the aforementioned solution, we can designate the matrix $e^{\mathbf{A}(t-t_0)}$ as the *transition matrix*. This matrix delineates the transition paths of both initial values and inputs from time t_0 to t. However, for a broader class encompassing general linear differential equations (including time-varying systems), a more comprehensive form of the transition matrix, denoted as $\Phi(t, t_0)$, is utilized. Thus, the general solution for any smooth linear differential equation modeled within the state-space framework is expressed as:

$$\mathbf{y}(t) = \mathbf{C}\Phi(t, t_0)\mathbf{z}(t_0) + \int_{t_0}^{t} \mathbf{C}\Phi(t,\lambda)\mathbf{B}\mathbf{x}(\lambda)d\lambda + \mathbf{D}\mathbf{x}(t). \tag{3.43}$$

The transition matrix has several important and useful properties, such as:
- $\frac{\partial\Phi(t,\lambda)}{\partial t} = \mathbf{A}(t)\Phi(t,\lambda)$;
- $\frac{\partial\Phi(t,\lambda)}{\partial \lambda} = -\Phi(t,\lambda)\mathbf{A}(t)$;
- $\Phi(t,\lambda) = \Phi(t,\tau)\Phi(\tau,\lambda)\ \forall\, t \le \tau \le \lambda$;
- $\Phi(t,\lambda) = \Phi(\lambda, t)^{-1}$;
- $\Phi(t,t) = \mathbf{I}$.

In general, obtaining a closed-form solution for the transition matrix in most linear time-varying systems is challenging. Nevertheless, numerical solutions are always feasible and can be employed for accurate results.

Example 98. For the linear dynamic system given in Example 93, find the state-space representation of the system from the differential equation. Find the transfer function and the solution in the time domain based on the state space matrices.

Using the same procedure discussed before, we find the state-space representation as:

$$\dot{\mathbf{z}} = \begin{bmatrix} \dot{z}_1 \\ \dot{z}_2 \end{bmatrix} = \begin{bmatrix} 0 & -2 \\ 1 & -\frac{1}{2} \end{bmatrix} \begin{bmatrix} z_1 \\ z_2 \end{bmatrix} + \begin{bmatrix} \frac{1}{8} \\ \frac{1}{2} \end{bmatrix} x(t)$$

$$y = [0\ 1] \begin{bmatrix} z_1 \\ z_2 \end{bmatrix}.$$

The transfer function is derived based on (3.36) as:

$$H(s) = \begin{bmatrix} 0 & 1 \end{bmatrix} \begin{bmatrix} s & 2 \\ -1 & s + \frac{1}{2} \end{bmatrix}^{-1} \begin{bmatrix} \frac{1}{8} \\ \frac{1}{2} \end{bmatrix}.$$

Since:

$$\begin{bmatrix} s & 2 \\ -1 & s + \frac{1}{2} \end{bmatrix}^{-1} = \frac{1}{s^2 + 0.5s + 2} \begin{bmatrix} s + \frac{1}{2} & -2 \\ 1 & s \end{bmatrix}.$$

Therefore, the transfer function becomes:

$$H(s) = \frac{s + 0.25}{2s^2 + s + 4}. \tag{3.44}$$

Which, naturally, is the same as the one obtained directly from the differential equation. The state transition matrix of LTI system is the inverse Laplace transform of

$$(s\mathbf{I} - \mathbf{A})^{-1}.$$

Therefore, using the Laplace transform tables, the state transition matrix is:

$$\Phi(t) = \begin{bmatrix} a_{11}(t) & a_{12}(t) \\ a_{21}(t) & a_{22}(t) \end{bmatrix},$$

where

$$a_{11}(t) = e^{-t/4}\left(\frac{1}{31}\cos(\sqrt{31}/4t) + \frac{1}{\sqrt{31}}\sin(\sqrt{31}/4t)\right)$$

$$a_{12}(t) = \frac{8}{\sqrt{31}} e^{-t/4} \sin(\sqrt{31}/4t)$$

$$a_{21}(t) = \frac{4}{\sqrt{31}} e^{-t/4} \sin(\sqrt{31}/4t)$$

$$a_{22}(t) = e^{-t/4}\left(\frac{1}{31}\cos(\sqrt{31}/4t) - \frac{1}{\sqrt{31}}\sin(\sqrt{31}/4t)\right).$$

Even for a straightforward Linear Time-Invariant (LTI) dynamic system, manually deriving the transition matrix can be a time-consuming and laborious task. Nevertheless, once the transition matrix is successfully derived, it provides comprehensive insights into the system's behavior, or more precisely, into the model of the system.

In general, for nonlinear time-varying dynamic systems, we may have the following state-space representation form:

$$\dot{\mathbf{z}} = \mathbf{f}(\mathbf{z}, \mathbf{x}, t), \tag{3.45}$$

$$\mathbf{y} = \mathbf{g}(\mathbf{z}, \mathbf{x}, t), \tag{3.46}$$

where $\mathbf{f}(\bullet)$ is the state nonlinear equation and $\mathbf{g}(\bullet)$ is the output equation.

Example 99. Construct the state-space model for the following nonlinear differential equation: $\frac{d^2y}{dt^2} + \sin(y)\frac{dy}{dt} + e^{-y} - x = 0$

Solution. We proceed with the same steps employed in the linear case, such as:

$$\dot{z}_1 = x - e^{-y} = \frac{d^2y}{dt^2} + \sin(y)\frac{dy}{dt} \tag{3.47}$$

$$\therefore z_1 = \frac{dy}{dt} - \cos(y) \tag{3.48}$$

$$\dot{z}_2 = z_1 + \cos(y) = \frac{dy}{dt} \tag{3.49}$$

$$\therefore z_2 = y. \tag{3.50}$$

Therefore

$$\dot{\mathbf{z}} = \begin{bmatrix} \dot{z}_1 \\ \dot{z}_2 \end{bmatrix} = \begin{bmatrix} x - e^{-z_2} \\ z_1 + \cos(z_2) \end{bmatrix} \tag{3.51}$$

$$y = z_2.$$

There exists no universal solution for nonlinear differential equations, as each nonlinear system typically demands a distinct approach. General techniques such as convolution with the impulse response, transfer function concepts, and transition matrix concepts are not universally applicable. However, if the system's operating point is known, it is often feasible to linearize the differential equation. The key prerequisite for linearization is the smoothness of the system. Fortunately, the deterministic component of most real systems exhibits smooth behavior. Consequently, linearization is a powerful method for effectively modeling nonlinear systems.

In the field of system continuity, two fundamental categories exist: smooth continuous systems and discrete systems in time. For instance, consider a system's state represented by the number of customers; such a state is updated at discrete time intervals. However, even within smooth, continuous systems, the prevalent practice in the digital age involves sampling observations and measurements. This practice results in the treatment of most systems in a practically discrete manner.

Within the sampling theorem, a collection of robust theorems delineates the impact of sampled signals. One fundamental theory asserts that by sampling a smooth signal at a rate twice its bandwidth, it's theoretically possible to restore all information within that signal. However, this hinges on the assumption of having non-causal filters to perfectly reconstruct the original signal. In practical terms, though, while perfect restoration might not be achievable, we can still reconstruct a significant portion of the original signal and even estimate the missing components.

Thus, when constructing the state-space representation of a smooth, continuous system, it becomes pivotal to transform it into its discrete counterpart. Numerous methods exist for discretizing continuous-time state-space systems, often involving the conversion of differential equations into difference equations.

Similarly, this transformation can be carried out in the complex frequency domain (Laplace) by converting to the *z-Transform*, which is analogous to a unit time shift in the time domain, expressed as $z^{-1}X(z) \rightleftharpoons x(k-1)$. A common method for transitioning from the *s-domain* to the *z-domain*, i. e., from continuous-time differential equations to difference equations, is through the *Bilinear* approximation:

$$s = \frac{2}{T_s}\frac{1-z^{-1}}{1+z^{-1}}. \tag{3.52}$$

Where T_s denotes the sampling time.

Moreover, an alternative approach involves a direct transformation to the state-space representation, exemplified as:

$$\mathbf{z}[k+1] = \mathbf{A}_d[k]\mathbf{z}[k] + \mathbf{B}_d[k]\mathbf{x}_d[k] \tag{3.53}$$

$$\mathbf{y}[k] = \mathbf{C}_d[k]\mathbf{z}[k] + \mathbf{D}_d[k]\mathbf{x}_d[k]. \tag{3.54}$$

Where we use the subscript d to refer to the discrete version. For the LTI system, we may formulate the discrete matrices as:

$$\begin{aligned} \mathbf{A}_d &= e^{\mathbf{A}T_s} \\ \mathbf{B}_d &= \left[\int_0^{T_s} e^{\mathbf{A}\tau}d\tau\right]\mathbf{B} \\ \mathbf{C}_d &= \mathbf{C} \quad \text{and finally} \quad \mathbf{D}_d = \mathbf{D}. \end{aligned} \tag{3.55}$$

It is also possible to have a discrete form by using the derivative (Euler) approximation:

$$\dot{\mathbf{z}}(t) \approx \frac{\mathbf{z}((k+1)T_s) - \mathbf{z}(kT_s)}{T_s}. \tag{3.56}$$

This approximation is accurate only for extremely small T_s compared to the fluctuations of the states $\mathbf{z}$. Otherwise it is not accurate enough. However, it offers much easier representation for the discrete form of the state-space equations. By substituting (3.56) in (3.32) we obtain the following transformed matrices:

$$\begin{aligned} \mathbf{A}_d &= T_s\mathbf{A} + \mathbf{I} \\ \mathbf{B}_d &= T_s\mathbf{B} \\ \mathbf{C}_d &= \mathbf{C} \quad \text{and} \quad \mathbf{D}_d = \mathbf{D}. \end{aligned} \tag{3.57}$$

However, remember that the transformation in (3.57) is less accurate than the transformation in (3.55). Since,

$$e^{\mathbf{A}T_s} = \mathbf{I} + T_s\mathbf{A} + \frac{1}{2!}T_s^2\mathbf{A}^2 + \cdots \tag{3.58}$$

Therefore, it is clear that Euler approximation considers only the first 2 terms of this expansion. Both Octave and Scilab (like *Matlab*) have very strong functions to handle and simulate continuous and discrete linear systems. For LTI systems (3.53) and (3.54) can be further simplified as

$$\mathbf{z}[k+1] = \mathbf{A}_d\mathbf{z}[k] + \mathbf{B}_d\mathbf{x}_d[k] \tag{3.59}$$

$$\mathbf{y}[k] = \mathbf{C}_d\mathbf{z}[k] + \mathbf{D}_d\mathbf{x}_d[k]. \tag{3.60}$$

These equations have particular importance in the analysis and modeling of deterministic as well as stochastic linear systems.

Example 100. Assume a dynamic system that is modeled as second-order linear differential equation as

$$\ddot{y}(t) + 3\dot{y}(t) + 2y(t) = u(t). \tag{3.61}$$

Determine the transfer function of the system in the Laplace domain. Develop the continuous-time state-space representation for the system. Derive the discrete form of the state-space equation using a sampling time of $T_s = 0.1$ s. Illustrate the block diagram of the state-space in discrete time. Display the impulse response of the system, showing both continuous and discrete time representations. Use Octave to build the system response for the input of $x(t) = e^{-0.5t}$ for $0 \le t \le 5$ seconds, using both continuous and discrete models. Compare the results.

Solution. The transfer function assuming zero initial states is given by

$$\begin{aligned} s^2Y(s) + 3sY(s) + 2Y(s) &= U(s) \\ \therefore H(s) = \frac{Y(s)}{U(s)} &= \frac{1}{s^2 + 3s + 2}. \end{aligned}$$

As described before, the construction of a state-space model of this simple linear differential equation is a straightforward as: $x_1 = y$, $x_2 = \dot{x}_1 = \dot{y}$, and $\dot{x}_2 = \ddot{y}$, hence $\dot{x}_2 = -3\dot{x}_1 - 2x_1 + u$, therefore, we can construct the state-space representation as follows

$$\dot{\mathbf{x}} = \begin{bmatrix} \dot{x}_1 \\ \dot{x}_2 \end{bmatrix} = \begin{bmatrix} 0 & 1 \\ -2 & -3 \end{bmatrix} \begin{bmatrix} x_1 \\ x_2 \end{bmatrix} + \begin{bmatrix} 0 \\ 1 \end{bmatrix} u(t)$$

$$y = [1\,0] \begin{bmatrix} x_1 \\ x_2 \end{bmatrix}.$$

We may use different approximations to convert from continuous to discrete time. In Octave as well, as there is built-in function to perform such conversion easily. However, to have a simple hand calculation, we will use the simplest form:

$$\mathbf{A}_d = 0.1 \begin{bmatrix} 0 & 1 \\ -2 & -3 \end{bmatrix} + \begin{bmatrix} 1 & 0 \\ 0 & 1 \end{bmatrix} = \begin{bmatrix} 1 & 0.1 \\ -0.2 & 0.7 \end{bmatrix}$$

$$\mathbf{B}_d = 0.1 \begin{bmatrix} 0 \\ 1 \end{bmatrix} = \begin{bmatrix} 0 \\ 0.1 \end{bmatrix}$$

$$\mathbf{C}_d = \mathbf{C} \quad \text{and} \quad \mathbf{D}_d = \mathbf{D}.$$

Figure 3.7 shows a simple block diagram of the internal dynamics of this system in discrete time.

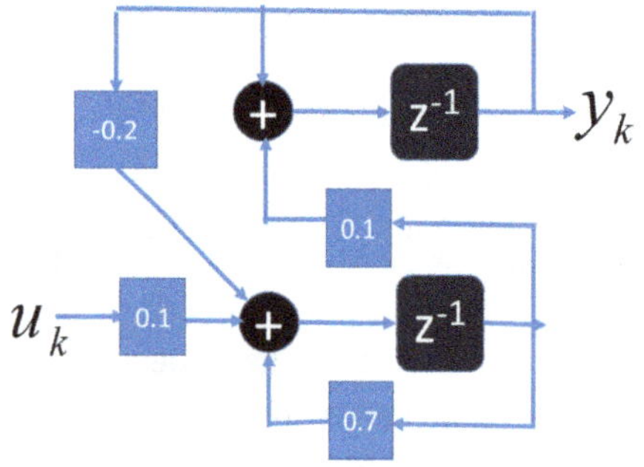

Figure 3.7: Block diagram of Example 100.

The simulation results in Figure 3.8 show the output of this dynamic system in both continuous and discrete cases. Such a linear simulation could be performed very easily in Octave using the command "LSIM" in control package.

The following code is used to simulate the example and obtain Figure 3.8.

```
clear all
% Continous States
A = [0 1; -2 -3];
B = [0;1];
C = [1 0];
D = [];
% Discrete States
```

```
Ad = [1 0.1; -0.2 0.7];
Bd = [0;0.1];
Cd = [1 0];
Dd = [];
Ts = 0.1; %sampling time of the discrete case
% Load Control package
pkg load control

SYSc = ss(A,B,C,D); % building continous ss model
SYSd = ss(Ad,Bd,Cd,Dd,Ts) % building discrete SS model

t=0:Ts:5; % Simulation time
u=exp(-0.5*t); % Input signal
lsim(SYSc,SYSd,u,t)
```

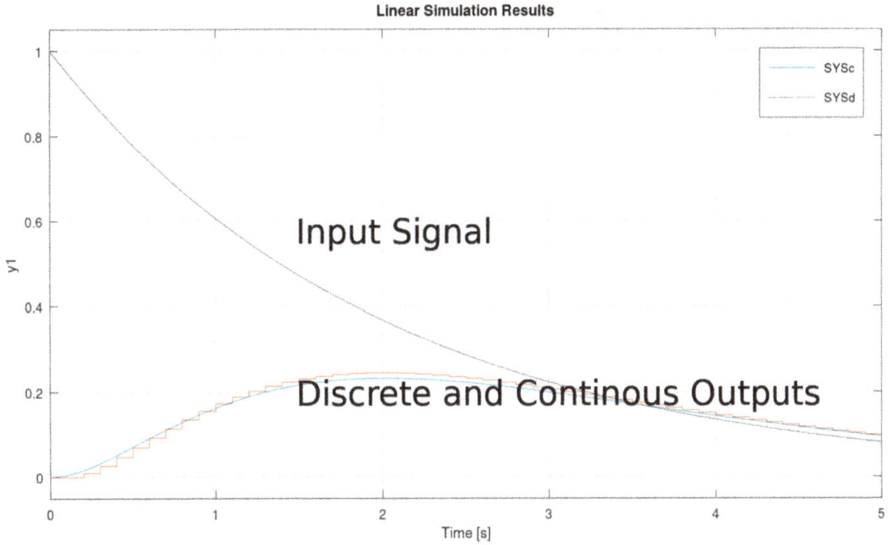

Figure 3.8: Linear simulation result.

3.3 Stochastic process and LTI systems

The system model establishes the relationship or mapping between system inputs and outputs. The accuracy of a model is measured by how closely its output aligns with the actual system output, a metric that holds true within specific classes of inputs and under defined conditions. Take, for instance, the conventional Ohm's law, which proves highly accurate under conditions of low frequency signals and constant temperature. Consequently, every model operates within a designated band of effectiveness.

The primary goal of system modeling is to address and minimize uncertainties. This objective finds application in various scenarios, including:

- Developing a system model provides the ability to anticipate how a system will behave under various conditions. Consider modeling electricity demand, factoring in anticipated weather conditions, time of day, public events, and more. Such models prove instrumental in predicting demand, enabling energy providers and generators to optimize their operations to meet these requirements efficiently. This supply-demand model finds application across diverse business sectors.
- After establishing the system model, we can forecast the output for a defined input. Additionally, we can estimate the input by analyzing the measured or observed output. This capability holds immense importance in various applications, such as in wireless communication. The transmitted signal, often unknown to us, is distorted while passing through the transmission media, particularly in a wireless channel. Consequently, what we receive is a corrupted and noisy rendition of the original signal. However, having a channel model allows us not only to recover but also to enhance the quality of the original signal. Another compelling application of this inverse modeling lies in restoring medical images from their corrupted versions.
- The constructed model enables in-depth study and optimization of system operations. Consequently, it becomes possible to minimize risks (and associated costs) or maximize rewards (yielding utilities), among other potential objectives.

Hence, one of the primary applications lies in predicting the system output for a specified input. However, practical scenarios seldom involve studying systems solely under well-known deterministic inputs. While this approach holds relevance in system calibration or adjustment, it finds limited application, primarily in fields such as radar or ultrasound imaging.

In many real-world applications, uncertainty envelopes the system input. Often, we possess only partial knowledge concerning the statistical properties of the input signal, driving our interest in understanding how this signal impacts the system output. Conversely, it's equally intriguing to decipher the statistical properties of the input signal based on the observed properties of the output signal.

Consider the study of reflected acoustic signals from ground layers—an endeavor that allows estimation of concealed structures such as groundwater or oil presence based on these signals. This exemplifies the quest to unveil hidden insights from signals, elucidating the characteristics of unknown elements within a system.

Assuming a random process denoted as $x(t)$ serves as an input to a deterministic Linear Time-Invariant (LTI) system characterized by its impulse response $h(t)$, the computation of the system output can be determined. Employing (3.5), we can assert that the output signal is given by:

$$y(t) = \int_{-\infty}^{\infty} x(\lambda)h(t-\lambda)d\lambda. \tag{3.62}$$

From a mathematical point of view, caution is warranted when delving into the analysis of dynamic systems with random signals. The question arises: Is this random process integrable, and can it be expressed as in (3.4)? This integrability condition is essential for employing the convolution integral. To circumvent unnecessary theoretical complexities, it is reasonable to assume that $x(t)$ is sufficiently smooth, ensuring the validity of both integration and differentiation. Further theoretical analysis is expounded upon in Section 2.10.2.

In practical systems, it often suffices to ascertain the first and second moments of the random output process. These moments are encapsulated by the mean and the autocorrelation function, respectively. Let's commence the analysis by delving into the mean properties of the output signal. The average is computed as:

$$\mu_y(t) = E_y[y(t)] = \int_{-\infty}^{\infty} E_x[x(\lambda)]h(t-\lambda)d\lambda. \tag{3.63}$$

Therefore, generally, the mean (average) of the output signal is the convolution of the input time-varying mean with the system impulse response. If the random input signal is stationary, i. e., $E_x[x(\lambda)] = \mu_x$, then the output signal mean is the mean of the input signal multiplied by the area under the curve of the impulse response of the system as

$$\mu_y(t) = E_y[y(t)] = \mu_x \int_{-\infty}^{\infty} h(\lambda)d\lambda. \tag{3.64}$$

Hence, for a zero mean input the mean of the output of the LTI will be always zero. Now, we extend the analysis to check the autocorrelation of the output $R_{yy}(t,\tau) = E_{yy}[y(t)y(\tau)]$:

$$R_{yy}(t,\tau) = \int_{-\infty}^{\infty}\int_{-\infty}^{\infty} E_{xx}[x(\lambda_1)x(\lambda_2)]h(t-\lambda_1)h(\tau-\lambda_2)d\lambda_1 d\lambda_2. \tag{3.65}$$

This shows that the output autocorrelation is just the double convolution of the input signal autocorrelation as:

$$R_{yy}(t,\tau) = \int_{-\infty}^{\infty}\int_{-\infty}^{\infty} R_{xx}(\lambda_1,\lambda_2)h(t-\lambda_1)h(\tau-\lambda_2)d\lambda_1 d\lambda_2. \tag{3.66}$$

In the case of a stationary input, the autocorrelation of the input can be expressed as: $R_{xx}(\lambda_1,\lambda_2) = R_{xx}(\lambda_1-\lambda_2)$. From the previous equation one can easily prove that the output will be stationary as well and $R_{yy}(t,\tau) = R_{yy}(t-\tau)$.

We know that the Fourier transform of the autocorrelation function gives the power spectral density. Therefore, from equation (3.66), the power spectral density of the output of a linear time-invariant system can be expressed as:

$$\Psi_{YY}(f) = |H(f)|^2 \Psi_{XX}(f). \tag{3.67}$$

White noise is characterized by a unit power spectral density, denoted as $\Psi_{XX}(f) = 1$. By employing white noise within a linear system, specifically tailored with desired properties as $\Psi_{YY}(f) = |H(f)|^2$, we create what is known as colored noise. This process enables the generation of noise with distinct spectral qualities. When simulating various systems, it becomes essential to generate random processes that exhibit specific statistical and spectral characteristics.

Example 101. Consider an LTI system with an impulse response function given as $h(t) = 0.1e^{-10^{-3}t}u(t)$, where $u(t) = 1$ for $t \geq 0$, and 0 otherwise. The input, if the system is a stationary white noise random signal with normal distribution, has zero mean and unite average power. Find the mean and the autocorrelation of the signal at the output. Moreover, find the power spectral density of the output signal.

Solution. As described before, applying a stationary signal in the LTI system will generate a stationary output. Furthermore, applying a normally distributed random process as input for any linear system, the output will be also normally distributed. For zero mean input, the output will also be zero mean as shown in equation (3.64). The autocorrelation of the output signal is:

$$R_{yy}(t,\tau) = \int_{-\infty}^{\infty}\int_{-\infty}^{\infty} R_{XX}(\lambda_1,\lambda_2)h(t-\lambda_1)h(\tau-\lambda_2)d\lambda_1 d\lambda_2$$

$$R_{yy}(t,\tau) = \int_{-\infty}^{\infty}\int_{-\infty}^{\infty} R_{XX}(t-\lambda_1,\tau-\lambda_2)h(\lambda_1)h(\lambda_2)d\lambda_1 d\lambda_2$$

$$R_{yy}(t,\tau) = 0.01\int_{0}^{\infty}\int_{0}^{\infty} \delta(t-\tau-\lambda_1+\lambda_2)e^{-10^{-3}\lambda_1}e^{-10^{-3}\lambda_2}d\lambda_1 d\lambda_2$$

$$R_{yy}(\tau-t) = 0.01\int_{0}^{\infty} e^{-10^{-3}(\tau-t+2\lambda_1)}d\lambda_1 = 0.01e^{-10^{-3}(\tau-t)}\int_{0}^{\infty} e^{-2\times10^{-3}\lambda_1}d\lambda_1$$

$$R_{yy}(\tau-t) = 5e^{-10^{-3}|\tau-t|}.$$

The power spectral density of the output can be derived by the taking the Fourier transform of the output autocorrelation function or by using equation (3.67). Let's do it both. Taking the Fourier transform of the autocorrelation function as:

$$\begin{aligned}\Psi_{YY}(f) &= 5\int_{-\infty}^{\infty} e^{-10^{-3}|\lambda|}e^{-j2\pi f\lambda}d\lambda\\ &= 5\int_{-\infty}^{0} e^{(10^{-3}-j2\pi f)\lambda}d\lambda + 5\int_{0}^{\infty} e^{-(10^{-3}+j2\pi f)\lambda}d\lambda\\ &= \frac{5}{10^{-3}-j2\pi f}+\frac{5}{10^{-3}+j2\pi f}\\ &= \frac{0.01}{10^{-6}+(2\pi f)^2}.\end{aligned}$$

Using equation (3.67), the output power spectral density should be $\Psi_{YY}(f) = |H(f)|^2$. It should be identical to the above result. The transfer function of the system is derived as:

$$\begin{aligned}H(f) &= 0.1\int_{-\infty}^{\infty} e^{-10^{-3}t}u(t)e^{-j2\pi ft}dt\\ &= 0.1\int_{0}^{\infty} e^{-10^{-3}t}e^{-j2\pi ft}dt\\ &= 0.1\int_{0}^{\infty} e^{-(10^{-3}+j2\pi f)t}dt\\ &= \frac{0.1}{10^{-3}+j2\pi f}.\end{aligned}$$

Based on the resultant transfer function, the output power spectral density should be

$$\Psi_{YY}(f) = |H(f)|^2 = \frac{0.01}{10^{-6}+(2\pi f)^2}.$$

Example 102. Consider the RL circuit depicted in Figure 3.9, featuring $R = 100\,\Omega$ and $L = 2\,\text{mH}$. Assuming that the input signal follows a white noise zero-mean normal distribution with unit variance, determine the power spectral density at the output. The output, in this case, refers to the voltage across resistor R. Additionally, compute the autocorrelation function of the output signal and derive its probability density function.

Solution. Using the voltage divider rule from circuit theory in the frequency domain, we obtain

$$\begin{aligned}V_o(f) &= V_i(f)\frac{R}{R+j2\pi fL}\\ \Rightarrow H(f) &= \frac{R}{R+j2\pi fL}\\ &= \frac{R/L}{R/L+j2\pi f}.\end{aligned}$$

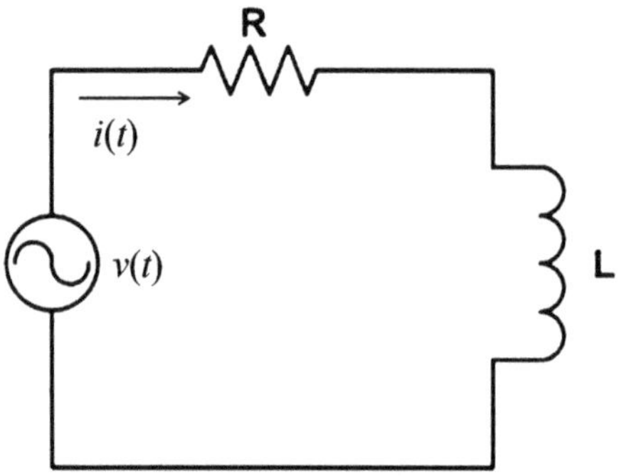

Figure 3.9: RL Circuit.

Since the input is white noise with unit average power, then the power spectral density of the output is

$$\Psi_{YY}(f) = |H(f)|^2 = \frac{(R/L)^2}{(R/L)^2 + (2\pi f)^2}.$$

The autocorrelation of the output signal can be calculated as the inverse Fourier transform of its power spectral density. Alternatively, it can also be obtained conveniently using Fourier transform tables such as:

$$R_{YY}(\tau) = \frac{1}{2}\left(\frac{R}{L}\right)e^{-\left(\frac{R}{L}\right)|\tau|}. \tag{3.68}$$

Since the input has zero mean, then the output will be also zero mean. Furthermore, the output variance will be $R_{XX}(0) = \frac{1}{2}(\frac{R}{L}) = 5 \times 10^4$. Since the system is linear and the input is a normal distribution, therefore, the output will be a normal distribution with zero mean and the given variance.

Example 103. Consider an LTI system with impulse response given by

$$h(t) = 0.1t \quad 0 \le t \le 10.$$

If the input signal $x(t)$ is stationary with a uniform distribution from 1 to 3 and an autocorrelation function given by

$$R_{XX}(\tau) = e^{-|\tau|}.$$

Find the mean, the autocorrelation function, and the power spectral density of the output signal $y(t)$.

Solution. Since the system is LTI and the input is stationary, hence, the output will be a stationary process as well. The mean of the output signal is

$$\mu_y = \mu_x \int_0^{10} 0.1t dt = 2 \times \frac{0.1t^2|_0^{10}}{2} = 10.$$

The output power spectral density is

$$\Psi_{YY}(f) = |H(f)|^2 \Psi_{XX}(f)$$

$$\Psi_{XX}(f) = \int_{-\infty}^{\infty} e^{-|\tau|} e^{-j2\pi f\tau} d\tau = \frac{2}{1 + (2\pi f)^2}$$

$$H(f) = 0.1 \int_0^{10} te^{-j2\pi ft} dt = \frac{-1 + e^{-j20\pi f}(1 + j20\pi f)}{40\pi^2 f^2}$$

$$\Psi_{YY}(f) = \frac{2| - 1 + e^{-j20\pi f}(1 + j20\pi f)|^2}{(1 + (2\pi f)^2)(40\pi^2 f^2)^2}.$$

Finally the autocorrelation function of the output signal is

$$R_{YY}(\tau) = \int_{-\infty}^{\infty} \frac{2| - 1 + e^{-j20\pi f}(1 + j20\pi f)|^2}{(1 + (2\pi f)^2)(40\pi^2 f^2)^2} e^{j2\pi f\tau} df. \quad (3.69)$$

3.4 General stochastic dynamic systems

We have seen in the previous section how deterministic linear time-invariant systems affect sufficiently smooth random signals. In many other modeling cases, we need to consider dynamic systems that are affected by random or unknown inputs that make the output different at each time realization. These systems cannot be expressed in a deterministic way.

In such systems, there are two types of uncertain inputs. The first is the input with inherent randomness, such as thermal or shot noise in electronic circuits and sensors. In this case, the random input is typically considered to be white noise.

The second type of uncertain input occurs when there are many unknown inputs that affect the dynamics of the system. In this case, it is possible to model the sum of all these system uncertainties as white or colored noise.

For example, the vibration of an electric generator can be modeled as a dynamic system. However, it is not possible to obtain a very accurate general model, because the vibration is affected by too many parameters that cannot be specified and parameterized accurately.

Another possible example is the fluctuations in stock markets. It is not possible to have a perfect model of such dynamics because there are too many parameters that can affect the prices. Moreover, most of these parameters are hidden and even unknown.

Another example is related to weather forecasting. The dynamic changes in the weather are not always perfectly predictable.

The wireless channel is another example that highlights the importance of dynamic uncertainties, such as the received signal strength, delay, distortion, and frequency shifts due to Doppler effects. By understanding these uncertainties and using accurate stochastic modeling of channel dynamics, we have been able to exploit the random fluctuations. For instance, opportunistic radio resource scheduling can be used to exploit the random fluctuations and achieve a certain degree of user diversity.

In fact, a deep understanding of the uncertainties of wireless channels has been one of the primary reasons for the current era of excellent performance in wireless communication.

To illustrate stochastic modeling, let's start with two elementary examples: one from basic electrical circuits and the other from finance.

Example 104. Figure 3.10 illustrates a basic circuit comprising only two components: a resistor R, representing all circuit losses, and a capacitor C. Thermal noise, caused by random electron fluctuations due to ambient temperature, is generated by each conductor. This noise is white, meaning it's uncorrelated. Additionally, the thermal noise voltage follows a normal distribution with zero mean and a variance of $E[V_R^2] = 4\kappa TBR$, where κ denotes Boltzmann's constant, B represents the system's bandwidth, T signifies the resistor's temperature in Kelvin, and R represents the resistance in ohms. In this simplified example, the temperature T represents the physical temperature. However, when dealing with more complex systems encompassing multiple components and noise sources, the temperature T may not directly correspond to the actual temperature (although it may be related). In such cases, it's referred to as the equivalent noise temperature (T_e). The maximum deliverable noise average power of the resistor R is $N_0 = \kappa TB$, independent of the resistance value. This is attributed to the fact that the maximum transmissible power is achieved when the load resistance is equal to R.

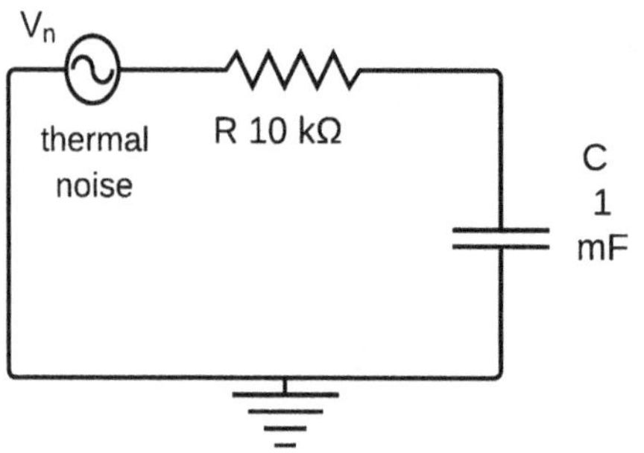

Figure 3.10: Circuit Example.

If we are interested in the voltage across the capacitance ($V_C(t)$), the dynamics of the circuit is

$$-V_n(t) + Ri(t) + V_C(t) = 0. \tag{3.70}$$

Since $i(t) = C\frac{dV_C(t)}{dt}$, the dynamics are governed by the differential equation

$$RC\frac{dV_C(t)}{dt} + V_C(t) - V_n(t) = 0. \tag{3.71}$$

While the differential equation may seem deceptively simple, it demands careful consideration, particularly regarding the white noise component $V_n(t)$. As demonstrated in Section 2.10.2, white noise can be mathematically modeled as the derivative of Brownian motion. In this section, we introduce a dedicated symbol for the Brownian process, denoted as β_n. As discussed in Section 2.10.2, the conventional notion of the derivative of the Brownian process does not exist. Nevertheless, we can express the relationship as $V_n(t)dt = d\beta_n(t)$. Consequently, the dynamics of the circuit can be described by:

$$dV_C(t) + \frac{1}{RC}V_C(t)dt - \frac{1}{RC}d\beta_n(t) = 0. \tag{3.72}$$

In the field of modeling stochastic dynamic systems, our objective is not to pinpoint the exact $V_C(t)$. In fact, if you were to measure the capacitor voltage multiple times within a given time period T, the probability of obtaining identical results is absolutely zero. Each iteration of a stochastic dynamic system yields distinct random outcomes. However, our focus lies in understanding the statistical behavior of the process paths over time. Further analysis of this aspect will be provided later in this section.

Example 105. The evolution of an initial investment of $x(0)$ over time t can be mathematically described as a dynamic system where the instantaneous rate of change (i. e., the derivative) at time t is directly proportional to the total investment at that time. This process can be represented mathematically as:

$$\frac{x(t)}{dt} = \alpha_t x(t). \tag{3.73}$$

Here, α_t represents the return rate, interest rate, or the investment rate. This simple yet versatile model finds application in numerous practical and theoretical scenarios, often referred to as the exponential growth model. It's employed in diverse fields such as biology to simulate the spread of microorganisms or human population growth, to understand nuclear chain reactions, and to describe exponential changes in electrical circuits.

While the model can accommodate time-varying α_t, the simplest form assumes an LTI system with a constant rate, denoted as $\alpha_t = \alpha_0$. Consequently, solving the differential equation yields an exponential function:

$$x(t) = x(0)e^{\alpha_0 t}. \tag{3.74}$$

A successful investment requires a positive rate, i. e., $\alpha_0 \geq 0$. While this model is intriguing and relatively straightforward for study and analysis, its accuracy in real-world applications is often compromised. In reality, precise knowledge of the future change rate of a_t is unattainable. It can fluctuate randomly due to various parameters, such as economic market collapses or unsuccessful bacterial cell divisions in biology. Acknowledging this inherent uncertainty, a more accurate model incorporates the definition $a_t = a_0 + n$, where n represents the uncertain component of the model. Given that the randomness (noise) arises from numerous parameters, it is reasonable to assert, with high confidence, that the noise component n follows a white and zero-mean normal distribution. Consequently, a_t can be characterized as a normal distribution with a mean of a_0. In this case, the above differential equation can be expanded as:

$$\begin{aligned} dx(t) &= a_0 x(t)dt + x(t)ndt \\ dx(t) &= a_0 x(t)dt + x(t)d\beta_t. \end{aligned} \tag{3.75}$$

One might assume that solving the problem with a random a_t follows a similar straightforward path as the deterministic solution, i. e., $x(t) = x(0)e^{a_t t}$. However, unfortunately, this approach leads to an incorrect solution.

Throughout this chapter, remember that $x(t)$ and x_t hold identical meanings, and we freely switch between these notations. Thus, to simplify expressions such as $g(x(t))$, which may appear cumbersome, we opt for the cleaner forms $g(x_t)$ or $g(x_t, t)$.

Both of the previous examples share a common feature: the presence of an uncertain component in the differential equation. However, they differ significantly from a philosophical perspective. In the RC circuit example, the uncertainty associated with the noise component is inherent and cannot be eliminated or resolved, at least according to the laws of thermodynamics. In contrast, the noise component in the investment example arises from our limited knowledge of the underlying market parameters. While this uncertainty cannot be completely eliminated, it can be partially reduced by acquiring more information. For instance, we may anticipate a 20 % decline in stock prices due to a planned strike in the future

Stochastic differential equations (SDEs) serve as valuable tools for modeling dynamic continuous-time uncertainties. These models provide insight into the statistical behavior of systems, enabling accurate estimation and prediction of future or unseen quantities such as market shares. Fortunately, most practical SDEs are linear (or can be approximated as linear), significantly simplifying the analysis process. Additionally, focusing on first-order SDEs is sufficient, as higher-order linear differential equations can be readily decomposed into a set of first-order equations, as evidenced by state-space representations.

Example 106. To illustrate the appearance of a solution to a stochastic differential equation, let's take a reverse approach. We'll begin with the solution itself and then proceed to construct the corresponding differential equation. Consider the following representation of a random process

$$x(t) = x_0 + at + \lambda\beta_t, \quad t \geq 0. \tag{3.76}$$

Where x_0 is a deterministic initial value of x at $t = 0$. Differentiating the equation leads to

$$dx(t) = adt + \lambda d\beta_t, \quad t \geq 0. \tag{3.77}$$

As we mentioned before, we usually avoid using the form $\frac{d\beta_t}{dt}$, because it is not a function in the strict sense, but it is interpreted as

$$\frac{d\beta_t}{dt} = n(t), \tag{3.78}$$

where $n(t)$ is white noise.

Therefore, we may consider the differential equation as follows:

$$\frac{dx(t)}{dt} = a + \lambda n(t). \tag{3.79}$$

Again equation (3.79) is not an accurate representation, and it is better to use the (3.77) form.

Example 107. For the basic stochastic differential equation given by (3.79) with $x_0 = 1$, $a = 1$ and $\lambda = 0.5$, find the mean, autocorrelation, and the auto-covariance of $x(t)$. What is the probability that $x \geq 5$ will occur during the time $t \leq 10$ seconds?

Solution. Since we know the solution as given in (3.76), we can solve for the mean and autocorrelation as follows:

$$\mu_X(t) = E[x(t)] = E[x_0 + at + \lambda\beta_t] = x_0 + at \tag{3.80}$$

$$\begin{aligned} R_{XX}(t,\tau) &= E[x(t)x(\tau)] \\ &= E[(x_0 + at + \lambda\beta_t)(x_0 + a\tau + \lambda\beta_\tau)] \\ \because E[\beta_t] &= E[\beta_\tau] = 0, \quad \text{and} \quad E[\beta_t\beta_\tau] = \min(t,\tau) \end{aligned}$$

$$\therefore R_{XX}(t,\tau) = x_0^2 + a\tau x_0 + atx_0 + a^2 t\tau + \lambda^2 \min(t,\tau) \tag{3.81}$$

$$\mathrm{Cov}_{XX}(t,\tau) = R_{XX}(t,\tau) - E[x(t)]E[x(\tau)] = \lambda^2 \min(t,\tau). \tag{3.82}$$

To ascertain the probability of the random process $x(t)$, our initial step involves determining its probability density function. As discussed in Section 2.10.2, the distribution of the Brownian motion β_t adheres to a zero-mean normal distribution, characterized by the probability density function given by equation (2.240). It's established that any linear operations performed on a random process following a normal distribution will yield a resulting distribution that is also normal, but could have a different mean and variance. Therefore, the distribution of $x(t)$ is as follows

$$f_{X_t}(x,t|x_0) = \frac{1}{\sqrt{2\pi\lambda^2 t}} e^{-\frac{(x-x_0-at)^2}{2\lambda^2 t}}. \tag{3.83}$$

With the parameters given in the example, the probability density function becomes:

$$f_{X_t}(x,t|x_0=1) = \frac{1}{\sqrt{0.5\pi t}} e^{-\frac{(x-1-t)^2}{0.5t}}. \tag{3.84}$$

The probability of $x(t) \geq 5$ is

$$P(x \geq 5, t|x_0=1) = \frac{1}{\sqrt{5\pi}} \int_5^\infty e^{-\frac{(x-1-t)^2}{5}} dx \tag{3.85}$$

$$= Q\left(\frac{4-t}{0.5\sqrt{t}}\right). \tag{3.86}$$

The probability of $x(t) \geq 5$ at each instance of time t is given by equation (3.86). Observe that the time is not a random variable, therefore, we can compute the average probability over time from t_1 to $t_2 > T_1$ as:

$$\frac{1}{t_2 - t_1} \int_{t_1}^{t_2} Q\left(\frac{4-t}{0.5\sqrt{t}}\right) dt. \tag{3.87}$$

If we are interested to compute the probability that $x(t) \geq 5$ for the time period $0 \leq t \leq 10$, we can compute it as:

$$\frac{1}{10} \int_0^{10} Q\left(\frac{4-t}{0.5\sqrt{t}}\right) dt = 0.5875. \tag{3.88}$$

Three distinct realizations of the stochastic differential equation are illustrated in Figure 3.11. Evidently, each realization exhibits unique trajectories, emphasizing the inherent randomness introduced by the stochastic nature of the equation. The straight line in the plot represents the solution of the corresponding deterministic differential equation, where noise is assumed to be zero.

It is essential to note that, in the realm of stochastic differential equations, our focus is not on pinpointing exact values of outcomes due to the inherent unpredictability. Instead, our interest lies in understanding the statistical behavior of the system. This is exemplified in our analysis of the simple scenario, where we computed the probability of the output trajectory falling within specific ranges over any time period.

Consider this scenario as a representation of the expected payoff for a project laden with uncertainties. In practical terms, this could mirror a decision-making process for investing in a project. For instance, you might opt to invest in the project if the probability of attaining a certain payoff exceeds a predefined threshold after a few months.

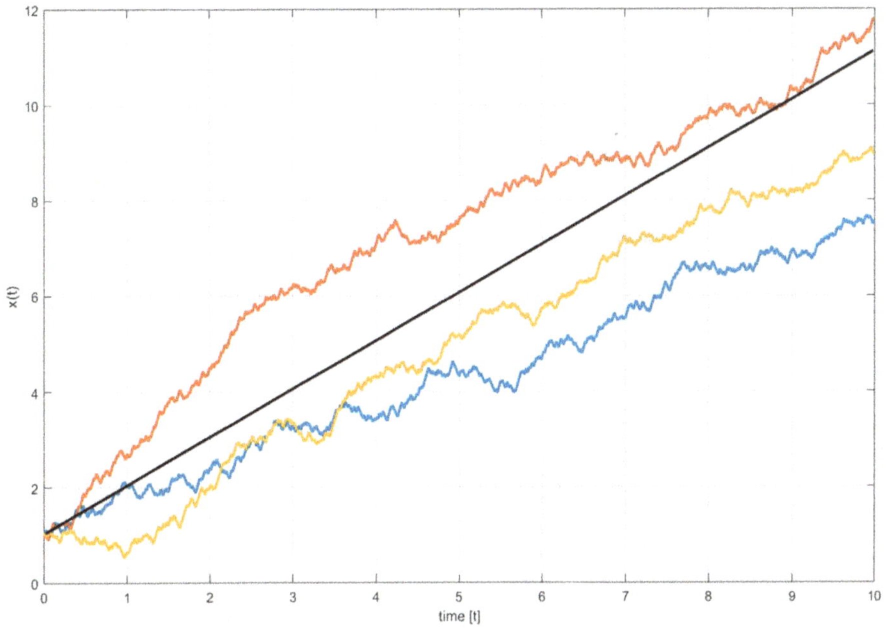

Figure 3.11: Different Realizations of the results.

Given that all future expectations are inherently uncertain, the study and simulation of stochastic differential equations emerge as a crucial tool for navigating and understanding complex systems with inherent uncertainties.

The same procedure in the previous example can be intended for any time function $g(t)$. Assuming the result of a stochastic differential equation is

$$x(t) = x_0 + g(t) + \lambda\beta_t, \quad t \geq 0. \tag{3.89}$$

Differentiating the equation leads to

$$dx(t) = \dot{g}(t)dt + \lambda d\beta_t, \quad t \geq 0, \tag{3.90}$$

where $\dot{g}(t) = \frac{dg(t)}{dt}$. You may try the previous example with $g(t) = \cos(\omega t)$. A second form of stochastic differential equation may take the following shape:

$$dx(t) = g(t)n(t)dt. \tag{3.91}$$

It is slightly more complected than the equation (3.79). As shown before, $n(t)dt = d\beta_t$, hence,

$$dx(t) = g(t)d\beta_t. \tag{3.92}$$

Since $d\beta_t$ is a normal distribution, therefore, $x(t)$ will be a normal distribution with zero mean, because

$$x(t) = \int_{t_0}^{t} g(\tau)d\beta_\tau \Rightarrow E[x(t)] = \int_{t_0}^{t} g(\tau)dE[\beta_\tau] = 0. \tag{3.93}$$

The variance of $x(t)$ can be proved to be

$$E[x(t)^2] = \sigma^2 \int_{t_0}^{t} g(\tau)^2 d\tau, \tag{3.94}$$

where σ^2 is the unit time variance of Brownian motion as described in the previous chapter.

The third possible form of stochastic differential equation may take the following shape:

$$dx(t) = g(x_t, t)dt + d\beta_t. \tag{3.95}$$

Where x_t is just another representation of $x(t)$. This stochastic differential equation has a more complicated form than the previous two. It can be a nonlinear and a stochastic differential equation. Again remember that we are looking for the statistical behavior of the trajectory $x(t)$. However, if the function $g(x_t, t)$ is nonlinear in the trajectory x_t, we cannot claim that x_t still has a normal distribution. Nevertheless, let's consider the following linear form as

$$dx(t) = -ax_t dt + d\beta_t. \tag{3.96}$$

Since it is linear, so that $x(t)$ will have a normal distribution, and then we just need to find the mean and the autocorrelation function. Let's define $\hat{X}_t$ as the average value of the random process $x(t)$, i. e., $\hat{X}_t = E[x(t)]$. Taking the expectation of the stochastic differential equation (3.96) as

$$\begin{aligned} E[dx(t)] &= E[-ax_t dt + d\beta_t] \\ \Rightarrow d\hat{X}_t &= -a\hat{X}_t + dE(\beta_t) = -a\hat{X}_t \\ \Rightarrow \hat{X}_t &= \hat{X}_0 e^{-at}. \end{aligned} \tag{3.97}$$

Where $\hat{X}_0$ is considered as the mean of the process at $t = 0$. Assuming the initial value process was a zero mean normal distribution with unit variance, hence $\hat{X}_0 = 0$, therefore, $\hat{X}_t = 0$. From (3.96), we may solve the process $x(t)$ assuming $t_0 = 0$ as

$$x(t) = x_0 e^{-at} + \int_{0}^{t} e^{-a(t-\tau)} d\beta_\tau. \tag{3.98}$$

We can proceed to the autocorrelation function and then the variance of the normal random process $x(t)$ as follows:

$$E[x_{t_1}x_{t_2}] = E\left[\left(x_0 e^{-at_1} + \int_0^{t_1} e^{-a(\tau-t_1)} d\beta_\tau\right)\left(x_0 e^{-at_2} + \int_0^{t_2} e^{-a(\tau_1-t_2)} d\beta_{\tau_1}\right)\right]$$

$$\begin{aligned} E[x_0] &= 0 \\ E[x_0^2] &= 1 \\ E[\beta_t] &= 0 \\ E[\beta_t\beta_\tau] &= \sigma^2 \min(t,\tau). \end{aligned}$$

Therefore, the autocorrelation function becomes:

$$E[x_{t_1}x_{t_2}] = E[x_0^2]e^{-a(t_1+t_2)} + \int_0^{t_1}\int_0^{t_2} e^{-a(t_1-\tau)}e^{-a(t_2-\tau_1)} E[d\beta_\tau d\beta_{\tau_1}].$$

Let's define $\lambda = \min(t, t_1)$, therefore

$$\begin{aligned} E[x_{t_1}x_{t_2}] &= e^{-a(t_1+t_2)} + \sigma^2 \int_0^{\lambda} e^{-2a(\lambda-\tau)} d\tau \\ &= e^{-a(t_1+t_2)} + \sigma^2 e^{-2a\lambda} \int_0^{\lambda} e^{2a\tau} d\tau \\ &= e^{-a(t_1+t_2)} + \sigma^2 \frac{e^{-2a\lambda}}{2a}(e^{2a\lambda} - 1). \end{aligned}$$

Finally, we can determine the variance of the random process $x(t)$ as:

$$\begin{aligned} E[x_t^2] &= e^{-2at} + \sigma^2 \frac{e^{-2at}}{2a}(e^{2at} - 1) \\ &= e^{-2at} + \frac{\sigma^2}{2a}(1 - e^{-2at}). \end{aligned} \tag{3.99}$$

To conclude, the solution of the linear stochastic differential equation given in equation (3.96) is a random process that has a zero mean normal distribution with time dependent variance given by equation (3.99).

Example 108. Consider a dynamic random process described by the differential equation:

$$\frac{dx(t)}{dt} = -2x(t) + 0.1n(t).$$

Here $n(t)$ represents white noise following a normal distribution with zero mean and unit variance. Additionally, the initial condition of the random process x_0 is a random variable following a normal distribution with zero mean and unit variance. We aim to derive a mathematical formula that provides the probability of the trajectory being within a specified interval at any time t. Additionally, we aim to create an Octave code to simulate $x(t)$ and visualize three different realizations of this process.

Solution. The stochastic differential equation in this example is identical with the one discussed before. The solution should have a zero mean normal distribution with variance given by equation (3.99) as

$$f_{X_t}(x,t) = \frac{1}{\sqrt{2\pi\theta_t}} e^{-\frac{x_t^2}{2\theta_t}} \tag{3.100}$$

$$\theta_t = e^{-4t} + \frac{0.01}{4}(1 - e^{-4t}). \tag{3.101}$$

Therefore, the probability of the trajectory to be within a certain interval at any time is given by:

$$P(x_0 \le x_t \le x_1, t) = \frac{1}{\sqrt{2\pi\theta_t}} \int_{x_0}^{x_1} e^{-\frac{x_t^2}{2\theta_t}} dx_t \tag{3.102}$$

$$= Q\left(\frac{x_0}{\sqrt{\theta_t}}\right) - Q\left(\frac{x_1}{\sqrt{\theta_t}}\right). \tag{3.103}$$

Figure 3.12 shows a one-time realization of the process $x(t)$ and the probability curve with time to be $0.2 \le x(t) \le 0.8$.

The Octave code is given next

```
% SDE Process
clear all
t=0:.001:5
theta=exp(-4*t)+0.01*(1-exp(-4*t))/4;
x=theta.^.5 .*randn(3,length(theta));
Pro = qfunc(0.2./sqrt(theta))-qfunc(0.8./sqrt(theta));
subplot(211); plot(t,x(1,:),'Linewidth',0.5)
xlabel('Time [s]','fontsize',18);
ylabel('Process x(t)','fontsize',18)
subplot(212); semilogy(t,Pro,'Linewidth',2)
xlabel('Time [s]','fontsize',18);
ylabel('Probability','fontsize',18)
```

The forth possible form of stochastic differential equation and probably the most unclear one is given by:

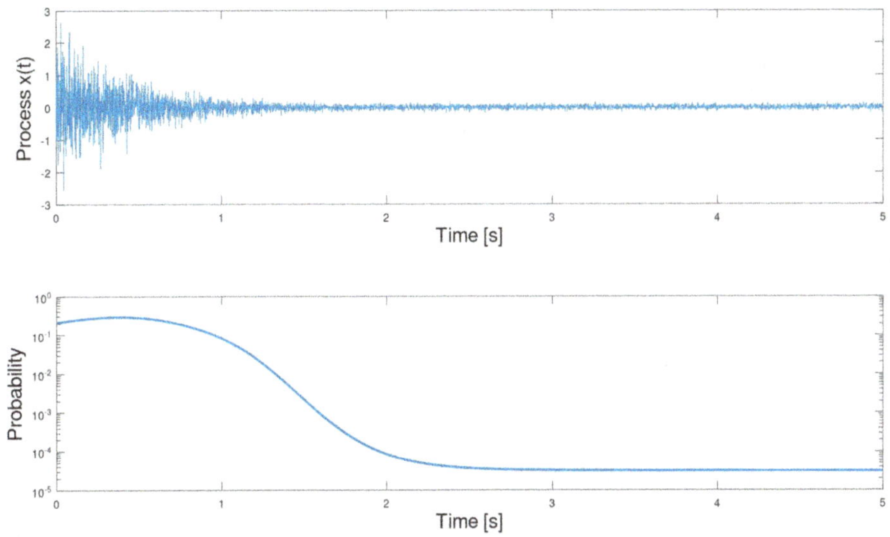

Figure 3.12: Stochastic Random Process.

$$dx(t) = g(x_t, t)d\beta_t. \tag{3.104}$$

In this differential equation, we need to address the rate of changes of a random process with respect to another random process given by Brownian motion. The integral form of this differential equation is given by

$$x(t) = \int_0^t g(x_\tau, \tau)d\beta_\tau. \tag{3.105}$$

The stochastic integral in (3.105) can be interpreted in various ways depending on the framework and the assumptions made about the function $g(x_t, t)$ and the stochastic process x_t. However, the most widely accepted interpretation in engineering is based on Ito calculus. This formulation of stochastic integration is rooted in the concept of approximating deterministic integration using sums of products of infinitesimal increments of Brownian motion and infinitesimal increments of $g(x_t, t)$. The Ito integral explicitly accounts for the path irregularities inherent in Brownian motion. Skipping extensive proofs, a fascinating outcome of the Ito integral formulation is:

$$\int_0^t g(x_\tau, \tau)d\beta_\tau^2 = \sigma^2 \int_0^t g(x_\tau, \tau)d\tau. \tag{3.106}$$

The equation above yields an intriguing result: integrating a random process with respect to the square of Brownian motion results in a deterministic integration. The inherent randomness of Brownian motion originates from infinitesimal ± increments. Consequently, the square of Brownian motion may be the underlying factor leading to the result in (3.106).

For a general stochastic differential equation given by:

$$dx_t = r(x_t, t)dt + g(x_t, t)d\beta_t. \tag{3.107}$$

Ito's lemma states that for a sufficiently smooth function, $h(x_t, t)$, the differential equation based on h can be expressed as:

$$dh(x_t, t) = \frac{\partial h}{\partial t}dt + \frac{\partial h}{\partial x_t}dx_t + \frac{\sigma^2}{2}g(x_t, t)^2\frac{\partial^2 h}{\partial x_t^2}dt. \tag{3.108}$$

Here, eliminating the term $d\beta_t$ allows us to maintain the same approach for handling statistical properties as previously employed.

Example 109. Let's revisit the investment problem in equation (3.75) and find the mean and variance of the equation, given that $E(\beta_t^2) = \sigma^2 = 10^{-3}$, $\alpha_0 = 0.01$, and x_0 has a zero mean unit variance normal distribution. The stochastic differential equation becomes:

$$dx_t = 0.01x_t dt + x_t d\beta_t. \tag{3.109}$$

Compare the results obtained here with those derived from models that disregard the stochastic aspects of dynamic systems.

Solution. To find the mean of the process, let's define $\hat{X}_t = E[x_t]$, therefore,

$$d\hat{X}_t = 0.01\hat{X}_t dt \Rightarrow \hat{X}_t = \hat{X}_0 e^{0.01t} = 0. \tag{3.110}$$

To find the mean square of the process let's use $h(x_t, t) = x_t^2$ in the above Ito formula. Here $\frac{\partial h}{\partial t} = 0$, $\frac{\partial h}{\partial x_t} = 2x_t$, and $\frac{\partial^2 h}{\partial t^2} = 2$, therefore, the Ito integral becomes:

$$dx_t^2 = 2x_t dx_t + 10^{-3}x_t^2 dt. \tag{3.111}$$

Substitute for x_t, we obtain

$$\begin{aligned} dx_t^2 &= 2x_t(0.01x_t dt + x_t d\beta_t) + 10^{-3}x_t^2 dt \\ &= 0.021x_t^2 dt + 2x_t^2 dB_t. \end{aligned}$$

Let's define $\hat{Y}_t = E[x_t^2]$, therefore taking the expectation of the last differential equations leads to (remember $E[d\beta_t] = 0$),

$$d\hat{Y}_t = 0.021\hat{Y}_t dt \Rightarrow \hat{Y}_t = \hat{Y}_0 e^{0.021t}. \tag{3.112}$$

Where $\hat{Y}_0 = E[x_0^2] = 1$.

4 Estimation theory

The purpose of this chapter is not to provide a complete and detailed explanation of estimation theory. Rather, the goal is to introduce the topic very briefly and place it in the context of the book. There are many excellent and comprehensive books that deal with the subject of estimation theory, such as [3, 4], and [5].

After completing this chapter and working through the provided examples, you will:

- Understand the fundamental concepts of estimation theory and its role in modeling systems with uncertainties.
- Grasp the criteria for evaluating estimator performance, including bias, consistency, and efficiency.
- Apply various estimation techniques, such as least mean square error (LMSE), minimum absolute error, and maximum a posteriori (MAP) estimation, to extract hidden parameters from data.
- Comprehend the principles of regression, classification, and clustering and their application in modeling relationships and categorizing data.

Estimation theory serves as a fundamental gateway to modeling systems, especially in the face of inherent uncertainties. This field encompasses a diverse set of statistical tools designed to extract valuable hidden insights from data, whether derived from measurements or observations.

Let's define the problem as

$$\mathbf{y} = h(\mathbf{x}, \theta) + \mathbf{n}. \tag{4.1}$$

Here, $\mathbf{x} \in \mathbf{R}^{M\times1}$ represents the input vector, which can be either known or unknown. The observable noisy output vector is denoted as $\mathbf{y} \in \mathbf{R}^{N\times1}$, while $h(.)$ signifies the mapping function linking the input and output, reliant upon mapping parameter θ. This mapping function can also exhibit time variability. Additionally, we have the random additive noise term, $\mathbf{n}$. In the domain of estimation theory, we often explore at least three intriguing scenarios:

- We possess (at least partial) knowledge about $\mathbf{x}$ and the observed data $\mathbf{y}$, and our objective is to identify the optimal set of functions, denoted as h, along with their associated parameters θ. This task is commonly referred to as a regression problem or curve fitting, and it serves as a fundamental tool in various data modeling and machine learning applications. In some cases, we may have prior knowledge of the general form of the mapping function h—such as the use of increasing exponential functions in population models. However, the challenge lies in determining or estimating the specific parameters of these functions. There are cases where the mapping function is not explicitly known, requiring exploration and adaptation of general shapes to align with the available $\mathbf{x}$ and $\mathbf{y}$ data. For more intricate mappings, the application of black-box modeling techniques such as neural networks becomes a viable approach.

https://doi.org/10.1515/9783111585055-004

- In the second scenario, we have observations $\mathbf{y}$ and knowledge of h (or at least validated assumptions about it) along with its parameters θ. Our aim here is to estimate $\mathbf{x}$. Since the vector $\mathbf{x}$ is either fully or partially unknown to us (uncertainty stemming from a lack of knowledge), it needs to be treated as a random process. This particular problem finds application in various domains and is recognized under different names. For instance, if the mapping function represents convolution, we refer to the task of determining the input as the deconvolution process. It is alternatively known as inverse filtering, a restoration problem, among other names.
- The third compelling scenario in estimation theory arises when only observations $\mathbf{y}$ are available, and neither the mapping functions h nor the inputs $\mathbf{x}$ are known. Nevertheless, it is possible to establish statistical relationships between the inputs and the outputs. The key question is: What constitutes the optimal estimate of the statistical parameters, leveraging the provided information (e. g., $\mathbf{y}$)?

While there may be numerous intriguing problems in estimation theory, our focus will primarily be on the aforementioned three cases, particularly the third one.

The literature is replete with various estimation techniques and algorithms. A crucial question that arises is how we can assess the performance of an estimation technique. This is somewhat challenging, given that we are estimating an unknown parameter. Therefore, we require objective tools to assess the reliability of our estimates.

We will now briefly discuss three performance criteria tools for estimators: bias, consistency, and efficiency.

4.1 Unbiased estimation

Consider our aim to estimate unseen or hidden signals or parameters denoted as $\mathbf{x}_{\text{est}}$. This estimation relies on measurements or observations captured by $\mathbf{y}$. Given the inherent uncertainty and potential noise in these measurements, the estimated parameters $\mathbf{x}_{\text{est}}$ also become random variables. For instance, if our objective is to estimate the mean (average) of a random variable based on N measured samples, the estimation process will be:

$$x_{\text{est}} = \frac{1}{N}\sum_{i=1}^{N} y_i. \tag{4.2}$$

The estimated value obtained is a single realization based on the available data. If we acquire another set of independent N measured samples from the same process, we will obtain a different x_{est}. Each repetition of the process with distinct measurements yields a new estimate. Consequently, x_{est} becomes another random variable, complete with its probability density function. If the probability density function governing the generated samples y_i is a normal distribution, then the distribution of x_{est} follows suit. Conversely, if the measurements y_i are exponential random variables, the sum corresponds to a gamma distribution. Regardless, the key point is that x_{est} itself is a random variable characterized by a specific distribution.

It's worth noting that if we have numerous independent samples, according to the *central limit theorem*, the estimated mean x_{est} tends to exhibit a normal distribution, irrespective of the original probability distributions of the measurements. In any case, our estimate is unbiased if and only if:

$$E[\mathbf{x}_{est}] = \mathbf{x}. \tag{4.3}$$

This means that the expected value of our estimated parameter will be equal to the parameter itself. To make this clear, let's see if our estimate of the mean discussed before is unbiased or not. If we take the expected value of the estimated mean to be

$$\begin{aligned} E[x_{est}] &= \frac{1}{N} E\left[\sum_{i=1}^{N} y_i\right] \\ &= \frac{1}{N} \sum_{i=1}^{N} E[y_i] \\ &= \frac{1}{N} \sum_{i=1}^{N} x = \frac{Nx}{N} = x. \end{aligned}$$

Therefore, estimating the mean by averaging over samples is an unbiased estimation.

4.2 Consistent estimation

The estimator $\mathbf{x}_{est}$ is considered consistent if an increase in the number of measured samples (observations), denoted by N, results in a reduction of its error variance during the estimation process.

$$\lim_{N\to\infty} E[(\mathbf{x}_{est} - E[\mathbf{x}_{est}])^2] = 0. \tag{4.4}$$

An unbiased and consistent estimator, by definition, converges (in the mean-square sense) to the actual unknown parameter as the number of samples approaches infinity. As such, it is generally preferable to have an estimator that is both unbiased and consistent. However, if we have two different estimators that are both unbiased and consistent, which one should we choose? Given that we usually have a finite number of samples, we should select the estimator with the lower error variance for a given finite number of samples, N. This indicates that it is a more efficient estimator.

4.3 Efficient estimator

To ascertain the efficiency of an estimator, understanding the minimum variance for estimators in a specific problem is crucial. Fortunately, this minimum variance, or more precisely, the lower bound of the variance for unbiased estimators at a given sample size N, is provided by the Cramer–Rao (CR) lower bound. While the derivation is not covered here, it is not a complex process and can be found in various statistical estimation textbooks. For the model introduced earlier in this chapter (referenced as (4.1)), let's establish the joint probability density function for the inputs. $\mathbf{x} = [x_1, x_2, \ldots x_M]^T$ and outputs $\mathbf{y} = [y_1, y_2, \ldots y_N]^T$ as $f_{XY}(\mathbf{x}; \mathbf{y})$. The covariance matrix of the estimate vectors is

$$\mathbf{R}_{\text{est}} = E\left[(\mathbf{x}_{\text{est}} - \mathbf{x})(\mathbf{x}_{\text{est}} - \mathbf{x})^T\right]. \tag{4.5}$$

Where the ith diagonal of $\mathbf{R}_{\text{est}}$ is the variance of the ith estimated input, i. e.,

$$\text{Var}(x_{i,\text{est}}) = [\mathbf{R}_{\text{est}}]_{i,i}. \tag{4.6}$$

The element (i, j) of the Fisher information matrix is given by

$$[\mathbf{I}(\mathbf{x})]_{ij} = -E\left[\frac{\partial^2}{\partial x_i \partial x_j} \log f_{XY}(\mathbf{y}; \mathbf{x})\right], \quad i, j = 1, 2, \ldots M. \tag{4.7}$$

Where $f_{XY}(\mathbf{y}; \mathbf{x})$ is the joint probability density function between the observations $\mathbf{y}$ and the input/parameters $\mathbf{x}$. Under the following regularity condition:

$$E\left[\frac{\partial}{\partial x_i} \log f_{XY}(\mathbf{y}; \mathbf{x})\right] = 0, \quad i = 1, 2, \ldots M \tag{4.8}$$

the CR lower bound of the estimate variance could be expressed as

$$\text{Var}([\mathbf{x}_{\text{est}}]_i) \geq [\mathbf{I}(\mathbf{x})^{-1}]_{ii}. \tag{4.9}$$

This implies that the variance of the ith parameter estimate is constrained by the diagonal elements of the inverse Fisher information matrix, assuming the regularity condition holds.

Example 110. Find the Fisher information matrix for the univariate normal distribution.

Solution. We have two parameters for the normal distribution random variables represented by it as mean and variance: $f(y) = \frac{1}{\sqrt{2\pi\sigma^2}} e^{-\frac{(y-\mu)^2}{2\sigma^2}}$. Let's call $x_1 = \mu$ and $x_2 = \sigma^2$. Now, we may find all elements of the Fisher matrix as presented in (4.7). Taking the natural logarithm of the normal distribution results in:

$$\log(f(y)) = -\frac{1}{2}\log(2\pi) - \frac{1}{2}\log(x_2) - \frac{(y-x_1)^2}{2x_2}. \tag{4.10}$$

Taking all partial derivatives as:

$$\begin{aligned}
\frac{\partial}{\partial x_1}\log(f(y)) &= \frac{y-x_1}{x_2}\\
\frac{\partial^2}{\partial x_1^2}\log(f(y)) &= -\frac{1}{x_2}\\
\frac{\partial^2}{\partial x_1 \partial x_2}\log(f(y)) &= -\frac{y-x_1}{x_2^2}\\
\frac{\partial^2}{\partial x_2^2}\log(f(y)) &= \frac{1}{2x_2^2} - \frac{(y-x_1)^2}{x_2^3}.
\end{aligned}$$

The final step to compute the Fisher matrix is to take the expectation operation of each element as in (4.7). Hence, the main diagonal of Fisher's matrix is computed as:

$$[\mathbf{I}]_{1,1} = -E\left[-\frac{1}{x_2}\right] = \frac{1}{x_2} = \frac{1}{\sigma^2}. \tag{4.11}$$

The other diagonal element

$$\begin{aligned}
[\mathbf{I}]_{2,2} &= -E\left[\frac{1}{2x_2^2} - \frac{(y-x_1)^2}{x_2^3}\right] \\
&= -\frac{1}{2x_2^2} + \frac{E(y-x_1)^2}{x_2^3} = -\frac{1}{2x_2^2} + \frac{x_2}{x_2^3} = \frac{1}{2\sigma^4}.
\end{aligned} \tag{4.12}$$

Since $E[y] = x_1$, the other two elements of Fisher's matrix are:

$$[\mathbf{I}]_{1,2} = [\mathbf{I}]_{2,1} = -E\left[-\frac{y-x_1}{x_2^2}\right] = 0. \tag{4.13}$$

Therefore, the Fisher matrix for univariate normal distribution with two parameters (mean and variance) is given by:

$$\mathbf{I} = \begin{bmatrix} \frac{1}{\sigma^2} & 0 \\ 0 & \frac{1}{2\sigma^4} \end{bmatrix}. \tag{4.14}$$

Building on the previous example's findings, computing the lower bound of the estimation error variance becomes achievable through the inverse of the Fisher matrix. This applies universally, independent of the estimation technique used.

When estimating the mean of a random variable with a normal distribution from a single observation y, the lower bound of the error variance remains σ^2. Additionally, when estimating the variance of a normal distribution based on a single observation sample y, the lower bound on the error variance equates to $2\sigma^4$. What happens to these lower bounds when there are N independent observations? $[y_1, y_2, \ldots y_N]$. Let's explore this in the next example.

Example 111. Find the Fisher information matrix for the independent multivariate normal distribution.

Solution. The joint normal distribution of N independent identical random variables is given by:

$$f(y_1, \ldots, y_N) = \frac{1}{(2\pi\sigma^2)^{N/2}} e^{-\frac{\sum_{i=1}^{N}(y_i-\mu)^2}{2\sigma^2}}.$$

Again, let's call $x_1 = \mu$ and $x_2 = \sigma^2$. Taking the natural logarithm of the normal distribution results in:

$$-\frac{N}{2}\log(2\pi) - \frac{N}{2}\log(x_2) - \frac{\sum_{i=1}^{N}(y_i - x_1)^2}{2x_2}.$$

Taking all partial derivatives and then the expectations as have been shown in the previous example we obtain the following results:

$$\begin{aligned}
[\mathbf{I}]_{1,1} &= -\frac{N}{\sigma^2} \\
[\mathbf{I}]_{1,2} &= [\mathbf{I}]_{2,1} = 0 \\
[\mathbf{I}]_{2,2} &= -\frac{N}{2\sigma^4}.
\end{aligned}$$

Therefore, the lower bound on the error variance when estimating the mean of a normally distributed process from N independent samples, irrespective of the estimation method used, is $\frac{\sigma^2}{N}$. Similarly, when estimating the variance of a normally distributed process from N independent samples, the lower bound on the error variance remains $\frac{\sigma^4}{2N}$.

Hence, an estimation technique is deemed efficient if its error variance is equal to the lower bound.

4.4 Estimation techniques

Having familiarized ourselves with the key performance measure criteria in estimation theory, we will now delve into some fundamental estimation techniques.

4.4.1 Least mean square error estimate

In this approach, the estimate is derived by minimizing the squared error between the estimated parameter and its true value. The Least Mean Square Error (LMSE) estimate holds paramount importance in estimation theory for several compelling reasons. Specifically, LMSE emerges as the optimal criterion when dealing with parameters affected by normally distributed noise, a proof we'll delve into later.

Moreover, the squared error signal aligns with its power, carrying tangible implications for electrical engineers. Thus, the power minimization holds a meaningful interpretation in this context. For the sake of simplicity, we derive the estimation technique based on a single parameter, with the straightforward extension to multidimensional parameters.

Consider the scenario where the goal is to estimate the unknown parameters, denoted as x, utilizing a set of samples y_i within the confines of the following mathematical model:

$$y_i = h(x) + n_i. \tag{4.15}$$

In this context, $h(\bullet)$ represents an unknown corruption or mapping associated with the unknown parameter x, while n_i denotes the ith additive noise sample. The estimation of the concealed parameter x can be achieved by different error criteria. In the case of the Least Mean Square Error (LMSE), the optimal estimate x_{est} minimizes the mean squared error, defined as:

$$\min E\left[(x - x_{\text{est}}(\mathbf{y}))^2\right]. \tag{4.16}$$

Note that in the case of an unbiased estimator, where $x_{\text{est}} = E[x]$, the LMSE aligns with minimizing the variance of x. This correlation is logical, since reducing the variance implies diminishing our uncertainty regarding x given the observations $\mathbf{y}$. The expected value of the squared error is expressed as:

$$E\left[(x - x_{\text{est}}(\mathbf{y}))^2\right] = \int_x \int_{\mathbf{y}} \cdots \int (x - x_{\text{est}}(\mathbf{y}))^2 f_{X\mathbf{Y}}(x;\mathbf{y})\,dx\,dy_1 \ldots dy_N. \tag{4.17}$$

Where $f_{X\mathbf{Y}}(x;\mathbf{y})$ is the joint distribution between the measurements $\mathbf{y} = [y_1, \ldots, y_N]^T$ and the parameter we want to estimate, x. The joint distribution can be expanded as $f_{X\mathbf{Y}}(x;\mathbf{y}) = f_{X|\mathbf{Y}}(x \mid \mathbf{y}) f_{\mathbf{Y}}(\mathbf{y})$. The conditional probability density function $f_{X|\mathbf{Y}}(x \mid \mathbf{y})$ represents the distribution of the parameter x conditioned on the observations $\mathbf{y}$, and $f_{\mathbf{Y}}(\mathbf{y})$ is the joint distribution of the observations regardless of the parameter x. This computation is straightforward, achieved by averaging the observation over all conceivable values of x. In the case of continuous x, for instance, this can be expressed as:

$$f_{\mathbf{Y}}(\mathbf{y}) = \int_{-\infty}^{\infty} f_{X\mathbf{Y}}(x;\mathbf{y})\,dx.$$

Equation (4.17) can be reformulated as

$$E[(x - x_{\text{est}}(\mathbf{y}))^2] = \int_x \int_{\mathbf{y}} (x - x_{\text{est}}(\mathbf{y}))^2 f_{X|\mathbf{Y}}(x \mid \mathbf{y}) f_{\mathbf{Y}}(\mathbf{y}) dx d\mathbf{y}. \tag{4.18}$$

To simplify, the multiple integrals have been condensed into a single integral symbol. The preceding equation can be further reformulated as:

$$E[(x - x_{\text{est}}(\mathbf{y}))^2] = \int_{\mathbf{y}} \left[\int_x (x - x_{\text{est}}(\mathbf{y}))^2 f_{X|\mathbf{Y}}(x \mid \mathbf{y}) dx \right] f_{\mathbf{Y}}(\mathbf{y}) d\mathbf{y}. \tag{4.19}$$

To identify the minimum error, it is necessary to differentiate equation (4.19) with respect to x, which yields:

$$\frac{d}{dx} E[(x - x_{\text{est}}(\mathbf{y}))^2] = \frac{d}{dx} \int_{\mathbf{y}} \left[\int_x (x - x_{\text{est}}(\mathbf{y}))^2 f_{X|\mathbf{Y}}(x \mid \mathbf{y}) dx \right] f_{\mathbf{Y}}(\mathbf{y}) d\mathbf{y} = 0. \tag{4.20}$$

Certainly, it's evident that $f_{\mathbf{Y}}(\mathbf{y}) > 0$, and it's not a function of x. Therefore, the differentiation can be executed solely for the inner integral, yielding:

$$\frac{d}{dx} \left[\int_x (x - x_{\text{est}}(\mathbf{y}))^2 f_{X|\mathbf{Y}}(x \mid \mathbf{y}) dx \right] = 0.$$

This differentiation is straightforward and its result is:

$$2 \int_x (x - x_{\text{est}}(\mathbf{y})) f_{X|\mathbf{Y}}(x \mid \mathbf{y}) dx = 0.$$

Since,

$$\int_x x f_{X|\mathbf{Y}}(x \mid \mathbf{y}) dx = E[x \mid \mathbf{y}]$$

and

$$\int_x x_{\text{est}} f_{X|\mathbf{Y}}(x \mid \mathbf{y}) dx = x_{\text{est}} \int_x f_{X|\mathbf{Y}}(x \mid \mathbf{y}) dx = x_{\text{est}}.$$

Hence, the optimal estimate derived from minimizing the mean square error is the conditional mean, represented as:

$$x_{\text{est}} = E[x \mid \mathbf{y}]. \tag{4.21}$$

While the obtained result is intriguing as it provides the optimal estimation of the hidden parameter regardless of the shape of the corruption mapping or the additive noise distribution, it's crucial to note that all statistical details related to the corruption mapping, the additive noise, and the estimated parameter are encapsulated in the conditional probability density function $f_{X|\mathbf{Y}}(x \mid \mathbf{y})$. However, practical challenges arise in determining this function. Its complexity stems from its dependence on the unknown parameter x and its impact on the given observations $\mathbf{y}$. Fortunately, the use of Bayes' rules allows us to express the conditional posterior distribution in terms of the likelihood distribution, a topic that will be explored in-depth later in the context of maximum likelihood estimators.

4.4.2 Minimum absolute error estimate

It's also feasible to derive the optimal estimate for the hidden parameter x in (4.15) by minimizing the absolute error, expressed as:

$$\min E[\mid x - x_{\text{est}}(\mathbf{y}) \mid]. \tag{4.22}$$

We can identify the optimal estimate that minimizes the absolute error by employing similar mathematical steps as in the case of LMSE. The optimal estimate is the one that achieves:

$$\frac{d}{dx_{\text{est}}}\left[\int_X \mid x - x_{\text{est}}(\mathbf{y}) \mid f_{X|\mathbf{Y}}(x \mid \mathbf{y})dx\right] = 0.$$

The absolute value could be expanded as:

$$\frac{d}{dx_{\text{est}}}\left[\int_{-\infty}^{x_{\text{est}}} (x - x_{\text{est}}(\mathbf{y}))f_{X|\mathbf{Y}}(x \mid \mathbf{y})dx - \int_{x_{\text{est}}}^{\infty} (x - x_{\text{est}}(\mathbf{y}))f_{X|\mathbf{Y}}(x \mid \mathbf{y})dx\right] = 0.$$

After differentiation, the formula for the optimal estimate is derived as follows:

$$\int_{-\infty}^{x_{\text{est}}} f_{X|\mathbf{Y}}(x \mid \mathbf{y})dx = \int_{x_{\text{est}}}^{\infty} f_{X|\mathbf{Y}}(x \mid \mathbf{y})dx. \tag{4.23}$$

In this scenario, the optimal estimate is evidently the *median* of the conditional distribution $f_{X|\mathbf{Y}}(x \mid \mathbf{y})$. Notably, the optimal estimate can be determined based on various norm criteria of the error. However, the focus now shifts to the infinite norm criterion.

4.4.3 MinMax error estimate criterion

The minimax error criterion involves minimizing the maximum error, that is:

$$\min \max E[\mid x - x_{\text{est}}(\mathbf{y}) \mid]. \tag{4.24}$$

It is also known as the infinite-norm criterion, because for an N-dimensional error vector $\mathbf{e} = [e_1, e_2, \ldots e_N]$, it is a straightforward to prove that

$$\max(\mathbf{e}) = \| \mathbf{e} \|_\infty . \tag{4.25}$$

Where the pth norm is given be

$$\| \mathbf{e} \|_p = \left(\sum_{i=1}^{N} \mid e_i \mid^p \right)^{\frac{1}{p}} \quad \forall p > 0. \tag{4.26}$$

Applying the same mathematical steps as before we obtain,

$$\frac{d}{dx_{\text{est}}} \left[\int_X \max \mid x - x_{\text{est}}(\mathbf{y}) \mid f_{X|\mathbf{Y}}(x \mid \mathbf{y}) dx \right] = 0. \tag{4.27}$$

In broad terms, the maximum error tends to occur around the mode of the conditional probability density function, namely:

$$x_{\text{est}} = \max[f_{X|\mathbf{Y}}(x \mid \mathbf{y})]. \tag{4.28}$$

Therefore, it is termed the maximum a posteriori estimate (MAP). So far, we've explored three distinct estimation criteria: minimum mean square error, minimum absolute error, and minimum maximum error. One might ask: which of these criteria is superior? In reality, there's no universally superior criterion! Each criterion yields an optimal solution. However, in practice, certain criteria may outperform others depending on the nature of uncertainties involved. When dealing with a unimodal and symmetric distribution of $f_{X|\mathbf{Y}}(x \mid \mathbf{y})$, all these criteria converge, resulting in identical outcomes where the mean equals the median equals the mode. An essential example of this is the conditional normal distribution. This characteristic holds immense significance in the principles underlying Kalman filters, which we'll explore in detail in the forthcoming chapter.

4.4.4 Maximum likelihood estimator

All of the above mentioned estimators are optimal in various normative senses, encompassing minimum mean square error (ℓ_2-norm), minimum absolute error (ℓ_1-norm), and minimum-maximum error (ℓ_∞-norm). Nevertheless, it's crucial to note that all these estimators rely on the conditional probability density function: $f_{X|\mathbf{Y}}(x \mid \mathbf{y})$ which is usually very difficult to know.

We can expand this distribution accordingly using Bayes' rule as follows:

$$f_{X|\mathbf{Y}}(x \mid \mathbf{y}) = \frac{f_{\mathbf{Y}|X}(\mathbf{y} \mid x) f_X(x)}{f_{\mathbf{Y}}(\mathbf{y})}. \tag{4.29}$$

The expression $f_{X|\mathbf{Y}}(x \mid \mathbf{y})$ denotes the probability distribution of an unknown parameter after observing the outcomes $\mathbf{y}$. This distribution is referred to as the 'posterior probability density function.' On the other hand, $f_{\mathbf{Y}|X}(\mathbf{y} \mid x)$ represents how observations are expected to behave given a certain parameter x; thus, it's termed the *likelihood function.*

In practical scenarios, determining the likelihood function is often more feasible. For instance, in wireless communications, sending a known signal (a *training sequence*) allows the receiver to estimate the likelihood distribution of the channel based on measurements $\mathbf{y}$. This estimated distribution is later employed to decode unknown transmitted signals. Additionally, estimating the likelihood distribution from available databases is a straightforward process.

The distribution $f_{\mathbf{Y}}(\mathbf{y})$ signifies the general joint distribution of measurements, independent of the parameter x.

To understand some of the philosophical concepts behind Bayes' rule, the distribution $f_X(x)$ holds special significance. In this chapter, our focus is on inferring the value of a concealed parameter x based on a series of observations or measurements $\mathbf{y}$. This parameter, denoted as x, could represent an unknown fixed or time-varying signal with unidentified parameters, or it may even embody a random process.

When x takes on a fixed but unknown value, the interpretation of $f_X(x)$ becomes intriguing. In this scenario, it reflects our uncertainty regarding the specific value of x. For instance, even though we do not know the exact value of x, we may assert with 60 % confidence that it falls within the range of 2 to 3. Our initial uncertainty stems from known physical behaviors or historical knowledge. Despite x being fixed, our imprecise knowledge transforms it into a random variable with a specific distribution function. Consider the extreme case where we have no information whatsoever about x. In this scenario, for us, x could assume any real number from $-\infty$ to $+\infty$ with identical uncertainty levels. The distribution of x in this case would follow a uniform distribution function, represented as $f_X(x) = \frac{1}{2a}$, where $-\infty \leq a \leq \infty$. However, there is no valid uniform probability density function that defined over all real numbers. A uniform distribution spanning from negative infinity to positive infinity doesn't align with the typical notion of a probability distribution. This is due to its infinite range, making it impossible to assign equal probabilities over an unbounded interval. In our discussion, this concept of a uniform distribution over the entire real number line is employed to represent a scenario of complete lack of information or even conjecture about the variable x prior to conducting measurements.

We have seen that the *mode* or MAP is one of the optimal estimators. Therefore, the best estimation of the parameter x is one that maximizes the left or the right side of (4.29). Since the denominator function $f_{\mathbf{Y}}(\mathbf{y})$ does not depend on x and it is the same for all x values, hence, it can be ignored. The optimal $x = x_{\text{est}}$ parameter is the one that maximizes the numerator $[f_{\mathbf{Y}|X}(\mathbf{y} \mid x) f_X(x)]$.

In estimation theory, it's often standard to assume that observations are independent. While this assumption holds true in scenarios such as independent entries in databases, in other cases, this assumption lacks justification but is still employed to streamline the analytical approach. However, this simplification can introduce bias into the final result, acknowledging the potential impact on the accuracy of the estimation. Hence, in the case of independence, the likelihood distribution is given by:

$$f_{\mathbf{Y}|X}(\mathbf{y} \mid x) = \prod_{i=1}^{N} f_{\mathbf{Y}|X}(y_i \mid x). \tag{4.30}$$

Finding the maximum point involves taking the derivative function with respect to x. However, due to the complexity introduced by this multiplicative function, obtaining closed-form mathematical solutions might prove challenging. Fortunately, logarithmic functions, being monotonically increasing functions, retain the location of the extreme points, simplifying the process. For $g(x) > 0\ \forall x$, then if $\hat{x} = \arg\max g(x)$, then it is always true that $\hat{x} = \arg\max \log[g(x)]$. Therefore, the optimal MAP estimate could be expressed as

$$\begin{aligned} x_{\text{MAP}} &= \arg\max \log[f_{X|\mathbf{Y}}(x \mid \mathbf{y})] \\ &= \arg\max \left[\sum_{i=1}^{N} \log f_{\mathbf{Y}|X}(y_i \mid x) + \log f_X(x) \right]. \end{aligned} \tag{4.31}$$

From the analytical expression above, it's evident that deriving the optimal MAP estimate relies on knowing both the likelihood and the prior distribution functions. However, if there's no prior statistical knowledge about x, the distribution conceptually becomes a fixed value, as previously explained. In this scenario, the optimal estimate is derived by maximizing the following expression:

$$\sum_{i=1}^{N} \log f_{\mathbf{Y}|X}(y_i \mid x) - \log(2a). \tag{4.32}$$

The second term remains constant and evaluates to zero when differentiated to find the optimal estimate. In this context, the optimal estimate coincides with maximizing the likelihood function, hence the term maximum likelihood estimate (ML). How does ML estimation differ from MAP? When there's no prior knowledge about the parameter, ML provides the same optimal solution as a MAP estimate. However, if there's any prior knowledge about x that isn't utilized in the estimation process, ML will deliver a suboptimal solution. We will explain these concepts in more detail with some examples.

Example 112. Consider a scenario where we're examining a process with two potential outcomes: success or failure. This setup finds application in various practical contexts.

A few illustrative examples include:
- Hitting or missing a targets.
- Correctly or incorrectly receiving a transmitted symbol or message.
- Positive or negative revenue.
- Spam email or not.
- Positive or negative diagnosis of a particular disease.

Using historical independent observations, our aim is to estimate the process parameter—specifically, the probability of success, denoted by $x = p$. Initially, let's assume that we have no prior knowledge about the process. Hence, x can potentially take any value between 0 and 1. Consider that from a set of N historical observations, we've recorded M instances of success and $N - M$ instances of failure. It is clear that:

$$f_{\mathbf{Y}|X}(\mathbf{y} \mid x = p) = x^M (1 - x)^{N-M}. \tag{4.33}$$

Since we lack prior knowledge about the values of x, its distribution is assumed to be uniform between 0 and 1. In this case, the optimal estimate is

$$x_{\mathrm{MAP}} = x_{\mathrm{ML}} = \arg\max\left[M \log(x) + (N - M) \log(1 - x)\right]. \tag{4.34}$$

Observe that $\log(1) = 0$. Differentiating the equation in order to find the extreme point, we have

$$\frac{d}{dx}\left[M \log(x) + (N - M) \log(1 - x)\right]\Big|_{x=x_{\mathrm{ML}}} = 0. \tag{4.35}$$

Differentiating the above equation yields

$$\frac{M}{x} - \frac{N - M}{1 - x} = 0 \rightarrow x_{\mathrm{ML}} = \frac{M}{N}. \tag{4.36}$$

This estimation is optimal as it satisfies both the MAP (Maximum a Posteriori) and ML (Maximum Likelihood) criteria.

Is the optimal estimate of the probability of success in the previous example unbiased, consistent, and efficient? To address this crucial query, let's introduce the indicator parameter z_i, where $z_i = 1$ if the ith outcome was a success and $z_i = 0$ otherwise. Hence, we can express the optimal estimate as:

$$x_{\mathrm{ML}} = \frac{\sum_{i=1}^{N} z_i}{N}. \tag{4.37}$$

It is possible to compute the expected value of z_i as:

$$E(z_i) = 1 \times P(z_i = 1) + 0 \times P(z_i = 0) = p.$$

Here, p is the hidden parameter that we are interested in estimating. Applying these results we obtain:

$$E[x_{\mathrm{ML}}] = \frac{E[\sum_{i=1}^{N} z_i]}{N} = \frac{Np}{N} = p. \tag{4.38}$$

Therefore, the estimate is unbiased.

To test the consistency of the estimate, one should check whether the error variance vanishes with the number of observations N or not. The error variance of the estimate can be computed as in the following detailed steps:

$$\begin{aligned}
\operatorname{Var}(x_{\mathrm{ML}}) &= E[(x_{\mathrm{ML}} - E[x_{\mathrm{ML}}])^2] \\
&= E\left[\left(\frac{\sum_{i=1}^{N} z_i}{N} - p\right)^2\right] \\
&= E\left[\left(\frac{\sum_{i=1}^{N} z_i}{N}\right)^2\right] - p^2 \\
&= E\left[\frac{\sum_{j=1}^{N}\sum_{i=1}^{N} z_i z_j}{N^2}\right] - p^2 \\
&= E\left[\frac{\sum_{j=1}^{N} z_j^2 + \sum_{j=1}^{N}\sum_{i=1, i\neq j}^{N} z_i z_j}{N^2}\right] - p^2 \\
&= \frac{\sum_{j=1}^{N} E[z_j^2] + \sum_{j=1}^{N}\sum_{i=1, i\neq j}^{N} E[z_i z_j]}{N^2} - p^2 \\
&= \frac{\sum_{j=1}^{N} p + \sum_{j=1}^{N}\sum_{i=1, i\neq j}^{N} p^2}{N^2} - p^2 \\
&= \frac{Np + N(N-1)p^2}{N^2} - p^2 \\
&= \frac{p(1-p)}{N} \leq \frac{1}{4N}.
\end{aligned} \tag{4.39}$$

The final equation (4.39) demonstrates the estimator's consistency. With an increasing number of measurements approaching infinity, the error variance decreases to zero.

Can we enhance this estimator further for greater efficiency? In other words, does the estimator achieve optimal efficiency? Computing the Fisher information element for $x = p$ sheds light on this:

$$I = -E_M\left[\frac{d^2}{dx^2}(M\log(x) + (N-M)\log(1-x))\right] \tag{4.40}$$

$$= E_M\left[\frac{M}{x^2} + \frac{N-M}{(1-x)^2}\right]. \tag{4.41}$$

The expected number of success is $E_M[M] = Np$, therefore,

$$I = \frac{Np}{p^2} + \frac{N-Np}{(1-p)^2} = \frac{N}{p(1-p)}. \tag{4.42}$$

Furthermore, it is straightforward to prove the regularity condition in (4.8). Hence, the minimum possible variance of the estimate of x is

$$I^{-1} = \frac{p(1-p)}{N}. \tag{4.43}$$

Therefore, the algorithm for estimating the probability of occurrence in the independent Bernoulli problem is efficient.

Previously, we established that the ML estimate is optimal when no prior information about the estimated parameter exists. However, when information is available—derived from thoughts, past observations, or other sources—it's advantageous to leverage it to refine the estimation process. This concept is elucidated in the following example.

Example 113. In the preceding example, consider a scenario where our uncertainty regarding the probability of success is not uniform. For instance, it may follow a truncated exponential distribution, such as:

$$f_X(x) = 4e^{-ax}, \quad 0 \le x \le 1. \tag{4.44}$$

Find the value of a. How will this information affect the optimal estimation?

Solution. Since the total area under the probability density function must be one, i. e.,

$$4\int_0^1 e^{-ax}dx = 1. \tag{4.45}$$

Hence, we can determine that $a \approx 3.92$.

Neglecting this prior information, perhaps for the sake of simplicity, renders the ML estimate, expressed as $x_{\text{ML}} = \frac{M}{N}$, as a suboptimal solution especially for a small numbers of observations N. However, to uncover the optimal solution in this scenario, we need to determine the MAP solution. Examining the prior distribution of x we observe that the probabilities of success and failure are not uniform. Now we have a specific uncertainty about the probability. Figure 4.1 shows the cumulative distribution of the probability p. From the CDF figure we can see that the probability of success being less than 0.3 is about 70 %. This kind of prior information should be useful to improve the estimation accuracy of the parameter x.

From (4.31), we can compute the optimal MAP estimate as:

$$\frac{d}{dx}[M\log(x) + (N-M)\log(1-x) + \log(4) - 3.92x]\Big|_{x=x_{\text{MAP}}} = 0. \tag{4.46}$$

Solving the above equation leads to the following quadratic equation:

$$3.92x_{\text{MAP}}^2 - (3.92 + N)x_{\text{MAP}} + M = 0, \tag{4.47}$$

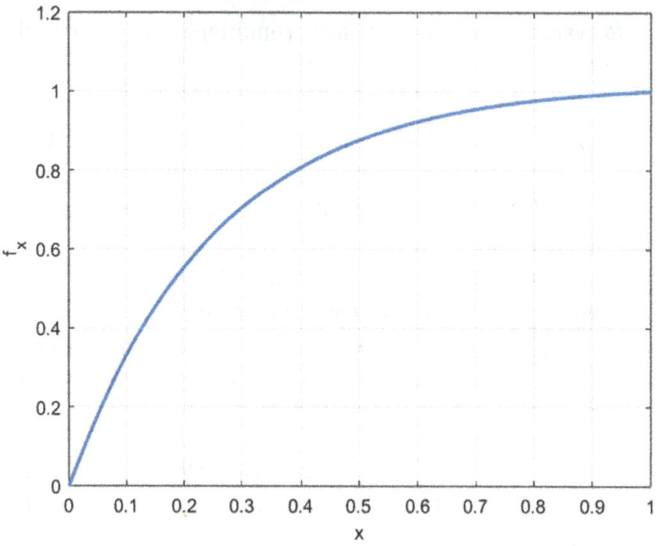

Figure 4.1: CDF of the probability of success.

where $0 \le x_{\text{MAP}} \le 1$. It is clear that this equation has two positive roots and at least one root should achieve the condition $0 \le x_{\text{MAP}} \le 1$ such as:

$$x_{\text{MAP}} = \frac{(3.92 + N) - \sqrt{(3.92 + N)^2 - 15.68M}}{7.84}. \tag{4.48}$$

Let's have the case that based on observations $\frac{M}{N} = \frac{1}{2}$, with the maximum likelihood estimation, $x_{\text{ML}} = \frac{1}{2}$. However, with MAP, this depends on the total number of observations N. Figure 4.2 shows the optimal estimation of the probability of success with a number of observations for $\frac{M}{N} = \frac{1}{2}$.

For a limited number of observations, the Maximum a Posteriori (MAP) estimate depends heavily on the prior information of the parameter. As illustrated in Figure 4.2 for the case of $M = 1$ and $N = 2$, the optimized estimate is $x_{\text{MAP}} \approx 0.19$, notably lower than the maximum likelihood (ML) estimate of 0.5. However, the figure demonstrates that as the number of observations N increases, the MAP estimate gradually converges towards the ML estimate. This indicates a shift in reliance from prior information to observational data. Is this trend universal? To explore this, let's consider a scenario where our prior information about x follows a uniform distribution between 0.6 and 0.9. For an enlightening exercise, repeat the analysis based on this updated prior distribution.

The previous example shows the optimal estimation of the probability of the Bernoulli distribution based on N observations. The next example shows more general cases known as multinomial density. Here we have $K > 2$ uncertain classes. Assuming

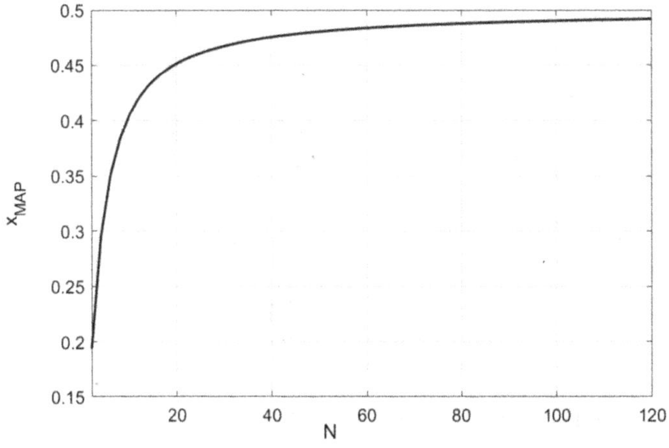

Figure 4.2: The optimal MAP estimate with the number of observations.

independent outcomes, the probability for each outcome to be in class i is p_i, and hence,

$$\sum_{i=1}^{K} p_i = 1,$$

where $i = 1, 2, \ldots K$. Assume the outcomes in each class i are n_i, then $\sum_{i=1}^{K} n_i = N$, where N is the total number of outcomes. The probability for a certain number of events in each class can be easily proved to be (for independent events):

$$P(n_1, n_2, \ldots, n_K) = \prod_{i=1}^{K} p_i^{n_i}. \tag{4.49}$$

The multinomial distribution proves to be exceptionally valuable in applications ranging from survey assessments to machine learning implementations, due to its versatile and adaptable nature.

Example 114. In a study of mouse behavior involving the selection of three different colored balls—red, green, and black—the mouse chooses one ball to hit during each trial. The criteria guiding the mouse's choice remain unknown, prompting the need for statistical inference about its behavior. Given a set of N observations, what would be the maximum likelihood estimate of the probabilities, assuming independence among the ball selections?

Solution. Assume that the ith estimated parameter is x_i, where x_i should be the optimal estimate of the probability of the ith class p_i. Since we have no prior information, the optimal estimate will be the maximum likelihood estimator. We have N observations with distributions n_1, n_2, n_3, and $\sum_{i=1}^{3} n_i = N$. The ML estimation for the ith element is the argument maximum:

$$\sum_{i=1}^{K} n_i \log(x_i)$$

with the constraint $\sum_{i=1}^{K} x_i = 1$. The maximization problem can be addressed using the method of Lagrange multipliers, like this:

$$\frac{\partial}{\partial x_m}\left[\sum_{i=1}^{K} n_i \log(x_i) + \lambda\left(\sum_{i=1}^{K} x_i - 1\right)\right]_{x_m = x_{m,\mathrm{ML}}} = 0, \tag{4.50}$$

$$\frac{\partial}{\partial \lambda}\left[\sum_{i=1}^{K} n_i \log(x_i) + \lambda\left(\sum_{i=1}^{K} x_i - 1\right)\right] = 0, \tag{4.51}$$

where λ is the Lagrange multiplier. Solve the above two equations:

$$\frac{n_m}{x_m} + \lambda = 0 \to x_m = -\frac{n_m}{\lambda} \tag{4.52}$$

$$\because \sum_{i=1}^{K} x_i = 1 \to -\frac{1}{\lambda}\sum_{i=1}^{K} n_i = 1 \tag{4.53}$$

$$\because \sum_{i=1}^{K} n_i = N \to \lambda = -N \tag{4.54}$$

$$\therefore x_{m,\mathrm{ML}} = \frac{n_m}{N}. \tag{4.55}$$

The outcome aligns logically; the maximum likelihood estimation of the probability for the mth class corresponds to the frequency of observations for that class divided by the total number of observations. For instance, if the mouse hits the red ball 7 times, the green ball 5 times, and the black ball 10 times, the maximum likelihood probabilities would be $\frac{7}{22}$, $\frac{5}{22}$, and $\frac{10}{22}$, respectively. However, this calculation assumes that the events are independent. Considering the potential interdependence between events might reflect reality more accurately. The mouse could follow certain criteria when hitting balls in a sequence. To explore such possibilities, a more sophisticated model like Markov chains could be employed.

Let's revisit equation (4.1) and simplify it into its most basic uncertain model form as:

$$y_i = x + n_i, \quad i = 1, 2, \ldots N. \tag{4.56}$$

We've explored various scenarios related to this estimation problem. In this context, let's consider the parameter to be estimated, denoted by x, as an unknown constant, while n_i represents independent samples with a zero-mean normal distribution and variance σ^2. It's evident that the measurements y_i also follow an independent normal distribution, centered around x. The variance of the additive noise n_i might be known or unknown. If unknown, it can be estimated from the measurements y_i.

If there's no prior knowledge available about x, it can be assumed to have a uniform distribution within the range of interest, the maximum likelihood (ML) estimation stands out as the most optimal estimate for the mean value of x. Therefore:

$$\frac{d}{dx}\left[\sum_{i=1}^{N}\log f_{Y|X}(y_i \mid x)\right]_{x=x_{\mathrm{ML}}} = 0 \tag{4.57}$$

$$\because f_{Y|X}(y_i \mid x) = \frac{1}{\sqrt{2\pi\sigma^2}}e^{\frac{-(y_i-x)^2}{2\sigma^2}} \tag{4.58}$$

$$\therefore \frac{d}{dx}\left[-\frac{N}{2}\log(2\pi\sigma^2) - \frac{\sum_{i=1}^{N}(y_i-x)^2}{2\sigma^2}\right]_{x=x_{\mathrm{ML}}} = 0 \tag{4.59}$$

$$\Rightarrow x_{\mathrm{ML}} = \frac{\sum_{i=1}^{N}y_i}{N}. \tag{4.60}$$

Hence, the arithmetic mean stands out as the optimal method for computing the mean of an unknown normal distribution based on its samples. As we've observed in previous chapters, the arithmetic mean converges to the true mean of the random variable (assuming it exists) as the number of samples ($N \to \infty$) increases, irrespective of the distribution type. However, it's important to note that while the arithmetic mean often serves as an optimal approach, this isn't universally true, as later proofs will demonstrate.

Example 115. Given that the arithmetic mean (or average) is the optimal method for computing the mean of a normal distribution in the absence of prior information about its mean and when dealing with independent samples, let's establish the proof that this maximum likelihood (ML) estimate is unbiased, consistent, and efficient.

Solution. Since

$$E[x_{\mathrm{ML}}] = \frac{\sum_{i=1}^{N}E[y_i]}{N} = \frac{Nx}{N} = x.$$

Therefore, the ML estimation is unbiased.

For the consistence, we need to compute the error variance $E[(x_{\mathrm{ML}} - x)^2]$ and then check if it will converge to 0 as $N \to \infty$. We may derive the error variance as:

$$E[(x_{\mathrm{ML}} - x)^2] = E[x_{\mathrm{ML}}^2] - x^2. \tag{4.61}$$

The expectation equation of $E[x_{\mathrm{ML}}^2]$ can be derived as

$$\begin{aligned} E[x_{\mathrm{ML}}^2] &= \frac{1}{N^2}E\left[\left(\sum_{i=1}^{N}y_i\right)^2\right] \\ &= \frac{1}{N^2}E\left[\sum_{i=1}^{N}\sum_{j=1}^{N}y_iy_j\right] \end{aligned}$$

$$= \frac{1}{N^2} E\left[\sum_{i=1}^{N} y_i^2 + \sum_{i=1}^{N}\sum_{j\neq i}^{N} y_i y_j\right]$$
$$= \frac{1}{N^2}\left[\sum_{i=1}^{N} E[y_i^2] + \sum_{i=1}^{N}\sum_{j\neq i}^{N} E[y_i y_j]\right],$$
$$= \frac{1}{N^2}[N(\sigma^2 + x^2) + N(N-1)x^2],$$
$$= \frac{\sigma^2}{N} + x^2$$
$$\Rightarrow E[(x_{\mathrm{ML}} - x)^2] = \frac{\sigma^2}{N} + x^2 - x^2 = \frac{\sigma^2}{N}. \tag{4.62}$$

As the sample size N approaches infinity, it becomes evident that the error variance converges to zero, establishing the consistency of this maximum likelihood (ML) estimate. Now, the question arises: is this estimate also efficient? To determine efficiency, one must ascertain the Cramer–Rao lower bound of the multivariate normal distribution, a calculation previously undertaken in Example 111. The analysis reveals that the error variance derived from this ML estimation for the mean aligns precisely with the lower bound. Consequently, we can conclude that the estimate is not only consistent, but also efficient.

The optimal estimate for x in (4.56) has been derived in the previous example. To extend this analysis and determine the optimal estimate for the variance of $x_2 = \sigma^2$, we can follow similar steps as those undertaken in the previous example. The optimal estimate, in this case, remains the maximum likelihood (ML) estimate, expressed as:

$$\frac{d}{dx_2}\left[-\frac{N}{2}\log(2\pi x_2) - \frac{\sum_{i=1}^{N}(y_i - x)^2}{2x_2}\right]_{x_2 = s_{\mathrm{ML}}} = 0$$
$$\Rightarrow -\frac{N}{2x_2} + \frac{\sum_{i=1}^{N}(y_i - x)^2}{2x_2^2}\bigg|_{x_2 = s_{\mathrm{ML}}} = 0.$$

Therefore, the maximum likelihood estimation of the variance is

$$s_{\mathrm{ML}} = \frac{\sum_{i=1}^{N}(y_i - x)^2}{N}. \tag{4.63}$$

Practically, the exact mean x is unknown, and hence the ML estimate of the data variance should be changed as follows:

$$s_{\mathrm{ML}} = \frac{\sum_{i=1}^{N}(y_i - x_{\mathrm{ML}})^2}{N}. \tag{4.64}$$

Is the maximum likelihood (ML) estimate of the variance unbiased, consistent, and efficient? Let's see:

$$\begin{aligned}
E[s_{\mathrm{ML}}] &= \frac{\sum_{i=1}^{N} E[(y_i - x_{\mathrm{ML}})^2]}{N} \\
&= \frac{\sum_{i=1}^{N} (E[y_i^2] - 2E[y_i x_{\mathrm{ML}}] + E[x_{\mathrm{ML}}^2])}{N} \\
\because E[x_{\mathrm{ML}}^2] &= \frac{\sigma^2}{N} + x^2 \\
\because E[y_i x_{\mathrm{ML}}] &= E\left[y_i \frac{\sum_{k=1}^{N} y_k}{N}\right] = \frac{\sigma^2 + Nx^2}{N} \\
\therefore E[s_{\mathrm{ML}}] &= \frac{N(\sigma^2 + x^2) - 2\sigma^2 - 2Nx^2 + \sigma^2 + Nx^2}{N} \\
&= \frac{(N-1)\sigma^2}{N}.
\end{aligned} \tag{4.65}$$

Therefore, the ML estimate of the variance is biased! To make the estimate unbiased, one should multiply s_{ML} by the factor $\frac{N}{N-1}$; therefore, the unbiased version of the ML estimate of the variance is

$$s_{\mathrm{ML}} = \frac{\sum_{i=1}^{N} (y_i - x_{\mathrm{ML}})^2}{N-1}. \tag{4.66}$$

When dealing with a substantial number of samples N, the distinction between dividing by N or $N-1$ becomes negligible.

For the consistency check, one needs to derive the error variance, i. e.,

$$E[(s_{\mathrm{ML}} - \sigma^2)^2].$$

The derivation of the variance follows a straightforward procedure similar to the one employed for the bias. Therefore, I leave it to the reader to verify the consistency and provide a proof of its consistency and efficiency.

What if there is prior knowledge available about the parameter x? How does non-uniform uncertainty impact the optimal estimate? This poses an intriguing problem. In such scenarios, the arithmetic mean may no longer be the optimal estimate. Let's consider a case where our uncertainty about the parameter x follows a normal distribution with a known mean μ_X and a known variance σ_X^2. In this case, the maximum likelihood (ML) estimate is not necessarily the optimal choice. To determine the optimal estimate, we can turn to the Maximum A Posteriori (MAP) criterion. As demonstrated in (4.31), the MAP estimate can be derived as follows:

$$\frac{d}{dx}\left[\sum_{i=1}^{N} \log f_{Y|X}(y_i \mid x) + \log f_X(x)\right]_{x=x_{\mathrm{MAP}}} = 0. \tag{4.67}$$

Substituting the likelihood and prior functions leads to

$$\frac{d}{dx}\left[-\frac{N\log(2\pi\sigma^2)}{2} - \frac{\sum_{i=1}^{N}(y_i - x)^2}{2\sigma^2} - \frac{\log(2\pi\sigma_X^2)}{2} - \frac{(x-\mu_X)^2}{2\sigma_X^2}\right] = 0. \tag{4.68}$$

Solving the differentiation leads to

$$\frac{\sum_{i=1}^{N}(y_i - x_{\text{MAP}})}{\sigma^2} - \frac{(x_{\text{MAP}} - \mu_X)}{\sigma_X^2} = 0. \tag{4.69}$$

Let's define

$$x_{\text{ML}} = \frac{\sum_{i=1}^{N} y_i}{N},$$
$$\beta = \frac{\sigma^2}{\sigma_X^2}.$$

Therefore, the optimal estimate could be easily formulated as:

$$x_{\text{MAP}} = \frac{1}{(1+\beta/N)} x_{\text{ML}} + \frac{\beta/N}{(1+\beta/N)} \mu_X. \tag{4.70}$$

The MAP estimate provided above is notably elegant. Smaller values of σ_X signify higher confidence that the parameter x is in close proximity to μ_X. The parameter β assumes larger values when σ_X is considerably smaller than σ. Therefore, when we have a small number of samples, such that β/N is very large, then we will rely more on our prior knowledge about x, i. e., $x_{\text{MAP}} \approx \mu_X$. On the other hand, when we have a very large number of samples, in the way that β/N is very small, then we rely more on the measurements, i. e.,

$$x_{\text{MAP}} \approx x_{\text{ML}} = \frac{\sum_{i=1}^{N} y_i}{N}.$$

In the spectrum between these two extremes, the optimal estimate of x emerges as a weighted sum that balances the influence of prior information and the available measurements.

Example 116. For the identical measurement model outlined in (4.56), when characterizing our uncertainty regarding x, it is represented by a χ^2 model with parameters s and degrees of freedom n, as follows:

$$f_X(x) = \frac{1}{2^{n/2}\Gamma(\frac{n}{2})} x^{\frac{n}{2}-1} e^{-x/2s}, \quad x \geq 0. \tag{4.71}$$

Find the MAP estimation of x based on the measurements y_i.

Solution. The approach to solving this issue mirrors the methodology applied in the example immediately above. Nevertheless, it is crucial to note that, in this particular case, considering the underlying uncertainty model, the variable x cannot assume negative values. While the specifics of the solution are left for the reader to explore, the solution for the case of $n = 2$ and $s = 1$ is provided as follows:

$$x_{\mathrm{MAP}} = \max\left(0, x_{\mathrm{ML}} - \frac{\sigma^2}{2N}\right). \tag{4.72}$$

In the preceding analysis, we made the assumption of uncorrelated measurements. However, this assumption may not hold true in all practical scenarios. Measurements or observations can exhibit correlation. Estimating unknown parameters under the assumption of uncorrelated measurements could result in significant errors when dealing with highly correlated measurements. Now, let's derive the maximum likelihood estimate for the measurement model in (4.56), taking into account correlated noise. The likelihood function can be formulated in the case of correlated noise, with a fixed unknown parameter x as:

$$f_{\mathbf{y}|x}(\mathbf{y} \mid x) = \frac{1}{(2\pi)^{N/2} \mid \mathbf{R}_{nn} \mid^{0.5}} e^{-\frac{1}{2}(\mathbf{y}-\mathbf{1}x)^T \mathbf{R}_{nn}^{-1}(\mathbf{y}-\mathbf{1}x)}, \tag{4.73}$$

where $\mathbf{R}_{nn}$ is the covariance matrix of the noise samples, $\mid \mathbf{R}_{nn} \mid$ is the determinant of the covariance matrix, N is the number of measurement samples, and $\mathbf{1} = [1, 1, \ldots 1]^T$. The maximum likelihood estimation could be found in the same way as shown before as:

$$\frac{d}{dx}\left[-\frac{N\log(2\pi)}{2} - \frac{\log(\mid \mathbf{R}_{nn} \mid)}{2} - \frac{(\mathbf{y} - \mathbf{1}x)^T \mathbf{R}_{nn}^{-1}(\mathbf{y} - \mathbf{1}x)}{2}\right]_{x_{\mathrm{ML}}} = 0. \tag{4.74}$$

Through basic manipulations employing matrix calculus, the optimal estimate is determined as:

$$x_{\mathrm{ML}} = \frac{\mathbf{1}^T \mathbf{R}_{nn}^{-1} \mathbf{y}}{\mathbf{1}^T \mathbf{R}_{nn}^{-1} \mathbf{1}}. \tag{4.75}$$

In the scenario of independent and identical noise samples, the covariance matrix takes on a diagonal structure, with the noise variance occupying the diagonal elements. Consequently, (4.75) simplifies to the conventional form $\frac{\sum_{i=1}^{N} y_i}{N}$. Thus, (4.75) stands as a generalized expression for the maximum likelihood estimator of the model presented in (4.56). It also addresses the optimal estimation scenario when we have independent noise samples with different variances, i. e., not identical noise samples. This form can be further extended to accommodate the case of time-varying x, where the objective is to find the optimal vector $\mathbf{x} = [x_1, x_2, \ldots, x_N]$. Additional insights into optimal vector estimation are provided in the *Kalman filters* chapter.

Example 117. Consider two measured samples of the measurement model described in (4.56), denoted as y_1 and y_2. Determine the optimal estimate of x given the provided covariance matrices:

– Independent and identically distributed noise samples:

$$\mathbf{R}_{nn} = \begin{bmatrix} \sigma^2 & 0 \\ 0 & \sigma^2 \end{bmatrix}.$$

- Independent noise samples but with different variances:

$$\mathbf{R}_{nn} = \begin{bmatrix} \sigma_1^2 & 0 \\ 0 & \sigma_2^2 \end{bmatrix}.$$

- Non-identical and correlated noise samples:

$$\mathbf{R}_{nn} = \begin{bmatrix} \sigma_1^2 & \alpha \\ \alpha & \sigma_2^2 \end{bmatrix}.$$

In all the aforementioned scenarios, we can consistently apply the same optimal algorithm in (4.75).

For the first scenario where we have independent and identical distributed measurements,

$$\begin{aligned} \mathbf{1}^T \mathbf{R}_{nn}^{-1} \mathbf{y} &= \frac{1}{\sigma^2}(y_1 + y_2) \\ \mathbf{1}^T \mathbf{R}_{nn}^{-1} \mathbf{1} &= \frac{2}{\sigma^2} \\ \Rightarrow x_{\mathrm{ML}} &= \frac{y_1 + y_2}{2}. \end{aligned} \tag{4.76}$$

This is the expected solution in this case.

For the case of independent but with different variances, the optimal estimation is computed as:

$$x_{\mathrm{ML}} = \frac{\sigma_2^2}{\sigma_1^2 + \sigma_2^2} y_1 + \frac{\sigma_1^2}{\sigma_1^2 + \sigma_2^2} y_2. \tag{4.77}$$

This result is intriguing. Optimal estimation involves a weighted summation of the measurements, assigning greater weight to those with lower error variance.

Finally for the case of correlated noise samples, the optimal estimation could be easily derived as:

$$x_{\mathrm{ML}} = \frac{\sigma_2^2 - \alpha}{\sigma_1^2 + \sigma_2^2 - 2\alpha} y_1 + \frac{\sigma_1^2 - \alpha}{\sigma_1^2 + \sigma_2^2 - 2\alpha} y_2. \tag{4.78}$$

It is also a weighted sum, but affected by the cross-covariance α.

We have delved into the realm of stochastic estimation, revealing that the optimal estimate for the hidden vector $\mathbf{x}$ in (4.1) aligns with one of the modes within the posterior probability density function $f_{\mathbf{X}|\mathbf{Y}}(\mathbf{X} \mid \mathbf{Y})$. This optimal estimation holds true for various probability density functions. However, in practical scenarios, we often encounter a series of measurements or observations, and the precise posterior probability density distribution remains elusive. Formulating assumptions prematurely can introduce significant bias errors into the estimation process. Nevertheless, it is feasible to estimate the distribution of measurements by leveraging techniques such as data histogram analysis or computing the cumulative distribution function (CDF) with the aid of smoothing windows.

4.5 Concepts of regression

In estimation theory, a fundamental requirement is to identify the optimal mathematical model that accurately links the inputs $\mathbf{x}$ to the noisy measurements $\mathbf{y}$. This involves finding the most suitable model for the unknown mapping function $h(\mathbf{x}, \theta)$ in (4.1). Constructing an accurate model using available data sources, such as measurements, observations, and databases, is a key aspect of machine learning. A reliable model, if accurately built, can predict output results for new input datasets, encompassing both extrapolation and interpolation.

In certain cases, we may have insight into the form of $h(\mathbf{x}, \theta)$ based on the nature of the problem at hand. For example, knowing that the growth rate of certain bacteria in a laboratory follows an exponential pattern allows us to anticipate the form of the function. However, the challenge lies in determining the optimal parameters that align with the available measurements.

Conversely, in many cases, the mathematical form of the function is unknown. Consider the scenario of collecting data on used car sale prices, including variables such as car model, date, and mileage. In such cases, one may hypothesize a certain form, such as an nth order polynomial, and then compute the optimal parameters to align with the measurements—a process commonly referred to as *curve fitting*.

A crucial question arises: how can we assess the adequacy of the proposed mathematical form? This evaluation can be achieved by partitioning the available measurements into two sets. The first set, known as the training set, is employed to compute the optimal parameters.

The second set, known as the validation set, plays a crucial role in testing the validity of the model. The proportion of training and validation data allocation depends on factors such as the overall dataset size and model complexity. Using a more complex model than required can lead to overfitting issues, where the average error in the training phase may be very small (even zero), but in the validation phase experiences higher average errors. This discrepancy occurs because the high complexity allows the model to store noise samples than the actual mapping, a phenomenon termed memorization. In practice, the goal is not to memorize data but to construct an accurate model that closely reflects the real mapping $h(\mathbf{x}, \theta)$ generating the observations.

Conversely, if the proposed model's complexity is significantly lower than the actual model, an underfitting problem arises. For instance, employing a simple linear model for highly nonlinear relationships between inputs and outputs will result in high average errors in both the training and validation phases.

Given that, in most practical cases, the exact mapping function $h(\mathbf{x}, \theta)$ and its parameters θ are unknown, the approach involves proposing a model $\hat{h}(\mathbf{x}, \theta)$ in the hope that it will approximate the true mapping closely. The optimization of model parameters is performed using the input/output measurements in the training set. Once optimized, the model is validated using the validation data. If the average error exceeds the defined requirements, adjustments to the model are necessary, prompting a repetition of the process. This process is illustrated in Figure 4.3. The estimation problem could be formulated as the best parameters of the model assumption $\hat{h}(\mathbf{x}, \theta)$ that minimizes the error. The error is defined as some norm of the difference between the actual observations and the model output. In the same context of this chapter, let's build the analytical formula to find the best parameters of the suggested model. We have input-output pairs $\{\mathbf{x}_i, y_i\}$, where $\mathbf{x}_i$ is the ith input vector, y_i is the ith corresponding output, and $i = 1, \ldots, N$ is the number of available measurements. Without loss of generality, we have assumed a single output. We may define the general log-likelihood function as:

$$L(\mathbf{x}; \theta) = \log[f_{\mathbf{X}|Y}(\mathbf{x} \mid y_i)]. \tag{4.79}$$

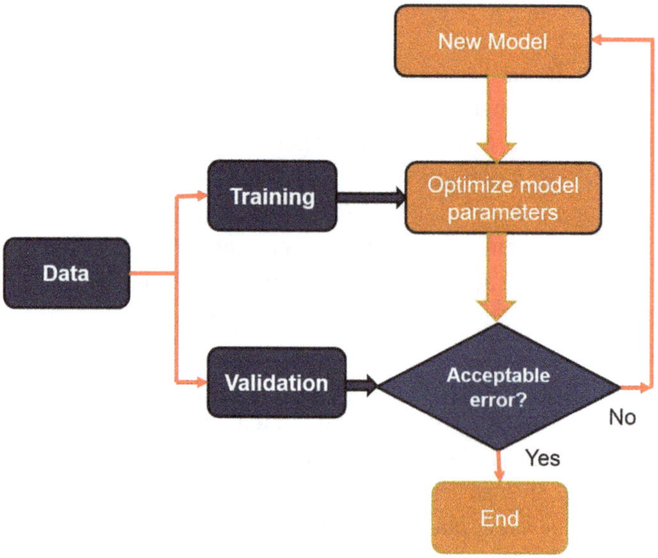

Figure 4.3: Regression process.

We assume that the inputs $\mathbf{x}_i$ represent known samples drawn from a random process $f_{\mathbf{X}}(\mathbf{x})$, and θ denotes the set of parameters for the proposed model $\hat{h}(\mathbf{x}, \theta)$. A straightforward and justifiable assumption is that the noise samples are independent and identically distributed as a zero-mean normal random process. In situations where no prior information is available regarding the random input process, we can formulate the likelihood function as follows:

$$L(\mathbf{x};\theta) = -N\log(\sqrt{2\pi}\sigma) - \frac{1}{2\sigma^2}\sum_{i=1}^{N}[y_i - \hat{h}(\mathbf{x}_i \mid \theta)]^2. \tag{4.80}$$

The first term is independent of the θ parameter. Therefore, to maximize the likelihood function, we aim to minimize the second term, which represents the sum of squared errors. Consequently, the optimal parameters, obtained by minimizing the least square error, correspond to the maximum likelihood (ML) estimate.

Therefore, we can find the optimum parameters by minimizing the sum of error squares formula such as:

$$\theta_{\text{opt}} = \arg\min\sum_{i=1}^{N}[y_i - \hat{h}(\mathbf{x}_i \mid \theta)]^2. \tag{4.81}$$

The next step is to derive the optimal parameters for the linear equations. However, before doing so, it's crucial to highlight the initial step in addressing the inherent uncertainty of available data: data preprocessing. This step is essential for refining the data by eliminating redundancy, linearly related inputs, estimating or correcting partially missing data, and reducing noise using techniques such as projection methods or the removal of rare and anomalous data (e. g., caused by spiky noise). There are both systematic and unsystematic (heuristic) data preprocessing approaches.

A widely employed systematic data preprocessing technique is Principal Component Analysis (PCA). However, for the purpose of this discussion, we assume that the available data has already undergone preprocessing and is now clean.

4.5.1 Optimal linear model regression

In a linear regression with $\mathbf{x}_i = [x_{i,1},\dots,x_{i,M}]$ we may define the M-order linear equations as

$$\hat{y}_i = \hat{h}(\mathbf{x}_i \mid \theta) = a_M x_{i,M} + a_{M-1}x_{i,M-1} + \cdots + a_1 x_1 + a_0, \tag{4.82}$$

where the parameter set is $\theta = \{a_M, a_{M-1},\dots,a_1,a_0\}$. In linear regression, each parameter a_k signifies the magnitude and direction of the influence of each input x_k on the model's output. With a total of N measurement points, the model can be expressed as follows:

$$\begin{bmatrix}\hat{y}_1\\ \hat{y}_2\\ \vdots\\ \hat{y}_N\end{bmatrix} = \begin{bmatrix} x_{1,1} & x_{1,2} & \cdots & x_{1,M}\\ x_{2,1} & x_{2,2} & \cdots & x_{2,M}\\ \vdots & \vdots & \cdots & \vdots\\ x_{N,1} & x_{N,2} & \cdots & x_{N,M}\end{bmatrix}\begin{bmatrix}a_1\\ a_2\\ \vdots\\ a_M\end{bmatrix} + \begin{bmatrix}1\\ 1\\ \vdots\\ 1\end{bmatrix}a_0. \tag{4.83}$$

We can augment the second term of the Ones vector in the main matrix as follows:

$$\begin{bmatrix} \hat{y}_1 \\ \hat{y}_2 \\ \vdots \\ \hat{y}_N \end{bmatrix} = \begin{bmatrix} x_{1,1} & x_{1,2} & \dots & x_{1,M} & 1 \\ x_{2,1} & x_{2,2} & \dots & x_{2,M} & 1 \\ \vdots & \vdots & \dots & \vdots & 1 \\ x_{N,1} & x_{N,2} & \dots & x_{N,M} & 1 \end{bmatrix} \begin{bmatrix} a_1 \\ a_2 \\ \vdots \\ a_M \\ a_0 \end{bmatrix}. \tag{4.84}$$

The previous set of equations could be represented in more compact form as

$$\hat{\mathbf{y}} = \mathbf{X}\mathbf{a}. \tag{4.85}$$

The optimization formula in (4.81) can be reformulated as:

$$\theta_{\text{opt}} = \arg\min\{(\mathbf{y} - \hat{\mathbf{y}})^T(\mathbf{y} - \hat{\mathbf{y}})\} \tag{4.86}$$

$$= \arg\min\{(\mathbf{y} - \mathbf{X}\mathbf{a})^T(\mathbf{y} - \mathbf{X}\mathbf{a})\}. \tag{4.87}$$

Differentiating the above equation with respect to the parameter vector $\mathbf{a}$ as

$$\frac{d}{d\mathbf{a}}[(\mathbf{y} - \mathbf{X}\mathbf{a})^T(\mathbf{y} - \mathbf{X}\mathbf{a})]_{\mathbf{a}=\mathbf{a}_{\text{opt}}} = 0 \tag{4.88}$$

$$\Rightarrow \frac{d}{d\mathbf{a}}[\mathbf{y}^T\mathbf{y} - \mathbf{y}^T\mathbf{X}\mathbf{a} - \mathbf{a}^T\mathbf{X}^T\mathbf{y} + \mathbf{a}^T\mathbf{X}^T\mathbf{X}\mathbf{a}]_{\mathbf{a}=\mathbf{a}_{\text{opt}}} = 0 \tag{4.89}$$

$$\Rightarrow -2\mathbf{X}^T\mathbf{y} + 2\mathbf{X}^T\mathbf{X}\mathbf{a}_{\text{opt}} = \mathbf{0}. \tag{4.90}$$

Therefore, the optimum parameters based on the available measurements are

$$\mathbf{a}_{\text{opt}} = (\mathbf{X}^T\mathbf{X})^{-1}\mathbf{X}^T\mathbf{y}. \tag{4.91}$$

Note that the formula above provides the optimal parameters for the linear regression model, subject to the necessary condition that the rank of the square matrix is satisfied, i. e., $\text{rank}(\mathbf{X}^T\mathbf{X}) = M$. To fulfill this requirement, the number of linearly independent training samples must exceed the number of parameters, denoted by M.

However, a numerical challenge may arise during matrix inversion, even in cases where the matrix is full rank but ill-conditioned. This situation can result in significant numerical errors when performing matrix inversion. An indicator used to assess the ill-conditioning of matrices is known as the condition number. The condition number of a square matrix is defined as the absolute value of the maximum singular value divided by the minimum singular value. Matrices characterized by low condition numbers are termed well-conditioned matrices, indicating stability and reliable numerical behavior. Conversely, matrices with large condition numbers are labeled as ill-conditioned. In the case of matrices with significantly large condition numbers (much larger than 1), the inversion process is susceptible to substantial numerical errors. Singular matrices, which lack invertibility without preprocessing, exhibit infinite condition numbers. Despite the challenges posed by ill-conditioned matrices, there are robust methods for matrix inversion that strike a balance between accuracy and stability. Interested readers can delve deeper into this topic by consulting specialized books on matrix calculus.

If the number of training samples is fewer than the number of parameters, the computation of the optimal parameters becomes:

$$\mathbf{a}_{\text{opt}} = \mathbf{X}^T(\mathbf{X}\mathbf{X}^T)^{-1}\mathbf{y}. \tag{4.92}$$

In both Octave and Scilab, you can compute the optimal inverse of the matrix **X** by using the command *pinv*, which means *pseudo-inverse*.

Example 118. A car dealer collected some data about the used car business. For a particular car type, they collected the information presented in Table 4.1.

Table 4.1: Table for Example 118.

Mileage	Age	Status	Gear	Price
50	8	1	A	8000
120	7	3	M	10500
636	12	2	M	4000
437	14	2	A	4750
383	16	2	M	5000
443	6	1	M	8200
477	3	3	A	28200

The data attributes include mileage (in 1000 s of km), age (in years at the selling time), status (rated as 3 for excellent, 2 for good, 1 for acceptable, and 0 for some damage), gear (A for automatic, M for manual), and price (in Euros)—representing the selling price that serves as the model output. The objective is to develop a mathematical model to predict the selling price based on these attributes. Although various models can be employed, we will start with a linear model for simplicity. Linear models are advantageous due to their low computational requirements, ease of generalization, and straightforward interpretability. Despite its simplicity, it is not guaranteed that a linear model is the optimal choice for expressing this relationship. Nevertheless, starting with a linear model provides a basis for assessing its validity. To facilitate modeling, the non-numeric Gear column can be mapped to real numbers, with 1 representing automatic gears and −1 representing manual gears. It's crucial to note that dividing the data into training and validation sets is a standard practice. For this demonstration, the first five samples will constitute the training set, while the last two samples will be used for validation. This division helps evaluate the model's performance on new, unseen data and ensures its reliability. Nevertheless, in a real-world scenario, a significantly larger dataset would be essential. Table 4.1 shows that:

$$\mathbf{X} = \begin{bmatrix} 50 & 8 & 1 & 1 & 1 \\ 120 & 7 & 3 & -1 & 1 \\ 636 & 12 & 2 & -1 & 1 \\ 437 & 14 & 2 & 1 & 1 \\ 383 & 16 & 2 & -1 & 1 \end{bmatrix}. \tag{4.93}$$

The desired output is $\mathbf{y} = [8000, 10500, 4000, 4750, 5000]^T$. Calculating the optimal linear parameters is a straightforward process, since:

$$\mathbf{a}_{\text{opt}} = (\mathbf{X}^T\mathbf{X})^{-1}\mathbf{X}^T\mathbf{y} = [-7.8, -244.5, 1243, -158.4, 9262.3]^T. \tag{4.94}$$

As anticipated, the mileage parameter exhibits a negative sign, reflecting the expected inverse relationship with the price, a pattern similar to that observed for the age parameter. The status parameter aligns positively with the car price, indicating a proportional relationship. However, directly quantifying the relative importance of each attribute based on the parameter values proves challenging due to the disparate ranges and units of these attributes. The optimal offset found is $a_0 = 9262.3$. Notably, the Gear parameter is found to be negative, suggesting that, according to our training data, manual gear positively influences the selling price—a somewhat uncommon scenario, but consistent with the available samples. To assess the accuracy of the model, validation data, represented by the last two samples in the table, can be employed (Table 4.1). The expected price for the 6th sample is:

$$\hat{y}_6 = 443(-7.8) + 6(-244.5) + 1243 - (-158.4) + 9262.3 = 5733. \tag{4.95}$$

Similarly, computing the value for the last sample yields 8370 Euros. It's evident that our model has significantly underestimated the selling prices for the last two samples, particularly in the case of the final sample where the disparity is substantial. Consequently, based on the validation data, it becomes apparent that this model struggles to effectively capture the intricate relationship between the attributes and the price of such used cars. The model's performance does not indicate a thorough understanding of the underlying data.

While one might suggest incorporating the last two samples in the training set to enhance the model's accuracy, a critical question arises: How can we effectively assess or validate the accuracy of the updated model in such a scenario?

The inability to model the data in the previous example could stem from various factors. The following are some possible explanations for this failure:

- The inadequate size of the dataset may have hindered an accurate representation of the true relationship or mapping between attributes and price.
- Key attributes that are crucial to determining the price might be absent, such as the car's engine type (diesel, gasoline, electric, gas), color, advanced electronics like

driving or parking assist, and whether it's a dealer-owned or privately-owned car, among other factors.
- The relationship is nonlinear, leading to a considerable modeling error. For instance, the connection could be exponential in relation to the car status. Alternatively, the price may be significantly influenced by multiplying both the status and age of the car, rather than considering each factor separately!
- Certain data samples exhibit bias, particularly if there is inaccurate or erroneous information about the sales price is present. Such discrepancies can significantly impact the accuracy of the model.

While it is theoretically possible to use any nonlinear analytical model for regression, doing so is highly undesirable for several reasons, including:
- Nonlinear models often demonstrate nonconvexity, which presents a challenge and, in certain cases (especially for high-dimensional problems), rendering it difficult or even impossible to efficiently identify the global optimal parameters using conventional computing devices and within reasonable time.
- Nonlinear models lack a consistent approach to problem solving. Therefore, a unique solution should be identified for each specific problem.
- Nonlinear models can be very complex, making it difficult to extend, generalize, or interpret the results.

Therefore, it is highly preferable to employ linear models whenever feasible. However, some exceptionally complex relationships cannot be accurately modeled by any linear approach. In such cases, black-box nonlinear models such as neural networks are used to skilfully learn the intricate nonlinear connections between inputs and outputs. Additionally, there is a clever strategy to address certain nonlinear relationships using linear models, albeit at the expense of expanding the input dimension of the problem.

4.5.2 Optimal nonlinear model with linear regression

Let's illustrate this approach by considering a dataset with two attributes, x_1 and x_2, and one output, y. One option is to use the model:

$$\hat{y} = a_2 x_2 + a_1 x_1 + a_0.$$

The optimal parameters are determined from available observations, as previously demonstrated. However, in many cases, this linear model may fail to capture the relationship with acceptable accuracy due to its inherent nonlinearity. In such cases, the regression error remains high no matter how abundant our data samples are. Nevertheless, in practical scenarios, it is still possible to address the non-linear relationship using the same linear model. For example,

$$\hat{y} = a_3 x_1 x_2 + a_2 x_2 + a_1 x_1 + a_0.$$

Despite this, we maintain linearity in the parameters, and we can employ the same approach as in (4.91) to solve it. However, the model is nonlinear in the inputs. This concept can be extended to accommodate more complex shapes, such as:

$$\hat{y} = a_6 e^{-\frac{x_2}{x_1}} + a_5 x_1^2 + a_4 x_2^2 + a_3 x_1 x_2 + a_2 x_2 + a_1 x_1 + a_0.$$

Regardless of the relationship's complexity, it remains solvable as a system of linear equations because it retains linearity in the parameters. Nevertheless, we are now faced with a higher dimensional problem. For an M-input problem $\mathbf{x}$, we can generalize the linear regression model as:

$$\hat{y}_i = \sum_{k=0}^{L} a_k \Psi_k(\mathbf{x}_i). \tag{4.96}$$

A notable advantage of this formulation is its convex nature. Consequently, the parameters acquired through methods like (4.91) ensure the global optimum. Additionally, there are robust, efficient, and reliable mathematical techniques for solving linear equations.

Example 119. In this example, we have a single input and a single output sample of certain measurements, as shown in Table 4.2. Find the optimal linear regression to model the data, $\hat{y} = a_1 x + a_0$. Use the first 7 samples for training and the last 3 for validation.

Table 4.2: Table for Example 119.

x	y
−1.55	−8.53
2.69	17.80
0.10	5.59
−0.32	3.56
−0.13	4.77
−1.08	−1.69
1.11	7.63
−0.82	0.19
−2.29	−24.31
1.56	8.43

Solution, we can find the optimal parameters using (4.91) by constructing matrix $\mathbf{X}$ and vector $\mathbf{y}$ as

$$\mathbf{X} = \begin{bmatrix} -1.55 & 1 \\ 2.69 & 1 \\ 0.10 & 1 \\ -0.32 & 1 \\ -0.13 & 1 \\ -1.08 & 1 \\ 1.11 & 1 \end{bmatrix}. \tag{4.97}$$

The desired output vector is $\mathbf{y} = [-8.53, 17.80, 5.59, 3.56, 4.77, -1.69, 7.63]^T$. Computing the optimal linear parameters is a straightforward task, since:

$$\mathbf{a}_{\text{opt}} = (\mathbf{X}^T\mathbf{X})^{-1}\mathbf{X}^T\mathbf{y} = [5.53, 3.51]^T. \tag{4.98}$$

Figure 4.4 displays both the original measurements and the linear model optimized by training on the available data. However, it's evident that the model doesn't perform well at certain points. For instance, at the validation point −2.29, the model predicts −9.15, while the actual measurement is −24.31. An alternative is to explore a higher order linear model, such as:

$$\hat{y} = a_2x^2 + a_1x + a_0.$$

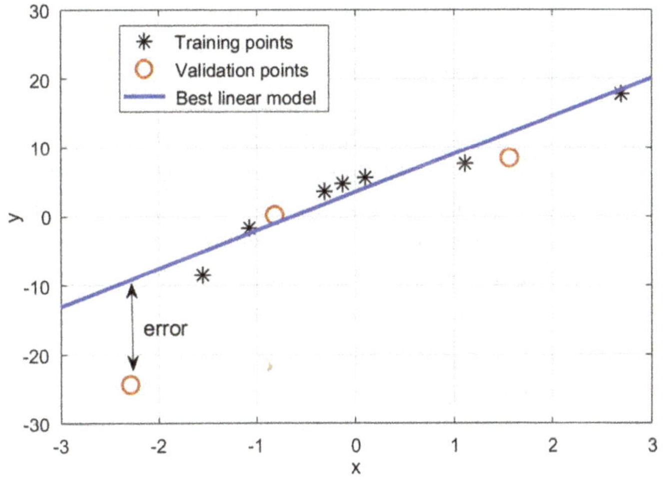

Figure 4.4: Best linear regression.

Despite the need for squared input values x_i, the model remains linear in parameters. In this case, the matrix **X** becomes:

$$\mathbf{X} = \begin{bmatrix} 2.41 & -1.55 & 1 \\ 7.23 & 2.69 & 1 \\ 0.01 & 0.10 & 1 \\ 0.1 & -0.32 & 1 \\ 0.02 & -0.13 & 1 \\ 1.16 & -1.08 & 1 \\ 1.23 & 1.11 & 1 \end{bmatrix}. \tag{4.99}$$

The vector $\mathbf{y}$ is the same. Solving the system of linear equations in (4.91), we get the following set of optimal parameters

$$\mathbf{a}_{\text{opt}} = (\mathbf{X}^T\mathbf{X})^{-1}\mathbf{X}^T\mathbf{y} = [-0.61, 6.27, 4.5]^T. \tag{4.100}$$

The updated model is now expressed as:

$$\hat{y} = -0.61x^2 + 6.27x + 4.5.$$

We can iterate through the same regression process for higher order linear equations, such as the following third order equation:

$$\hat{y} = a_3x^3 + a_2x^2 + a_1x + a_0$$

or the forth order equation:

$$\hat{y} = a_4x^4 + a_3x^3 + a_2x^2 + a_1x + a_0$$

or perhaps higher. However, there's a risk of overfitting if the model's complexity is greater than necessary. The model's accuracy should be assessed based on validation (test) data. Once the model complexity is determined, it's crucial to repeat the regression using all available data (training + validation). Figure 4.5 illustrates the modeling results for the same problem with different levels of complexity. It appears that the third order model provides the best fit for the given training sequence.

Example 120. Consider a database containing N samples of single-input, single-output measurements. Below is an Octave code that reads the data, allocates 70 % for training, and reserves the remaining 30 % for validation. The code iterates through polynomial orders from the first to the fifteenth and uses the validation data to identify the best model. Finally, it returns the optimal parameters of the best model.

```
clear all
load Data1 % Data1 is data saved in [xi y] format
N=length(xi);
K=15; % maximum modelling order
M=fix(0.7*N); %data length of training
a1=xi(1:M); %for training
```

```
y1=y(1:M); %for training
a2=xi(M+1:N); %for validation
y2=y(M+1:N); %for validation
L=N-M-1; %Length of validating data
for i=1:K,
z(:,i)=a1.^i;
A=[z ones(size(a1))];
x{i}=pinv(A)*y1;
err_sum(i)=sum((y1-A*x{i}).^2);
end
% Validation Part
for i=1:K,
z2(:,i)=a2.^i;
A1=[z2 ones(size(a2))];
err_sum2(i)=sum((y2-A1*x{i}).^2)/L;
end
Best_Ord=find(err_sum2==min(err_sum2));
Parameters=x{Best_Ord};
```

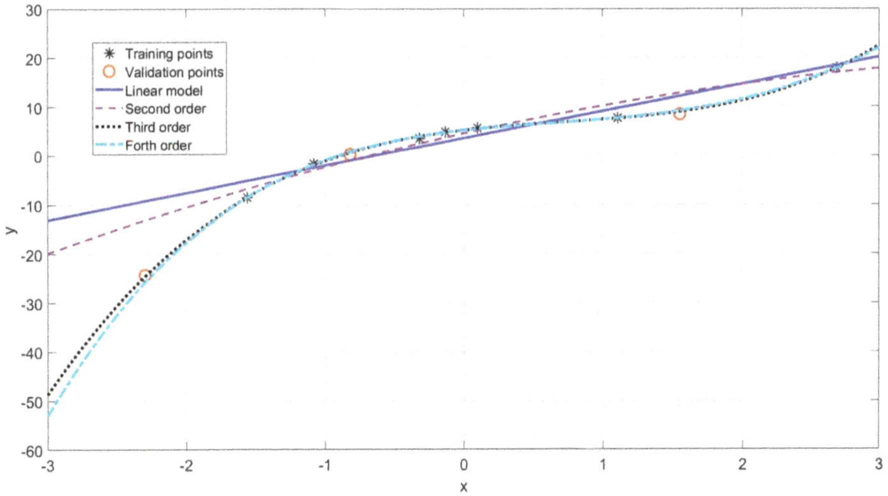

Figure 4.5: Regression with different orders.

Note that as model complexity increases, there is a tendency for a minimal average error during the training phase, possibly even reaching zero when the model's order matches the number of training samples. However, this seemingly impressive performance doesn't necessarily indicate a robust model. In such cases, the model may have effectively memorized the available training data without truly understanding the underlying relationship between input and output variables. This phenomenon is commonly referred to as overfitting. To address this issue, it is crucial to employ an independent validation set for evaluating the model's accuracy in representing the actual unknown mapping. Relying solely on the training phase results can lead to a misleading assessment of the model's generalization capabilities. The relationship between average errors on the training and validation sets is illustrated in Figure 4.6.

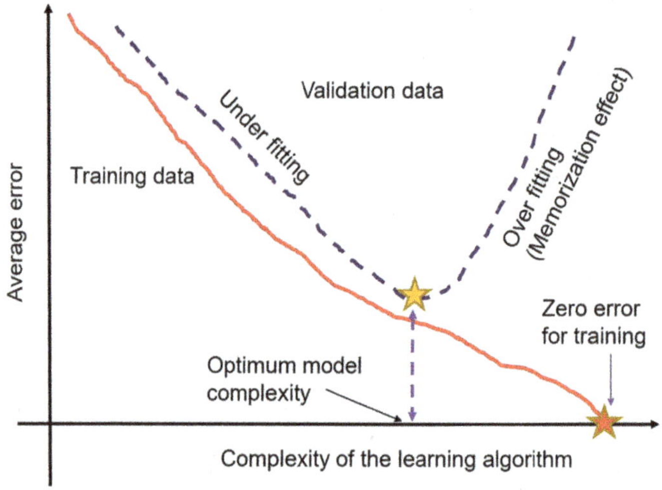

Figure 4.6: Training vs. validation vs. algorithm complexity.

Some nonlinear relationships can be reformulated as a set of linear equations without increasing the input dimensions. For example, if certain measurements could be represented as

$$y_i = a_0 e^{a_1 x_i}.$$

It is necessary to find the optimal parameters a_0 and a_1 based on the available measurements. Taking the natural logarithm of the relation, we obtain

$$\log(y_i) = \log(a_0) + a_1 x_i.$$

This could be defined as:

$$\hat{y}_i = \hat{a}_0 + a_1 x_i,$$

where $\hat{y}_i = \log(y_i)$ and $\hat{a}_0 = \log(a_0)$. Subsequently, we can apply the fundamental form of first order linear regression by taking the logarithm of the output values y_i.

4.6 Concepts of classification

In the previous section, we introduced several concepts related to the regression process, emphasizing its effectiveness as a method for constructing a mathematical model from data obtained through measurements or observations. A desirable model is characterized by its ability to minimize uncertainties in the data by consolidating diverse measurements into a deterministic analytical framework. However, it's important to note that not all data outputs are numerical. Numerous applications involve categorical data organized into finite classes. Examples of such applications include:
- Spam email or not a spam.
- Fraud or trustworthy transaction or customer.
- Faulty or healthy device.
- Based on the diagnosis, the patient may have disease A or B or C.
- The location of the smart grid fault is in part A or B or C.
- Any other application where the output needs to be categorized into two or more finite classes.

Classification is often treated as a special case of regression. While data with linearly separable classes is a possibility, it tends to be uncommon in practical scenarios. In the majority of real-world problems, data exhibit nonlinear class distributions. This prevalence can be attributed to at least three key reasons:
- The inherent nature of the problem is nonlinearly separable.
- Some crucial attributes may be missing from the dataset.
- Presence of noise or biased data in the dataset.

Figure 4.7 shows a simple example of linearly separable and non-separable two-dimensional data.

There are at least two methods for addressing uncertainty in classification. One approach involves leveraging the statistical information or assumptions inherent in the available data. The alternative method relies solely on the actual available data, without regard to their statistical properties. In the following sections, we will explore both of these approaches.

4.6.1 Stochastic classification

Consider each set of data within distinct classes as outcomes generated by distinct random processes. These processes can be described by their respective statistical characteristics, which are determined by a two-step approach. Firstly, identify the overall structure of the probability distribution for each class. Secondly, employ the existing data along with their corresponding labels (indicating class assignments for each data input) to estimate the parameters of each class's distribution, typically

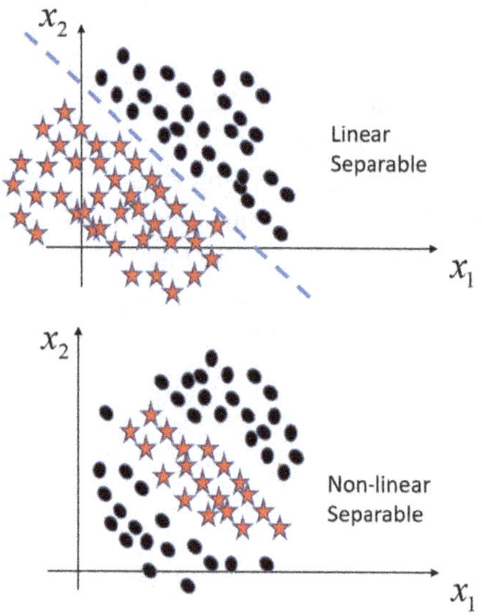

Figure 4.7: Linear vs. non-linear separable classes.

through methods such as maximum likelihood estimation (MLE) in the learning process.

The determination of the probability distribution for each class can be accomplished by three main methods:

- Determine the distribution by relying on the inherent data characteristics, such as employing the histogram method of estimation.
- Using existing prior knowledge regarding the process, for instance, based on certain physical properties.
- Based on assumptions, particularly in the context of multivariate normal distribution where the general assumption is justified in many cases. This assertion is easily supported by leveraging the distinctive properties of the normal distribution, such as the central limit theorem.

The learning process involves converting existing data (from databases) into a collection of probability distributions and their corresponding parameters, with the number of distributions corresponding to the number of data classes. After completing the learning process, it is possible to assign new data inputs to the class that maximizes the probability of inclusion. The new data input $\mathbf{x}_i$ will be assigned to class j (assuming equal risk across all classes) when:

$$P(C_j \mid \mathbf{x}_i) \geq P(C_k \mid \mathbf{x}_i) \quad \forall k = 1, \ldots, K$$

Where, K is the total number of classes.

Assuming that the data in each class follows a multivariate normal distribution

$$f_{\mathbf{X}|C}(\mathbf{x} \mid C_i) = \frac{1}{(2\pi)^{n/2}\det(\mathbf{R}_{XX,i})^{0.5}} e^{-\frac{1}{2}(\mathbf{x}-\mu_{\mathbf{X}_i})^T \mathbf{R}_{XX,i}^{-1}(\mathbf{x}-\mu_{\mathbf{X}_i})}, \tag{4.101}$$

where $\mathbf{R}_{XX,i}$ is the covariance matrix computed from all data that belong to the ith class, $\mu_{\mathbf{X}_i}$ is the mean vector of the ith class.

From the Bayes formula,

$$P(C_j \mid \mathbf{x}_i) = \frac{f_{\mathbf{X}|C_j}(\mathbf{x}_i \mid C_j)P(C_j)}{f_{\mathbf{X}}(\mathbf{x}_i)}. \tag{4.102}$$

It is evident that $f_{\mathbf{X}}(\mathbf{x}_i)$ represents the overall distribution of the input data, irrespective of its classes. Consequently, it exerts no influence on the classification decision and can be excluded. Therefore, the log-likelihood formula for the ith input belonging to the jth class is as follows:

$$L(\mathbf{x}_i, C_j) = \log(f_{\mathbf{X}|C_j}(\mathbf{x}_i \mid C_j)) + \log(P(C_j)). \tag{4.103}$$

Thus, maximizing the probability in (4.102) leads to maximizing the log-likelihood formula (4.103). The uncertainty associated with classified data can be addressed through a learning process that follows three key steps based on available historical data:

1. Define the likelihood function $f_{\mathbf{X}|C_j}(\mathbf{x}_i \mid C_j)$ for each class, as discussed earlier.
2. Utilize the training set from the available database to estimate the parameters of the likelihood function. For instance, if the likelihood function is modeled as a multivariate normal distribution, employ the data for estimating the mean vector and covariance matrix of each class, which is often achieved by methods like maximum likelihood (ML) estimation.
3. Estimate the probability of each class $P(C_j)$. This is calculated as the ratio of the number of inputs belonging to class j to the total number of inputs.

As mentioned earlier in this section, we assume that our database has undergone preprocessing. Once the aforementioned three steps are completed, the accuracy of the learned algorithm can be assessed using the validation set. If classification errors are high during the validation test, several factors could contribute to this outcome. Potential reasons include an insufficient training set size to capture all hidden relationships, missing key attributes in the data, significant bias and noise, or an inadequacy in the defined likelihood function to represent the uncertainty of the problem. In such cases, addressing the accuracy issue can involve various methods, such as increasing the size of the training set, searching for additional relevant attributes, or defining alternative likelihood functions. The log-likelihood function of the multivariate normal distribution can be easily derived from (4.101) as shown (disregarding the common factor function $f_{\mathbf{X}}(\mathbf{x}_i)$):

$$L(\mathbf{x}_i, C_j) = \log(\det(\mathbf{R}_{XX,j})) - (\mathbf{x}_i - \mu_{\mathbf{X}_j})^T \mathbf{R}_{XX,j}^{-1}(\mathbf{x}_i - \mu_{\mathbf{X}_j}) + \log(P(C_j)). \tag{4.104}$$

The optimal class to assign to the input data is determined by maximizing (4.104). Generally, the function used to classify the inputs is referred to as the *discriminant function*. To comprehend the functioning of the stochastic classification algorithm, let's delve into a straightforward numerical example. We will solve it meticulously, step by step.

Example 121. Consider a database from specific hospitals containing information about three types of diseases and associated attributes, as illustrated in Table 4.3. The table includes the patient's age, gender, and hemoglobin levels in g/dl. We have assumed the existence of three distinct diseases (classes). For numerical computations, we'll represent M as +1 and F as −1. All data presented here are used for training purposes.

Table 4.3: Table for Example 121.

Age [Years]	HB [g/dl]	Gender	Disease Class
52	8	M	A
74	12	M	C
24	15.5	F	A
48	11	M	B
19	17.5	F	B
82	10	F	A
33	9	M	C
28	13	M	B
66	15	F	A
16	9	F	C
90	12.5	M	C
56	14	F	B

Given the presence of three classes, we need to formulate three log-likelihood functions. Assuming a multivariate normal distribution, the task involves estimating the mean and covariance matrix for each class. Commencing with Class A, the initial matrix is as follows:

$$\mathbf{M}_A = \begin{bmatrix} 52 & 8 & 1 \\ 24 & 15.5 & -1 \\ 82 & 10 & -1 \\ 66 & 15 & -1 \end{bmatrix}. \tag{4.105}$$

Therefore, from this matrix we can easily calculate the maximum likelihood (ML) estimates for the mean vector and the covariance matrix:

$$\mu_A = [56,\ 12.125,\ -0.5]^T \tag{4.106}$$

$$\mathbf{R}_{XX,A} = \begin{bmatrix} 605.333 & -39.333 & -2.667 \\ -39.333 & 13.729 & -2.750 \\ -2.667 & -2.750 & 1.000 \end{bmatrix}. \tag{4.107}$$

The inverse of the covariance matrix is:

$$\mathbf{R}_{XX,A}^{-1} = \begin{bmatrix} 0.004 & 0.031 & 0.096 \\ 0.031 & 0.396 & 1.171 \\ 0.096 & 1.171 & 4.475 \end{bmatrix}. \tag{4.108}$$

Furthermore, the determinant of the covariance matrix is:

$$\det(\mathbf{R}_{XX,A}) = 1511. \tag{4.109}$$

The probability of class A is $P(C_A) = \frac{4}{12}$. Since all classes have the same probability, we can drop them from the likelihood function. Therefore, the log-likelihood based discriminant function of class A is formulated as:

$$L(\mathbf{x}_i, C_A) = 7.32 - (\mathbf{x}_i - \mu_A)^T \begin{bmatrix} 0.004 & 0.031 & 0.096 \\ 0.031 & 0.396 & 1.171 \\ 0.096 & 1.171 & 4.475 \end{bmatrix} (\mathbf{x}_i - \mu_A). \tag{4.110}$$

In the same way, we should construct the log-likelihood based discriminant functions for classes B and C. The results are:

$$\mu_B = [37.750,\ 13.875,\ 0]^T \tag{4.111}$$

$$\mathbf{R}_{XX,B} = \begin{bmatrix} 294.917 & -28.875 & 0.333 \\ -28.875 & 7.395 & -2.500 \\ 0.333 & -2.500 & 1.333 \end{bmatrix} \tag{4.112}$$

$$\mu_C = [53.250,\ 10.625,\ 0.500]^T \tag{4.113}$$

$$\mathbf{R}_{XX,C} = \begin{bmatrix} 1192.917 & 63.625 & 24.833 \\ 63.625 & 3.563 & 1.083 \\ 24.833 & 1.083 & 1.000 \end{bmatrix}. \tag{4.114}$$

Hence, the other two discriminant functions are

$$L(\mathbf{x}_i, C_B) = -0.523 - (\mathbf{x}_i - \mu_B)^T \begin{bmatrix} 6.093 & 63.562 & 117.656 \\ 63.562 & 663.375 & 1227.937 \\ 117.656 & 1227.937 & 2273.719 \end{bmatrix} (\mathbf{x}_i - \mu_B)$$

$$L(\mathbf{x}_i, C_C) = 3.332 - (\mathbf{x}_i - \mu_C)^T \begin{bmatrix} 0.085 & -1.311 & -0.698 \\ -1.311 & 20.572 & 10.271 \\ -0.698 & 10.271 & 7.198 \end{bmatrix} (\mathbf{x}_i - \mu_C).$$

The above three discriminant functions $L(\mathbf{x}_i, C_A)$, $L(\mathbf{x}_i, C_B)$, and $L(\mathbf{x}_i, C_C)$ are the result of the learning process over the available data. Now suppose we have a new patient record with attributes as $\mathbf{x} = [42,\ 9.6,\ F]$. To automatically classify the patient, we evaluate all three discriminant functions and identify the class associated with the highest value. In this case, the computed values are $L(\mathbf{x}_i, C_A) = -3.24$, $L(\mathbf{x}_i, C_B) = -20976$, and $L(\mathbf{x}_i, C_A) = -26.02$. Consequently, given these outcomes, the new patient is determined to belong to class A disease.

In the aforementioned Example 121, the reliability of the discriminant function for class B appears to be compromised. This is attributed to the ill-conditioned nature of the covariance matrix $\mathbf{R}_{XX,B}$. Confirmation of this condition can be obtained by computing the condition number. To address this issue, it is highly recommended to employ data preprocessing algorithms, such as Principal Component Analysis (PCA), prior to the training phase. These preprocessing techniques alleviate the ill-conditioned state of the matrix by reducing dimensionality and eliminating redundancies or nearly linearly dependent inputs. When the number of input data instances in at least one class is less than the number of attributes, the estimated covariance matrix becomes singular. In such cases, a pragmatic solution is to use a common covariance matrix for all discriminant functions. This matrix can be computed as the weighted sum of the covariance matrices across all classes, expressed as follows:

$$\mathbf{R}_{XX} = \sum_{k=1}^{K} \mathbf{R}_{XX,k} P(C_k). \tag{4.115}$$

In Example 121, the common covariance matrix will be:

$$\mathbf{R}_{XX} = \begin{bmatrix} 697.024 & -1.526 & 7.493 \\ -1.526 & 8.221 & -1.388 \\ 7.493 & -1.388 & 1.110 \end{bmatrix}. \tag{4.116}$$

This matrix has a more reliable inverse:

$$\mathbf{R}_{XX}^{-1} = \begin{bmatrix} 0.002 & -0.002 & -0.013 \\ -0.002 & 0.156 & 0.208 \\ -0.013 & 0.208 & 1.249 \end{bmatrix}. \tag{4.117}$$

Therefore, we can employ (4.104) to calculate the log-likelihood function for each class, assuming a shared covariance inverse. In essence, (4.104) can be conceptualized as the weighted sum of the Euclidean distance (commonly referred to as the *Mahalanobis* distance) between the new item and the mean of the previously learned stochastic zones for each class. The optimal class is determined by selecting the class with the smallest Mahalanobis distance from the tested input.

Example 122. Consider a database with 7 attributes, including income, age, weight, gender, education level, health class, and height. The database encompasses several thousand samples that are classified into 13 distinct classes denoted as [−2, −1, 0, 1, ..., 10]. Each class may represent different outcomes such as trust level, expected age, criminal profile, etc. A partial snapshot of the database is illustrated in Figure 4.8. To address

	A	B	C	D	E	F	G	H
1	Income	Age	Weight	Gender	Educ	Health	Hight	Class
2	7752	41	118	1	4	7	157	10
3	5597	59	85	-1	3	3	136	-1
4	8063	56	100	1	3	10	152	10
5	5314	53	88	-1	0	2	156	-1
6	7406	69	94	1	3	9	115	10
7	10009	35	79	1	4	10	101	10
8	9485	36	19	-1	3	10	133	4
9	10237	49	76	1	3	6	121	10
10	8318	39	37	-1	1	10	183	7
11	6554	76	71	1	1	0	164	9
12	7731	39	90	-1	4	8	189	10
13	4833	85	52	-1	0	5	153	-2
14	7646	68	45	-1	4	7	178	-2
15	9581	73	114	1	1	10	137	10
16	6690	20	79	1	4	10	154	10
17	10955	56	77	1	0	10	160	10
18	9976	31	41	1	5	10	137	10
19	5900	85	95	-1	2	4	173	-2
20	8360	16	61	-1	3	10	184	10
21	12134	40	46	1	1	10	169	10
22	7122	18	55	-1	1	7	104	10
23	13039	78	44	-1	2	10	156	-2
24	11623	44	26	-1	1	10	116	-2
25	[illegible]	44	90	1	5	10	165	10

Database1

Figure 4.8: Database of Example 122.

classification uncertainties, a straightforward Octave code is presented below. This code reads the database in Excel format, uses 5000 samples for training, and employs the subsequent 101 samples for validation. The computed probability of misclassification for the validated data offers insight into the model's performance. Feel free to enhance the accuracy of the model by adjusting the number of training samples.

The provided code is as follows.

```
clear all
% let's read the first 5000 inputs
N=5000;
AR=xlsread('Database2.xlsx','Database1','A2:H5001');
Class=AR(:,8);
s=0;
classNum=-2:10; %Available classes
m=length(classNum);
for i = classNum,
```

```
s=s+1;
D(s)={AR(find(Class==i),1:7)};
Smean(s)={mean(D{s})}; % Mean of each class
Scov(s)={cov(D{s})};   % Covariance of each class
n(s)=length(find(Class==i));
Ps(s)=n(s)/N; % A priori probability of each class
end
% validation set consists of 101 elements
x=xlsread('Database2.xlsx','Database1','A8400:G8500');

% the results of the validation set from the database
xx=xlsread('Database2.xlsx','Database1','H8400:H8500');

K=length(xx);
%building Discrimination Functions
for k=1:K,
s=0;
for i=classNum,
s=s+1;

Mean=Smean{s};
Cov=Scov{s};
g(s)=-0.5*log(det(Cov))-0.5*(x(k,:)-Mean)*pinv(Cov)*
    (x(k,:)-Mean)'+log(Ps(s));
end
% desciding the class of each input k
c=find(g==max(g));
R(k)=classNum(c);
end
% error probability in miss-classification
Pmc=length(find(R!=xx'))/101;
```

Using the computed covariance matrix for each class, it is possible to calculate the correlation between the attributes within each class. This analysis yields profound insights; for instance, it may reveal a stronger positive correlation between age and income in a specific class compared to another class.

In stochastic classification, the assumption is that the uncertainty in the data follows certain statistical properties, represented by specific multivariate distributions. As demonstrated previously, we addressed classification uncertainties using the multivariate normal distribution. However, it is equally feasible to employ alternative distributions that may better capture the data distribution. A notable example is the *multivariate t-distribution.*

4.6.2 Analytical classification

In the previous section, we delved into the realm of stochastic or statistical classification. However, this approach presents notable challenges, including:

- The need to make assumptions about the underlying distribution introduces a potential source of bias. If the actual data does not adhere to these assumed distributions, it can lead to skewed results.
- The computation of parameters for the distribution within each class relies on available data. In cases of a limited sample size, this may result in significant errors, undermining the robustness of the classification model.
- The flexibility of classifiers is constrained, particularly in terms of geometry. Take, for example, classifiers based on multivariate normal distribution, where only two parameters—covariance matrix and mean—shape the classifier for each class. This limitation can restrict the diversity of classification shapes.
- Handling vast datasets may demand substantial computational resources in stochastic classification. In contrast, deterministic methods, such as support vector machines (SVM), offer efficiency by focusing only on data proximal to classification boundaries, making them more resource-friendly for extensive datasets.

In summary, while stochastic classification provides valuable insights, its reliance on distributional assumptions, susceptibility to error with limited data, geometric limitations, and computational demands require a careful consideration of alternative approaches, such as deterministic methods like SVM. Analytical classifiers serve as a crucial alternative in certain applications, offering a straightforward approach to classifying data without relying on assumed distributions or statistical properties. Instead, these classifiers leverage analytical methods to construct discriminant functions that effectively distinguish between classes. The analytical classifier can be either linear or nonlinear.

In contrast to statistical classification, the analytical approach uses optimization methods to determine the parameters of the discriminant functions. In scenarios where the data is linearly separable, the task involves identifying the optimal parameters for a linear hyperplane. This hyperplane is designed to maximize the distance between the nearest points from different classes, ensuring a robust separation. In essence, optimization theory is employed to enhance the margin between points from different classes, resulting in a more reliable classification boundary.

However, the analytical classifier remains versatile, recognizing that real-world problems may not always be perfectly linearly separable due to inherent complexities, noise, or bias. Even in such cases, linear classification can be applied with the objective of minimizing the misclassification of points. This is achieved by strategically adjusting parameters to strike a balance between capturing the essence of the data and mitigating the impact of nonlinearity, noise, or bias, as illustrated in Figure 4.9. Furthermore, in cases where classes exhibit inherent nonlinearity and are challenging to separate, it

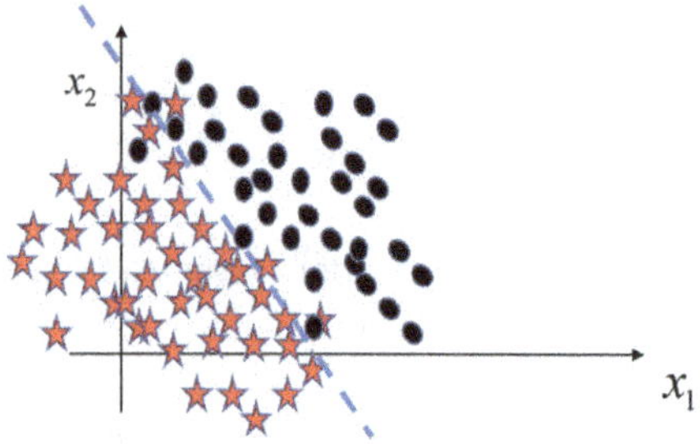

Figure 4.9: Linear separation to minimize errors.

becomes feasible to apply transformations to higher dimensions. This approach aims to reconfigure the data space, transforming it into a domain where the classes become linearly separable—an illustration of which will be presented later. The ultimate goal of the optimal classification algorithm is to minimize uncertainties associated with classifying new inputs, ensuring a robust and reliable categorization process.

In the case of the two-class problem, it is possible to classify inputs with two discriminant functions L_1 and L_2, where $L_1 > L_2$ for inputs belong to class 1 and vice versa. Hence, it would be possible to use one discriminant function as $L = L_1 - L_2$, where $L \geq 0$ for inputs belong to Class 1 and $L < 0$ for Class 2. In case we have $K > 2$ classes, we may build linear machine based on K linear discriminant functions such as, for $\mathbf{x}_n \in$ class k, we have $L_k(\mathbf{x}_n) \geq L_j(\mathbf{x}_n)\ \forall j = 1, 2, \ldots, K$, and $j \neq k$.

Consider a database comprising N samples, characterized by M input attributes, and categorized into $K \geq 2$ distinct classes. Our objective is to identify the optimal set of linear parameters, totaling $M + 1$ for each class. These parameters are instrumental in maximizing the separation distance between each input and the hyperplane boundary that encompasses all classes, ensuring accurate classification within the appropriate zones.

For the kth class, the linear separation hyperplane is determined by the following expression:

$$\begin{aligned} L_k(\mathbf{x}_n) &= a_{M,k}x_{M,n} + a_{M-1,k}x_{M-1,n} + \cdots + a_{1,k}x_{1,n} + a_{0,k} \\ &= \sum_{m=1}^{M} a_{m,k}x_{m,n} + a_{0,k}, \end{aligned} \tag{4.118}$$

where $n = 1, \ldots N$ is the number of samples, $L_k(\mathbf{x}_n)$ is the output of the linear discriminant function of class k for the nth sample, $a_{m,k}$ is the mth parameter of the linear model of the kth class, and $a_{0,n}$ is the model bias or threshold weight. To enhance the distinctiveness between classes, one strategy is to make the assumption that

$$L_k(\mathbf{x}_n) \geq +1. \tag{4.119}$$

This should be true for all inputs belonging to class k, and

$$L_{\hat{k}}(\mathbf{x}_n) \leq -1. \tag{4.120}$$

This should be true for all inputs belonging to the class $\hat{k}$ where $\hat{k}$ is the complement of k, i. e., $\hat{k} \neq k$. Therefore, the distance between different classes should be at least 2. In practical classification scenarios, it is often impractical to achieve perfect separation between classes. Consequently, determining the optimal parameters for the hyperplane requires meticulous construction of the optimization formula. To provide a concise overview, let's begin with the general formulation of the optimization problem before delving into a straightforward numerical example that illustrates key concepts. The general optimization problem, which focuses on minimizing specific cost functions while considering multiple inequalities and equality constraints, is expressed as follows:

$$\begin{aligned}
&\min_{\mathbf{x}\in\Omega} C(\mathbf{x}) \\
&\text{subject to} \\
&\qquad f_1(\mathbf{x}) \leq 0 \\
&\qquad f_2(\mathbf{x}) \leq 0 \\
&\qquad\quad \vdots \\
&\qquad f_M(\mathbf{x}) \leq 0 \\
&\qquad g_1(\mathbf{x}) = 0 \\
&\qquad g_2(\mathbf{x}) = 0 \\
&\qquad\quad \vdots \\
&\qquad g_N(\mathbf{x}) = 0.
\end{aligned} \tag{4.121}$$

Therefore, the task at hand is to identify the optimal vector $\mathbf{x}$, that lies within a specified feasible set Ω, while simultaneously satisfying all M inequalities ($f_i(\mathbf{x})$) and N equality functions ($g_k(\mathbf{x})$). This type of optimization problem is ubiquitous in various disciplines, including engineering, science, and business, with well-established theories for problem solving and analysis. While comprehensive discussions on these theories are beyond the scope here, one approach to convert constraint optimization problems into unconstrained ones is through the Lagrange multiplier method, as follows:

$$\min_{\mathbf{x}\in\Omega} L(\mathbf{x}, \lambda, \mu) = C(\mathbf{x}) + \sum_{m=1}^{M} \lambda_m f_m(\mathbf{x}) + \sum_{n=1}^{N} \mu_n g_n(\mathbf{x}). \tag{4.122}$$

Where λ_m and μ_n for all m and n are called Lagrange multipliers. If the cost function $C(\mathbf{x})$ and all constraints are convex functions within the feasible set Ω, there will be a single and unique optimal solution.

A function $f(\mathbf{x})$ is termed *convex* within the feasible set Ω if the following condition holds true for any two vectors $\mathbf{x}_1$ and $\mathbf{x}_2$ belonging to the feasible set Ω:

$$f(\lambda \mathbf{x}_1 + (1-\lambda)\mathbf{x}_2) \leq \lambda f(\mathbf{x}_1) + (1-\lambda) f(\mathbf{x}_2) \tag{4.123}$$

for any $0 \leq \lambda \leq 1$. The concept of convexity is visually illustrated in Figure 4.10 for a one-dimensional problem.

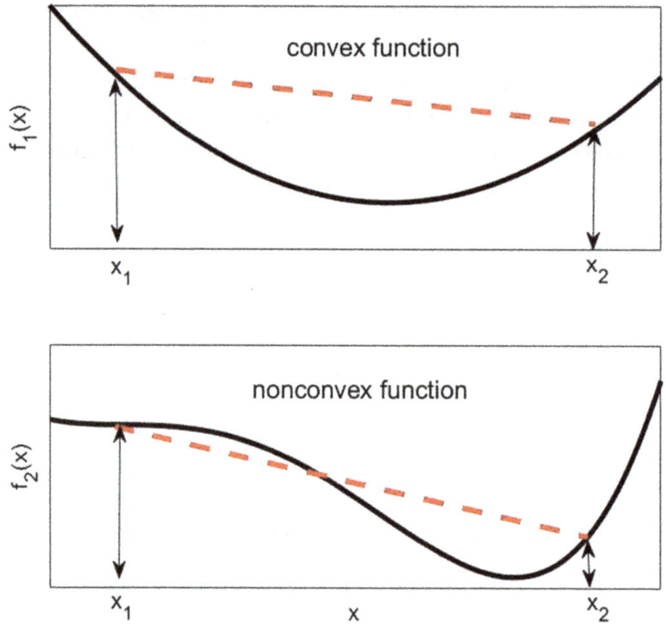

Figure 4.10: Convexity.

When dealing with continuous functions, solving unconstrained optimization involves differentiation with respect to all variables, namely $\mathbf{x}$, λ_m, and μ_n. These conditions are known as the Karush–Kuhn–Tucker (KKT) conditions for local optimality. It's crucial to note that to ensure the validity of the inequality condition, it is necessary to have $\lambda_m \geq 0$. If, by any chance, a particular inequality, say k, has $\lambda_k < 0$, it implies an inactive inequality. In such cases, the inactive inequality can be eliminated from the unconstrained problem by setting $\lambda_k = 0$. However, there are no validity conditions imposed on the Lagrange multipliers associated with the equality constraints. Finally, it is worth mentioning that we can keep the same formulation in (4.121) also for the maximization problem by using the fact:

$$\max C(\mathbf{x}) = \min[-C(\mathbf{x})]. \tag{4.124}$$

It's important to remember that if a constraint is expressed in the form $f_k(\mathbf{x}) \geq 0$, it can be equivalently represented as $-f_k(\mathbf{x}) \leq 0$. These concepts will be demonstrated more in the following simple example.

Example 123. Finding the optimal vector $\mathbf{x} = [x_1, x_2]^T$ satisfies the following optimization problem:

$$\begin{aligned} &\min_{\mathbf{x}\in\mathbb{R}} x_1^2 + x_2^2 \\ &\text{subject to} \\ &\qquad x_1 + x_2 \leq 4 \\ &\qquad 2x_1 + x_2 \geq 6. \end{aligned}$$

Initially, we transform the constrained optimization problem into an unconstrained form as follows:

$$L = x_1^2 + x_2^2 + \lambda_1(x_1 + x_2 - 4) - \lambda_2(2x_1 + x_2 - 6). \tag{4.125}$$

Apply KKT conditions as:

$$\begin{aligned} \frac{\partial L}{\partial x_1} &= 2x_1 + \lambda_1 - 2\lambda_2 = 0 \\ \frac{\partial L}{\partial x_2} &= 2x_2 + \lambda_1 - \lambda_2 = 0 \\ \frac{\partial L}{\partial \lambda_1} &= x_1 + x_2 - 4 = 0 \\ \frac{\partial L}{\partial \lambda_2} &= -2x_1 - x_2 + 6 = 0. \end{aligned}$$

Upon solving the system of linear equations outlined above, we obtain $x_1 = 2$ and $x_2 = 2$, leading to a minimum cost function value of 8. However, a critical step involves scrutinizing the Lagrange multipliers to ascertain their optimality. After solving for the multipliers, we find that $\lambda_1 = -4$ and $\lambda_2 = 0$. This indicates that the first constraint is inactive and can be disregarded. Consequently, let's readdress the problem without considering the first constraint:

$$\begin{aligned} \frac{\partial L}{\partial x_1} &= 2x_1 - 2\lambda_2 = 0 \\ \frac{\partial L}{\partial x_2} &= 2x_2 - \lambda_2 = 0 \\ \frac{\partial L}{\partial \lambda_2} &= -2x_1 - x_2 + 6 = 0. \end{aligned}$$

After solving these equations, we find that $x_1 = 2.4$, $x_2 = 1.2$, and $\lambda_2 = 2.4$. The positive value of λ_2 signifies that the second constraint is active. Substituting this solution into the first constraint yields $x_1 + x_2 = 3.6$, satisfying the initial constraint. Finally, substituting these values into the cost function, we get $x_1^2 + x_2^2 = 7.2$. This outcome is superior to the case where the first constraint was included. This observation is crucial because it demonstrates that by eliminating any inactive constraint (i. e., with its associated Lagrange multiplier $\lambda < 0$), it is possible to achieve better optimization results while still satisfying all active constraints.

Returning to the realm of optimal analytical classifiers, let's introduce an indicator function, denoted as $\xi_{n,k}$. Here, $\xi_{n,k} = +1$ signifies that the nth sample belongs to class k, while $\xi_{n,k} = -1$ indicates that the nth sample belongs to the complementary class $\hat{k}$ (the *k-complement*). Leveraging this indicator function, we can articulate for all input samples that:

$$\xi_{n,k}\left[\sum_{m=1}^{M} a_{m,k}x_{m,n} + a_{0,k}\right] \geq +1, \quad \forall n = 1, 2, \ldots, N. \tag{4.126}$$

The optimal weight vector of each linear model is the one that maximizes the distance margin between the inputs in different classes. Define the weight vector of the kth class $\mathbf{a}_k = [a_{M,k}, a_{M-1,k}, \ldots, a_{1,k}]^T$, and the nth input sample vector as $\mathbf{x}_n = [x_{M,n}, x_{M-1,n}, \ldots, x_{1,n}]^T$. The linear discriminant function becomes:

$$L_k(\mathbf{x}_n) = \mathbf{a}_k^T\mathbf{x}_n + a_{0,k}. \tag{4.127}$$

Hence, (4.126) could be rewritten in more compact form as:

$$\xi_{n,k}\left[\mathbf{a}_k^T\mathbf{x}_n + a_{0,k}\right] \geq +1, \quad \forall n = 1, 2, \ldots, N. \tag{4.128}$$

The perpendicular distance between input $\mathbf{x}_n$ and the kth hyperplane border is given by:

$$\frac{\mathbf{a}_k^T\mathbf{x}_n + a_{0,k}}{\|\mathbf{a}_k\|}. \tag{4.129}$$

Therefore, to maximize the distance we need to minimize the absolute value of the weight vector, i. e., $\|\mathbf{a}_k\|^2$. Hence, the optimization problem can be formulated as:

$$\begin{aligned} &\min \frac{1}{2}\|\mathbf{a}_k\|^2 \\ &\text{subject to} \\ &\quad \xi_{n,k}\left[\mathbf{a}_k^T\mathbf{x}_n + a_{0,k}\right] \geq +1, \quad \forall n = 1, 2, \ldots, N. \end{aligned} \tag{4.130}$$

The optimization problem stated above is convex and can be easily addressed using Lagrangian multipliers. However, a challenge arises when the actual problem is not linearly separable, meaning that the specified constraint cannot be satisfied. In fact, the majority of practical problems do not strictly adhere to linear separability. To address this common scenario and enhance the generality of the optimization problem, we will reframe it in two distinct ways. Initially, let's relax the requirement stipulating that the projection of the input vector $\mathbf{x}_n$ into the kth class must be greater than or equal to +1 when $\mathbf{x}_n$ belongs to it and less than or equal to −1 when it belongs to another class. Instead, if $\mathbf{x}_n \in \{k\}$, then:

$$L_k(\mathbf{x}_n) \geq L_j(\mathbf{x}_n), \quad \forall j = 1, 2, \ldots K, \text{ and } j \neq k. \tag{4.131}$$

The constraint $L_k(\mathbf{x}_n) \geq L_j(\mathbf{x}_n)$ implies that $L_k(\mathbf{x}_n) - L_j(\mathbf{x}_n) \geq 0$, or

$$(\mathbf{a}_k^T - \mathbf{a}_j^T)\mathbf{x}_n + a_{0,k} - a_{0,j} \geq 0 \quad \forall j, k, \text{ and } j \neq k. \tag{4.132}$$

Therefore, the formulation of the optimization problem can be expressed as follows:

$$\begin{aligned} &\min \frac{1}{2}\|\mathbf{a}_k\|^2 \\ &\text{Subject to} \\ &\quad (\mathbf{a}_k^T - \mathbf{a}_j^T)\mathbf{x}_n + a_{0,k} - a_{0,j} \geq 0, \quad \forall n, j = 1, \ldots, K \text{ and } j \neq k. \end{aligned} \tag{4.133}$$

The optimization problem presented above needs to be addressed simultaneously for all classes.

In the second formulation, we introduce a new positive slack variable, denoted as $\alpha_{n,k} \geq 0$, to relax the constraints in (4.130). This adjustment is expressed as:

$$\begin{aligned} &\min \frac{1}{2}\|\mathbf{a}_k\|^2 \\ &\text{Subject to} \\ &\quad \xi_{n,k}[\mathbf{a}_k^T\mathbf{x}_n + a_{0,k}] \geq 1 - \alpha_{n,k}, \quad \forall n = 1, 2, \ldots, N \\ &\quad \alpha_{n,k} \geq 0. \end{aligned} \tag{4.134}$$

Optimal conditions are evident when the slack variable $\alpha_{n,k}$ for each input reaches the value of 0. Additionally, when $0 < \alpha_{n,k} \leq 1$, the input point is situated in proximity to the class border. The most undesirable scenario occurs when $\alpha_{n,k} > 1$, leading to misclassification and incorrect localization of the input in the wrong class, i. e., where $\xi_{n,k} = -1$. Therefore, in the optimization problem formulation, minimizing the occurrence of such erroneous classifications is imperative. Employing the conventional Lagrange multiplier method, the optimization problem can be articulated as follows:

$$\begin{aligned} C_k = \frac{1}{2}\|\mathbf{a}_k\|^2 + \omega_k \sum_{n=1}^{N} \alpha_{n,k} - \sum_{n=1}^{N} \lambda_{k,n}[\xi_{n,k}(\mathbf{a}_k^T\mathbf{x}_n + a_{0,k}) - 1 + \alpha_{n,k}] \\ - \sum_{n=1}^{N} \mu_{k,n}\alpha_{n,k}, \end{aligned} \tag{4.135}$$

where C_k is the unconstrained cost function of class k, the second term in (4.135) is $\sum_{n=1}^{N} \alpha_{n,k}$ represents a penalty term that we want to minimize and ω_k is a compromise term of the kth class between achieving maximum distance separation and minimizing the number of erroneous classifications. Furthermore, $\lambda_{k,n} \geq 0$ is the Lagrange multiplier of the first constraint and $\mu_{k,n}$ is the Lagrange multiplier of the positivity constraint of the slack variable. We can determine the optimal k^{th} weight vector $\hat{\mathbf{a}}_k$ by differentiating C_k with respect to $\mathbf{a}_k$, ω_k, $\lambda_{k,n}$, and $\mu_{k,n}$, and setting them to zero. Remember that $\|\mathbf{a}_k\|^2 = \mathbf{a}_k^T\mathbf{a}_k$, and its derivative $d(\mathbf{a}_k^T\mathbf{a}_k) = 2\mathbf{a}_k^T d(\mathbf{a}_k)$. Therefore the derivatives of the unconstrained cost function are:

$$\frac{\partial C_k}{\partial \mathbf{a}_k} = \mathbf{a}_k^T - \sum_{n=1}^{N} \lambda_{k,n}\xi_{n,k}\mathbf{x}_n^T = \mathbf{0} \tag{4.136}$$

$$\frac{\partial C_k}{\partial a_{0,k}} = \sum_{n=1}^{N} \lambda_{k,n}\xi_{n,k} = 0 \tag{4.137}$$

$$\frac{\partial C_k}{\partial \alpha_{r,k}} = \omega_k - \lambda_{k,r} - \mu_{k,r} = 0 \quad \forall r = 1, \ldots N \tag{4.138}$$

$$\frac{\partial C_k}{\partial \lambda_{k,r}} = \xi_{r,k}(\mathbf{a}_k^T\mathbf{x}_r + a_{0,k}) - 1 + \alpha_{r,k} = 0 \quad \forall r = 1, \ldots N. \tag{4.139}$$

From (4.136) we can formulate the optimal weight vector as

$$\hat{\mathbf{a}}_k = \sum_{n=1}^{N} \lambda_{k,n}\xi_{n,k}\mathbf{x}_n. \tag{4.140}$$

The optimal weight vector is evidently a weighted sum of the inputs. However, we will exclusively focus on the inputs where their Lagrange multiplier $\lambda_{k,n}$ is greater than zero; otherwise, it is disregarded by setting $\lambda_{k,n} = 0$. Another constraint comes into play regarding the values of $\lambda_{k,n}$. Referring to (4.138) and considering that $\mu_{k,r} \geq 0$, the condition becomes $\omega_k \geq \lambda_{k,n} > 0$. Consequently, the inputs $\mathbf{x}_n$ satisfying this condition are termed support vectors. Upon substituting the optimal weight vector back into the unconstrained cost function in (4.135), while taking into account (4.138), we obtain:

$$D_k = -\frac{1}{2}\sum_{n=1}^{N}\sum_{r=1}^{N} \lambda_{k,n}\lambda_{k,r}\xi_{n,k}\xi_{r,k}\mathbf{x}_n^T\mathbf{x}_r + \sum_{n=1}^{N} \lambda_{k,n}. \tag{4.141}$$

Where D_k is the dual form of the original problem (4.135), where we solve for the Lagrange multipliers $\omega_k \geq \lambda_{k,n} > 0$ that maximize the dual problem with the constraint that:

$$\sum_{n=1}^{N} \lambda_{k,n}\xi_{n,k} = 0. \tag{4.142}$$

The classification problem can be effectively addressed by determining the optimal weight vector for each class, which is achieved either by minimizing the primal problem (4.135) or maximizing the dual problem (4.141). An advantage of the dual form lies in its complexity, which depends only on the number of samples N, and is independent of the problem's dimension. Consequently, using different projection methods for higher dimensions does not impact the process of identifying the optimal weight. Once the optimal Lagrange multipliers have been ascertained, they can be seamlessly substituted into (4.140) to obtain the optimal weights. Additionally, numerous numerical packages are available for efficiently solving such quadratic optimization problems.

In Octave, it is possible to use *quadprog()* function and in Scilab we have *qpsolve()* function.

As previously highlighted, a challenge arises when the data is not linearly separable by nature. However, in such cases, it may be feasible to transform the data into another space where linear separability can be achieved. A simple illustration of this concept is presented in Figure 4.11. In this example, one-dimensional data with two classes, denoted by *crosses* and *circles*, lacks a single linear boundary capable of accurately classifying both groups. Nevertheless, transforming the data into two dimensions, such as $[x, x^2]$, allows for straightforward classification using a linear discriminant function, as demonstrated in Figure 4.12. Generally, if different classes in the dataset $\mathbf{x}$ are not linearly separable, it is possible to employ a data transformation $\Psi(\mathbf{x})$ to higher dimensions where different classes exhibit linear separability with minimal classification error.

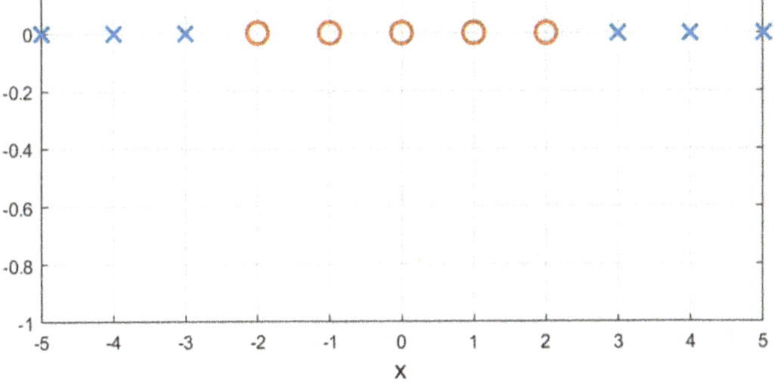

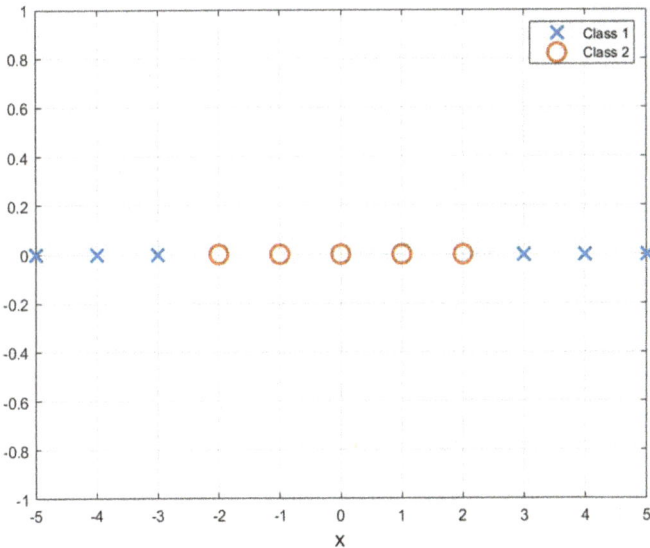

Figure 4.11: Nonlinearly separable one-dimensional data.

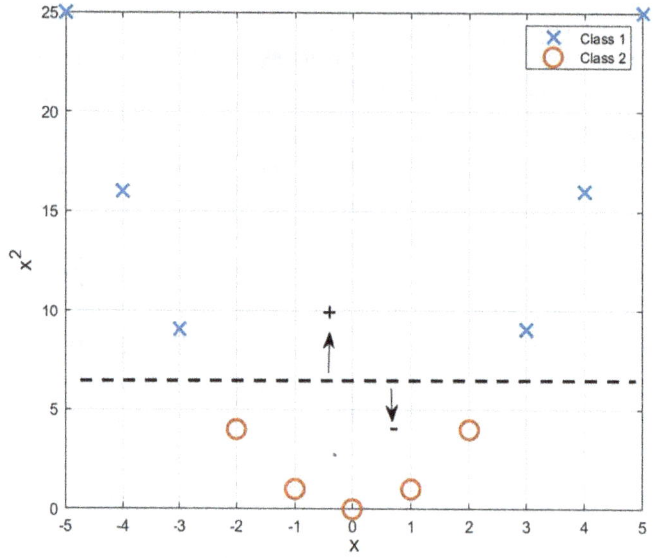

Figure 4.12: Linear separation in two dimensions.

4.7 Concepts of clustering

We've explored how to address uncertainty in classified data by building statistical or analytical classification models. However, the challenge becomes more complicated when learning data lacks classification. In essence, there is no teacher (associated labels) to guide us in assigning each data input to a specific class. In such scenarios, it becomes possible to infer labels by grouping *similar* inputs together through clustering. The challenge then lies in determining the criteria for similarity. Various similarity measures can yield different clustering shapes, and there may be hidden similarities or patterns in the data that require extraction.

Similar to classification, there are two main approaches to clustering: statistical clustering and analytical clustering. In statistical clustering, the process starts by making educated guesses about three parameters: the number of clusters, the initial mean of each cluster, and the probability distribution of the data within each cluster. Subsequently, data points are assigned to clusters based on maximizing the achieved probability. Mathematically, the input $\mathbf{x}_i$ will be assigned to cluster j when:

$$P(C_j \mid \mathbf{x}_i) \geq P(C_k \mid \mathbf{x}_i) \quad \forall k = 1, \ldots, K. \tag{4.143}$$

Here, K represents the initially guessed number of clusters. After this initial step, we proceed to re-estimate the mean of each cluster along with other statistical parameters, contingent on the assumed input distribution. For instance, if the input data is assumed to follow a multivariate normal distribution, the estimation also involves determining the covariance matrix for each cluster. Once the new mean and covariance matrix have been estimated for each cluster, we recalculate the conditional probabilities and redistribute the data among clusters based on their respective maximum probabilities. This iterative process continues until convergence is achieved, resulting in a discernible pattern distribution of the data, as shown in Figure 4.13.

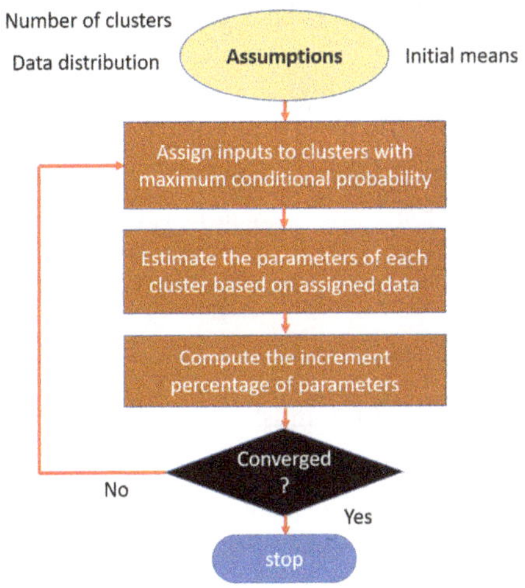

Figure 4.13: Flowchart of statistical clustering.

In analytical clustering, there's no requirement for assumptions about the statistical distribution or origins of the data. The key is to specify the number of clusters, establish similarity measures between the inputs, and set a threshold for accepting each input within a class. Intuitive insights into the problem can help experts estimate the number of clusters. It's worth noting that different clustering methods can produce different clustering shapes and data distributions based on initial assumptions and similarity criteria. Consequently, these methods have the potential to reveal different hidden relationships within the data.

Clustering proves to be a crucial method for addressing uncertainties in data, and it extends its utility even when labeled data is available. It enables a more insightful analysis of data without the constraints imposed by labels.

5 Kalman filters: statistical approach

This chapter does not aim to provide an exhaustive explanation of Kalman filters. Instead, it offers a brief introduction to the topic within the context of the book, focusing solely on the statistical approach. For a comprehensive understanding of Kalman filters, numerous excellent resources are available, such as [6] and [3].

After completing this chapter and working through the provided examples, you will:

- Understand the significance of Kalman filters in estimation, filtering, and prediction, particularly for linear systems with Gaussian noise.
- Comprehend the concept of Gauss-Markov processes and their statistical behavior.
- Master the derivation and implementation of the discrete Kalman filter algorithm for optimal state estimation in linear dynamic systems.
- Extend your knowledge to the extended Kalman filter for handling nonlinear systems through linear approximation.

When dealing with uncertainties in systems, Kalman filters must be mentioned. The Kalman filter is widely regarded as the most important algorithm in the field of estimation, filtering, and prediction. However, it is mainly used in one-dimensional time series signals. If you need to estimate or predict signals that have been distorted by linear operations and corrupted by Gaussian noise, then the Kalman filter is the best choice by far. Under these specific conditions—linearity and Gaussian noise—no other linear or nonlinear filter can outperform Kalman filters.

Despite its exceptional performance, the Kalman filter is surprisingly easy to implement. You can simulate it using a computer or apply it using DSPs/microcontrollers. Even students with basic knowledge of the Kalman filter can quickly write a simulation code and start using it for their projects.

There are already many textbooks and reference books available on Kalman filters, so you might be wondering why we're adding yet another chapter on this topic. Many existing books on Kalman filters tend to be either overly detailed or too superficial, lacking a balance between the two.

During one of my advanced courses on stochastic estimation and control, I asked students who had previously studied Kalman filters this question: 'Why is the Kalman filter the best estimator for parameters measured through a linear system and corrupted by Gaussian noise?' Surprisingly, most of them struggled to provide a clear answer, either relying on complex mathematical explanations or referring to the orthogonality principle.

This chapter introduces Kalman filters, emphasizing the key mathematical foundations while remaining accessible.

Let's focus our discussion in this chapter on time series signals, without loss of generality. Although the concepts presented can be extended to N dimensional data, we will specifically address time series signals. Time series signals typically exhibit correla-

https://doi.org/10.1515/9783111585055-005

tion, with the exception of white noise, which is uncorrelated. The correlation between samples in a time series can be modeled as a one-degree Markov process or through a general statistical correlation, as discussed previously in relation to the multivariate normal distribution. Additionally, there is another important scenario where the signal correlation in a time series arises from deterministic dynamics within the system. In other words, certain internal states of the system generate the time series signal. However, in practice, the observed signal is often affected by noise, whether originating from the measurement or from internal dynamic processes. This introduces uncertainty regarding the true underlying signal. Fortunately, the Kalman filter provides an exceptional solution for addressing this challenge with its remarkable performance. Next, we will introduce the derivation of the Kalman filter algorithm from a statistical perspective. It is worth noting that the same algorithm can also be derived using optimization and linear algebra. However, the statistical approach offers valuable insights into both the inherent strengths and limitations of Kalman filters. The operation of the Kalman filter could be explained for state estimation as shown in Figure 5.1. In this figure we have reused the result of Example 100. There are two internal states of this system. The Kalman filter is the optimal estimator for the internal states of a linear dynamic system given the Gaussian probability density function of the noise.

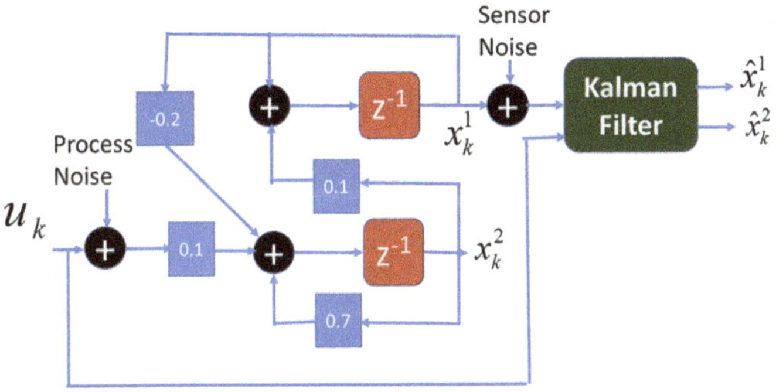

Figure 5.1: Operation structure of the Kalman filter.

In Chapter 3, we discussed several mathematical modeling methods for dynamic systems, emphasizing the significant role of state space representation. By using state-space representations, complex higher-order dynamic systems be decomposed into well-structured first-order dynamics, making them more manageable and comprehensible for control purposes. In control systems, it is necessary to measure the states of the system in order to compute the optimal control signal. However, these states are often subject to noise, and certain states may even be unobservable. As a result, applications involving such scenarios entail a high level of uncertainty regarding the internal states of the system. Hence, the Kalman filter is crucial in such situations.

Let's start by introducing the Gauss–Markov process. Studying the statistical behavior of this process will be useful for the detailed derivation of Kalman filters later.

5.1 Gauss–Markov processes

One of the interesting properties of the discrete-time Gauss–Markov process is that it has a similar structure as a state-space system representation as shown in equation (5.1)

$$\mathbf{x}_{k+1} = \mathbf{A}_k \mathbf{x}_k + \mathbf{w}_k. \tag{5.1}$$

Where $\mathbf{x}_k = [x_1, x_2, \ldots, x_N]^T$ is a vector representing the state of the linear system, $\mathbf{A}_k$ is an $N \times N$ known matrix, and $\mathbf{w}_k$ is an N-vector-valued uncorrelated Gaussian random noise sequence. The statistics of $\mathbf{w}_k$ can be expressed as follows:

$$E[\mathbf{w}_k] = \bar{\mathbf{w}}_k \tag{5.2}$$

$$E\left[(\mathbf{w}_k - \bar{\mathbf{w}}_k)(\mathbf{w}_k - \bar{\mathbf{w}}_k)^T\right] = \mathbf{W}_k \delta_{kl}, \tag{5.3}$$

where $\delta_{kl} = 1$ when $k = l$ and 0 otherwise. This condition indicates that the matrix $\mathbf{W}_k$ is a diagonal matrix. This is intuitive as we assumed that the noise samples are uncorrelated. If the mean of noise sequences is zero, then $\bar{\mathbf{w}}_k = \mathbf{0}$. We can study the statistical behavior of the system given in equation (5.1). However, we need to define the initial state of the system, i. e., $\mathbf{x}_0$ and if there is any relation or dependence between the states and the deriving noise sequence. Let's assume that the initial vector is a sample taken from a multivariate Gaussian distribution such as:

$$E[\mathbf{x}_0] = \bar{\mathbf{x}}_0 \tag{5.4}$$

$$E\left[(\mathbf{x}_0 - \bar{\mathbf{x}}_0)(\mathbf{x}_0 - \bar{\mathbf{x}}_0)^T\right] = \mathbf{P}_0. \tag{5.5}$$

Usually the noise samples are uncorrelated with the system states as follows:

$$E\left[(\mathbf{x}_k - \bar{\mathbf{x}}_k)(\mathbf{w}_j - \bar{\mathbf{w}}_j)^T\right] = \mathbf{0} \quad \forall j \geq k = \{0, 1, \ldots\}. \tag{5.6}$$

Since the system is derived from noise sequences $\{\mathbf{w}_k\}$ with the following Gaussian distribution, the state transition probability density function is given by

$$f_{x_{k+1}|x_k}(\mathbf{x}_{k+1} \mid \mathbf{x}_k) = \frac{1}{(2\pi)^{N/2} \mid \mathbf{W}_k \mid^{1/2}} e^{-\frac{1}{2}(\mathbf{x}_{k+1} - \mathbf{A}_k \mathbf{x}_k)^T \mathbf{W}_k^{-1} (\mathbf{x}_{k+1} - \mathbf{A}_k \mathbf{x}_k)}. \tag{5.7}$$

The process in equation (5.1) will be dynamic and adapt with time. The mean of the vector $\mathbf{x}_{k+1}$ could be derived as follows:

$$E[\mathbf{x}_{k+1}] = \bar{\mathbf{x}}_{k+1} = E[\mathbf{A}_k \mathbf{x}_k + \mathbf{w}_k] \tag{5.8}$$

$$= \mathbf{A}_k E[\mathbf{x}_k] + E[\mathbf{w}_k] = \mathbf{A}_k \bar{\mathbf{x}}_k + \bar{\mathbf{w}}_k. \tag{5.9}$$

The next intuitive question should be about the covariance matrix of $\mathbf{x}_{k+1}$, that can be evaluated as

$$\mathbf{P}_{k+1} = E[(\mathbf{x}_{k+1} - \bar{\mathbf{x}}_{k+1})(\mathbf{x}_{k+1} - \bar{\mathbf{x}}_{k+1})^T] \tag{5.10}$$

$$= E[(\mathbf{A}_k(\mathbf{x}_k - \bar{\mathbf{x}}_k) + \mathbf{w}_k - \bar{\mathbf{w}}_k)(\mathbf{A}_k(\mathbf{x}_k - \bar{\mathbf{x}}_k) + \mathbf{w}_k - \bar{\mathbf{w}}_k)^T] \tag{5.11}$$

$$= E[(\mathbf{A}_k(\mathbf{x}_k - \bar{\mathbf{x}}_k) + \mathbf{w}_k - \bar{\mathbf{w}}_k)((\mathbf{x}_k - \bar{\mathbf{x}}_k)^T\mathbf{A}_k^T + \mathbf{w}_k^T - \bar{\mathbf{w}}_k^T)] \tag{5.12}$$

$$= \mathbf{A}_k E[(\mathbf{x}_k - \bar{\mathbf{x}}_k)(\mathbf{x}_k - \bar{\mathbf{x}}_k)^T]\mathbf{A}_k^T + E[(\mathbf{w}_k - \bar{\mathbf{w}}_k)(\mathbf{w}_k - \bar{\mathbf{w}}_k)^T] \tag{5.13}$$

$$\therefore \mathbf{P}_{k+1} = \mathbf{A}_k\mathbf{P}_k\mathbf{A}_k^T + \mathbf{W}_k. \tag{5.14}$$

Denote from equation (5.6) that $E[(\mathbf{x}_k - \bar{\mathbf{x}}_k)(\mathbf{w}_k - \bar{\mathbf{w}}_k)^T] = \mathbf{0}$. In this section, the mathematical analysis demonstrates the behavior of the mean vector and covariance matrix in a multivariate linear stochastic process, as defined by equation (5.1). Understanding this analysis is crucial for comprehending the derivation of the discrete Kalman filters. Therefore, it is highly recommended for students to practice hand calculations with this analysis.

5.2 Optimal vector estimation

In this section, we present the methodology for obtaining the optimal estimate of a hidden vector of data that undergoes manipulation by a linear system and is affected by additive noise. This step serves as the foundation for developing the general framework of discrete Kalman filters. Assume $\mathbf{z} = [z_1, z_2, \ldots, z_M]^T$ as the column vector of measurements or observation signal. We have omitted the time index k for simplicity. We are interested in a hidden vector of the states $\mathbf{x} = [x_1, x_2, \ldots, x_N]^T$ which is related to the measurement vector as:

$$\mathbf{z} = \mathbf{H}\mathbf{x} + \mathbf{w} \tag{5.15}$$

The relationship between the measurement vector and the state vector is shown as a noisy linear projection. The measurement sensitivity matrix, denoted as $\mathbf{H}$, is expected to have dimensions of $M \times N$. Additionally, the noise vector follows a zero-mean multivariate Gaussian distribution and possesses dimensions of $M \times 1$. Currently, our objective is to ascertain the optimal estimation of the state vector, denoted as $\mathbf{x}$, utilizing the measurement vector $\mathbf{z}$ and statistical information pertaining to the additive noise vector $\mathbf{w}$.

As mentioned in earlier chapters, it is established that linear operations applied to a Gaussian-distributed process yield another Gaussian-distributed process, possibly with a different mean and covariance. The optimal estimation of such problems has been discussed in Chapter 4. It was granted that the optimal estimator can be any moment of the conditional distribution function $f_{\mathbf{X}|\mathbf{Z}}(\mathbf{x} \mid \mathbf{z})$. For instance, when considering the mean of $f_{\mathbf{X}|\mathbf{Z}}(\mathbf{x} \mid \mathbf{z})$, it corresponds to minimizing the mean square error, i. e., using L_2

norm. Similarly, the mode represents the minimum error in terms of the maximum deviation, known as the *minmax* criteria, or the L_∞ norm. Moreover, opting for the median of $f_{\mathbf{X}|\mathbf{Z}}(\mathbf{x} \mid \mathbf{z})$ is tantamount to minimizing the absolute error or utilizing the L_1 norm. In the context of a Gaussian distribution, all modes of the distribution—mean, median, and mode—are equivalent. Therefore, any of these measures could be considered as an optimal solution under a specific criterion. The conditional probability density function will also follow a Gaussian distribution. Once the conditional distribution is found, the optimal state vector estimate will be the mean of that distribution. This optimal vector will be the optimal under any criterion as long as the relation is linear as in (5.15) and the noise vector $\mathbf{w}$ has a Gaussian distribution.

The conditional probability density function could be formulated using the Bayesian formula as

$$f_{\mathbf{X}|\mathbf{Z}}(\mathbf{x} \mid \mathbf{z}) = \frac{f_{\mathbf{XZ}}(\mathbf{x}, \mathbf{z})}{f_{\mathbf{Z}}(\mathbf{z})} = \frac{f_{\mathbf{Z}|\mathbf{X}}(\mathbf{z} \mid \mathbf{x}) f_{\mathbf{X}}(\mathbf{x})}{f_{\mathbf{Z}}(\mathbf{z})}. \tag{5.16}$$

Let's define a new joint vector $\mathbf{v}$ of both $\mathbf{x}$ and $\mathbf{z}$, as follows

$$\mathbf{v} = \begin{bmatrix} \mathbf{x} \\ \mathbf{z} \end{bmatrix} = \begin{bmatrix} \mathbf{I} & \mathbf{0} \\ \mathbf{H} & \mathbf{I} \end{bmatrix} \begin{bmatrix} \mathbf{x} \\ \mathbf{w} \end{bmatrix}. \tag{5.17}$$

The vector mean of $\mathbf{v}$ is given by:

$$E[\mathbf{v}] = \bar{\mathbf{v}} = \begin{bmatrix} \bar{\mathbf{x}} \\ \bar{\mathbf{z}} \end{bmatrix} = \begin{bmatrix} \mathbf{I} & \mathbf{0} \\ \mathbf{H} & \mathbf{I} \end{bmatrix} \begin{bmatrix} \bar{\mathbf{x}} \\ \mathbf{0} \end{bmatrix} = \begin{bmatrix} \bar{\mathbf{x}} \\ \mathbf{H}\bar{\mathbf{x}} \end{bmatrix}. \tag{5.18}$$

The next step to find all statistics of the joint vector $\mathbf{v}$ is to determine its covariance matrix $\mathbf{R}_{vv} = E[(\mathbf{v} - \bar{\mathbf{v}})(\mathbf{v} - \bar{\mathbf{v}})^T]$ as follows:

$$\mathbf{R}_{vv} = E\left[\left(\begin{bmatrix} \mathbf{I} & \mathbf{0} \\ \mathbf{H} & \mathbf{I} \end{bmatrix} \begin{bmatrix} \mathbf{x} - \bar{\mathbf{x}} \\ \bar{\mathbf{w}} \end{bmatrix}\right)\left(\begin{bmatrix} \mathbf{x} - \bar{\mathbf{x}} \\ \bar{\mathbf{w}} \end{bmatrix}^T \begin{bmatrix} \mathbf{I} & \mathbf{0} \\ \mathbf{H} & \mathbf{I} \end{bmatrix}^T\right)\right]. \tag{5.19}$$

This leads to the following expression:

$$\mathbf{R}_{vv} = \begin{bmatrix} \mathbf{I} & \mathbf{0} \\ \mathbf{H} & \mathbf{I} \end{bmatrix} E\left[\begin{bmatrix} \mathbf{x} - \bar{\mathbf{x}} \\ \mathbf{w} \end{bmatrix} \begin{bmatrix} (\mathbf{x} - \bar{\mathbf{x}})^T & \mathbf{w}^T \end{bmatrix}\right] \begin{bmatrix} \mathbf{I} & \mathbf{H}^T \\ \mathbf{0} & \mathbf{I} \end{bmatrix}. \tag{5.20}$$

Define $\mathbf{R}_{xx} = E[(\mathbf{x}-\bar{\mathbf{x}})(\mathbf{x}-\bar{\mathbf{x}})^T]$, and $\mathbf{R}_{ww} = E[\mathbf{w}\mathbf{w}^T]$. Therefore the covariance matrix can be expressed as:

$$\mathbf{R}_{vv} = \begin{bmatrix} \mathbf{I} & \mathbf{0} \\ \mathbf{H} & \mathbf{I} \end{bmatrix} \begin{bmatrix} \mathbf{R}_{xx} & \mathbf{0} \\ \mathbf{0} & \mathbf{R}_{ww} \end{bmatrix} \begin{bmatrix} \mathbf{I} & \mathbf{H}^T \\ \mathbf{0} & \mathbf{I} \end{bmatrix}. \tag{5.21}$$

Next, we need to determine the inverse of this covariance matrix, which is given by:

$$\mathbf{R}_{vv}^{-1} = \begin{bmatrix} \mathbf{I} & -\mathbf{H}^T \\ \mathbf{0} & \mathbf{I} \end{bmatrix} \begin{bmatrix} \mathbf{R}_{xx}^{-1} & \mathbf{0} \\ \mathbf{0} & \mathbf{R}_{ww}^{-1} \end{bmatrix} \begin{bmatrix} \mathbf{I} & \mathbf{0} \\ -\mathbf{H} & \mathbf{I} \end{bmatrix}. \tag{5.22}$$

Performing matrix multiplication rules, we obtain the following result:

$$\mathbf{R}_{vv}^{-1} = \begin{bmatrix} \mathbf{R}_{xx}^{-1} + \mathbf{H}^T\mathbf{R}_{ww}^{-1}\mathbf{H} & -\mathbf{H}^T\mathbf{R}_{ww}^{-1} \\ -\mathbf{R}_{ww}^{-1}\mathbf{H} & \mathbf{R}_{ww}^{-1} \end{bmatrix}. \tag{5.23}$$

We may state the probability density function of the joint distribution as:

$$f_{\mathbf{V}}(\mathbf{v}) = f_{\mathbf{XZ}}(\mathbf{x}, \mathbf{z}) = \frac{1}{(2\pi)^{(M+N)/2} \mid \mathbf{R}_{vv} \mid^{1/2}} e^{-\frac{1}{2}(\mathbf{v}-\bar{\mathbf{v}})^T\mathbf{R}_{vv}^{-1}(\mathbf{v}-\bar{\mathbf{v}})}. \tag{5.24}$$

Example 124. In this example, we will demonstrate through simple numerical examples how a block matrix of submatrices can be manipulated.

As presented in (5.16), we should formulate $f_{\mathbf{z}}$ to be able to construct $f_{\mathbf{X}|\mathbf{Z}}$ formula. Actually equation (5.15) has a similar form as (5.1), hence, we can formulate the mean vector when the noise vector has zero mean as:

$$E[\mathbf{z}] = \bar{\mathbf{z}} = \mathbf{H}\bar{\mathbf{x}}. \tag{5.25}$$

Furthermore, the covariance matrix of $\mathbf{z}$ can be easily found as

$$\mathbf{R}_{zz} = \mathbf{H}\mathbf{R}_{\mathbf{xx}}\mathbf{H}^{\mathrm{T}} + \mathbf{R}_{\mathbf{ww}}. \tag{5.26}$$

Finally, the probability density function of the vector $\mathbf{z}$ is a multivariate Gaussian distribution as:

$$f_{\mathbf{Z}}(\mathbf{z}) = \frac{1}{(2\pi)^{M/2} \mid \mathbf{R}_{zz} \mid^{1/2}} e^{-\frac{1}{2}(\mathbf{z}-\bar{\mathbf{z}})^T\mathbf{R}_{zz}^{-1}(\mathbf{z}-\bar{\mathbf{z}})}. \tag{5.27}$$

We can now formulate our objected distribution $f_{\mathbf{X}|\mathbf{Z}}$ as:

$$f_{\mathbf{X}|\mathbf{Z}} = \frac{f_{\mathbf{XZ}}(\mathbf{xz})}{f_{\mathbf{Z}}(\mathbf{z})} = \frac{(2\pi)^{M/2} \mid \mathbf{R}_{zz} \mid^{1/2}}{(2\pi)^{(M+N)/2} \mid \mathbf{R}_{vv} \mid^{1/2}} e^{-\frac{1}{2}(\mathbf{v}-\bar{\mathbf{v}})^T\mathbf{R}_{vv}^{-1}(\mathbf{v}-\bar{\mathbf{v}})+\frac{1}{2}(\mathbf{z}-\bar{\mathbf{z}})^T\mathbf{R}_{zz}^{-1}(\mathbf{z}-\bar{\mathbf{z}})}. \tag{5.28}$$

Let's define the exponent part in the previous equation as follows

$$\Omega = (\mathbf{v} - \bar{\mathbf{v}})^T\mathbf{R}_{vv}^{-1}(\mathbf{v} - \bar{\mathbf{v}}) - (\mathbf{z} - \bar{\mathbf{z}})^T\mathbf{R}_{zz}^{-1}(\mathbf{z} - \bar{\mathbf{z}}). \tag{5.29}$$

What we need now is to put the previous formula in the following form:

$$\Omega = (\mathbf{x} - \hat{\mathbf{x}})^T\mathbf{P}^{-1}(\mathbf{x} - \hat{\mathbf{x}}). \tag{5.30}$$

Here $\hat{\mathbf{x}}$ is the conditional mean of the conditional probability distribution function and thus the optimal estimation of vector $\mathbf{x}$ under any criterion. Furthermore, $\mathbf{P}$ is the covariance matrix of the conditional mean. Let's start with the first part of (5.29),

$$(\mathbf{v}-\bar{\mathbf{v}})^T\mathbf{R}_{vv}^{-1}(\mathbf{v}-\bar{\mathbf{v}}) = \left[(\mathbf{x}-\bar{\mathbf{x}})^T\ (\mathbf{z}-\mathbf{H}\bar{\mathbf{z}})^T\right]$$
$$\begin{bmatrix} \mathbf{R}_{xx}^{-1}+\mathbf{H}^T\mathbf{R}_{ww}^{-1}\mathbf{H} & -\mathbf{H}^T\mathbf{R}_{ww}^{-1} \\ -\mathbf{R}_{ww}^{-1}\mathbf{H} & \mathbf{R}_{ww}^{-1} \end{bmatrix}\begin{bmatrix} \mathbf{x}-\bar{\mathbf{x}} \\ \mathbf{z}-\mathbf{H}\bar{\mathbf{z}} \end{bmatrix}. \tag{5.31}$$

The second part of (5.29) can be expressed as:

$$(\mathbf{z}-\bar{\mathbf{z}})^T\mathbf{R}_{zz}^{-1}(\mathbf{z}-\bar{\mathbf{z}}) = (\mathbf{z}-\mathbf{H}\bar{\mathbf{x}})^T\mathbf{R}_{zz}^{-1}(\mathbf{z}-\mathbf{H}\bar{\mathbf{x}}). \tag{5.32}$$

Therefore, the exponent factor in (5.29) can be expressed as:

$$\Omega = \left[(\mathbf{x}-\bar{\mathbf{x}})^T\ (\mathbf{z}-\mathbf{H}\bar{\mathbf{z}})^T\right]$$
$$\begin{bmatrix} \mathbf{R}_{xx}^{-1}+\mathbf{H}^T\mathbf{R}_{ww}^{-1}\mathbf{H} & -\mathbf{H}^T\mathbf{R}_{ww}^{-1} \\ -\mathbf{R}_{ww}^{-1}\mathbf{H} & \mathbf{R}_{ww}^{-1}-\mathbf{R}_{zz}^{-1} \end{bmatrix}\begin{bmatrix} \mathbf{x}-\bar{\mathbf{x}} \\ \mathbf{z}-\mathbf{H}\bar{\mathbf{z}} \end{bmatrix}. \tag{5.33}$$

Substitute $\mathbf{R}_{zz}$ using (5.26),

$$\Omega = \left[(\mathbf{x}-\bar{\mathbf{x}})^T\ (\mathbf{z}-\mathbf{H}\bar{\mathbf{z}})^T\right]$$
$$\begin{bmatrix} \mathbf{R}_{xx}^{-1}+\mathbf{H}^T\mathbf{R}_{ww}^{-1}\mathbf{H} & -\mathbf{H}^T\mathbf{R}_{ww}^{-1} \\ -\mathbf{R}_{ww}^{-1}\mathbf{H} & \boxed{\mathbf{R}_{ww}^{-1}-(\mathbf{H}\mathbf{R}_{\mathbf{xx}}\mathbf{H}^{\mathrm{T}}+\mathbf{R}_{\mathbf{ww}})^{-1}} \end{bmatrix}\begin{bmatrix} \mathbf{x}-\bar{\mathbf{x}} \\ \mathbf{z}-\mathbf{H}\bar{\mathbf{z}} \end{bmatrix}. \tag{5.34}$$

Unfortunately (5.34) is still not in the required form of (5.30). Several more steps are required. From Hemes inversion formula which states that

$$(\mathbf{A}+\mathbf{B}\mathbf{C}^{-1}\mathbf{D}^{\mathrm{T}})^{-1} = \mathbf{A}^{-1}-\mathbf{A}^{-1}\mathbf{B}(\mathbf{C}+\mathbf{D}^{\mathrm{T}}\mathbf{A}^{-1}\mathbf{B})^{-1}\mathbf{D}^{\mathrm{T}}\mathbf{A}^{-1}. \tag{5.35}$$

Applying *Hemes* formula for the boxed equation in (5.34) with:

$$\mathbf{R}_{ww} = \mathbf{A}$$
$$\mathbf{H} = \mathbf{B} = \mathbf{D}$$
$$\mathbf{R}_{xx} = \mathbf{C}^{-1}.$$

Therefore

$$(\mathbf{H}\mathbf{R}_{\mathbf{xx}}\mathbf{H}^{\mathrm{T}}+\mathbf{R}_{\mathbf{ww}})^{-1} = \mathbf{R}_{ww}^{-1}-\mathbf{R}_{ww}^{-1}\mathbf{H}(\mathbf{R}_{xx}^{-1}+\mathbf{H}^T\mathbf{R}_{ww}^{-1}\mathbf{H})^{-1}\mathbf{H}^T\mathbf{R}_{ww}^{-1}. \tag{5.36}$$

Substituting the last formula in the boxed equation we get:

$$\Omega = \left[(\mathbf{x}-\bar{\mathbf{x}})^T\ (\mathbf{z}-\mathbf{H}\bar{\mathbf{z}})^T\right]$$
$$\begin{bmatrix} \mathbf{R}_{xx}^{-1}+\mathbf{H}^T\mathbf{R}_{ww}^{-1}\mathbf{H} & -\mathbf{H}^T\mathbf{R}_{ww}^{-1} \\ -\mathbf{R}_{ww}^{-1}\mathbf{H} & \boxed{\mathbf{R}_{ww}^{-1}\mathbf{H}(\mathbf{R}_{xx}^{-1}+\mathbf{H}^T\mathbf{R}_{ww}^{-1}\mathbf{H})^{-1}\mathbf{H}^T\mathbf{R}_{ww}^{-1}} \end{bmatrix}\begin{bmatrix} \mathbf{x}-\bar{\mathbf{x}} \\ \mathbf{z}-\mathbf{H}\bar{\mathbf{z}} \end{bmatrix}. \tag{5.37}$$

You may observe the similarities between the terms in boxed equation and the other terms in the matrix.

Let's define:

$$\mathbf{A} \triangleq \mathbf{R}_{xx}^{-1} + \mathbf{H}^T \mathbf{R}_{ww}^{-1} \mathbf{H}.$$

Also

$$\mathbf{B} = \mathbf{H}^T \mathbf{R}_{ww}^{-1}.$$

Equation (5.34) becomes

$$\Omega = \left[(\mathbf{x} - \bar{\mathbf{x}})^T \ (\mathbf{z} - \mathbf{H}\bar{\mathbf{z}})^T \right] \begin{bmatrix} \mathbf{A} & -\mathbf{B} \\ -\mathbf{B}^T & \mathbf{B}^T \mathbf{A}^{-1} \mathbf{B} \end{bmatrix} \begin{bmatrix} \mathbf{x} - \bar{\mathbf{x}} \\ \mathbf{z} - \mathbf{H}\bar{\mathbf{z}} \end{bmatrix}. \tag{5.38}$$

By expanding (5.37) and put it in the form of (5.30), we obtain the optimal estimation as:

$$\hat{\mathbf{x}} = E[\mathbf{x} \mid \mathbf{z}] = \bar{\mathbf{x}} + \mathbf{A}^{-1} \mathbf{B}(\mathbf{z} - \mathbf{H}\bar{\mathbf{x}}). \tag{5.39}$$

In terms of the original formula, the optimum estimate is:

$$\hat{\mathbf{x}} = \bar{\mathbf{x}} + \left(\mathbf{R}_{xx}^{-1} + \mathbf{H}^T \mathbf{R}_{ww}^{-1} \mathbf{H} \right)^{-1} \mathbf{H}^T \mathbf{R}_{ww}^{-1} (\mathbf{z} - \mathbf{H}\bar{\mathbf{x}}). \tag{5.40}$$

Finally, the covariance matrix of this optimum estimate can be also found from (5.38) as $\mathbf{P} = \mathbf{A}^{-1} = (\mathbf{R}_{xx}^{-1} + \mathbf{H}^T \mathbf{R}_{ww}^{-1} \mathbf{H})^{-1}$, therefore the optimal vector estimate is:

$$\hat{\mathbf{x}} = \bar{\mathbf{x}} + \mathbf{P}\mathbf{H}^T \mathbf{R}_{ww}^{-1} (\mathbf{z} - \mathbf{H}\bar{\mathbf{x}}). \tag{5.41}$$

This optimal state vector $\hat{\mathbf{x}}$ is the optimal estimate by any criterion, as long as the relation between the measurement and the state vector is linear and the uncertainty follows a Gaussian distribution with zero mean.

In (5.15), we may know (or estimate) the matrix $\mathbf{H}$ and the statistics of the additive noise vector $\mathbf{w}$. But it is a common case that we have no idea about the hidden vector $\mathbf{x}$ as well as its statistics like the mean $\bar{\mathbf{x}}$ and the covariance matrix $\mathbf{R}_{xx}$. Nevertheless, fortunately it is possible to solve (5.40) in iterative way as:

- Start with initial vector $\hat{\mathbf{x}}_0$ and initial matrix $\mathbf{P}_0^{-1}$.
- For time slots $k = 1, 2, \ldots$
- $\mathbf{K}_k = (\mathbf{P}_{k-1} + \mathbf{H}_k^T \mathbf{R}_{ww}^{-1} \mathbf{H}_k)^{-1} \mathbf{H}_k^T \mathbf{R}_{ww}^{-1}$.
- $\hat{\mathbf{x}}_k = \hat{\mathbf{x}}_{k-1} + \mathbf{K}_k (\mathbf{z}_\mathbf{k} - \mathbf{H}_\mathbf{k} \hat{\mathbf{x}}_{k-1})$.
- $\mathbf{P}_k^{-1} = \mathbf{P}_{k-1}^{-1} + \mathbf{H}_k^T \mathbf{R}_{ww}^{-1} \mathbf{H}_k$.

In the above iterative procedure we assume that the measurement noise vector $\mathbf{w}$ is uncorrelated. It is interesting to see from the iterative algorithm that the measurement matrix $\mathbf{H}$ as well as the noise covariance matrix can be time-varying.

Example 125. Let's have this simulation example to evaluate the performance of the iterative optimum estimator. Assume that the relationship between the states and the output is given by

$$\mathbf{z} = \begin{bmatrix} z_1 \\ z_2 \end{bmatrix} = \begin{bmatrix} 1 & 2 & -1 \\ 5 & 1 & 3 \end{bmatrix} \begin{bmatrix} x_1 \\ x_2 \\ x_3 \end{bmatrix} + \begin{bmatrix} w_1 \\ w_2 \end{bmatrix}. \tag{5.42}$$

Our purpose is to track the hidden states based on the measurement vector **z**. The noise vector is uncorrelated with the covariance matrix given by $\mathbf{R}_{ww} = \left[\begin{smallmatrix} 0.2 & 0 \\ 0 & 0.4 \end{smallmatrix}\right]$. The code to perform the states estimation in Octave is shown next. It is clear that we have two measurements and we try to estimate all three hidden states. The results are shown in Figure 5.2.

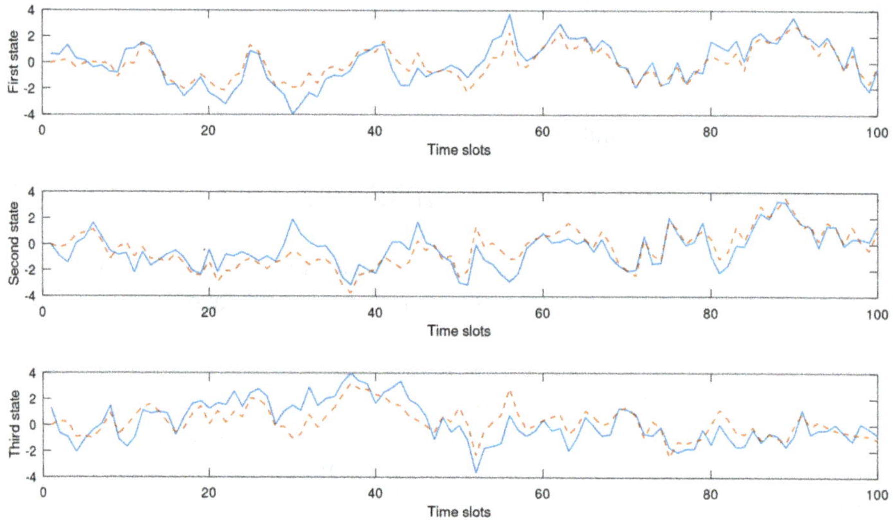

Figure 5.2: Optimal states estimation in Example 125.

```
% Octave code
# Example
clear all
T = 100; %Number of time samples
H= [1 2 -1;5 1 3];
Rww=[0.2 0;0 0.4];
w1=Rww(1,1)^.5*randn(1,T);
w2=Rww(2,2)^.5 * randn(1,T);
r=randn(3,T);
x=filter(1,[1 -0.8],r');
```

```
x=x';
z=H*x+[w1;w2];
xh=zeros(3,1);
Pi=eye(3); %initial estimate of the Covariance matrix inverse
for k=1:T-1,
K = inv(inv(Pi)+H'*inv(Rww)*H)*H'*inv(Rww)
xh(:,k+1)=xh(:,k)+K*(z(:,k+1)-H*xh(:,k));
Pi=Pi+H'*inv(Rww)*H;
end
t=1:T;
subplot (311);
plot(t,x(1,t),'linewidth', 1,t,xh(1,t),'linewidth', 1,'linestyle','--')
xlabel('Time slots');ylabel('First state')
subplot (312);
plot(t,x(2,t),'linewidth', 1,t,xh(2,t),'linewidth', 1,'linestyle','--')
xlabel('Time slots');ylabel('Second state')
subplot (313);
plot(t,x(3,t),'linewidth', 1,t,xh(3,t),'linewidth', 1,'linestyle','--')
xlabel('Time slots');ylabel('Third state')
```

The previous optimal vector estimation achieves about 75 % of the full discrete Kalman filter, as explained in the next section.

5.3 Discrete Kalman filter

The discrete Kalman filter algorithm is simply a one-step development of the previous optimal vector estimation. The concept here is that the state vector $\mathbf{x}$ is governed by certain system dynamics. The main concepts of state space representation was reviewed in Chapter 3. We use different variables here, but the concept is identical. Consider a discrete linear system that could be expressed as follows:

$$\mathbf{x}_{k+1} = \mathbf{A}_k\mathbf{x}_k + \mathbf{B}_k\mathbf{u}_k + \mathbf{v}_k \tag{5.43}$$

$$\mathbf{z}_k = \mathbf{H}_k\mathbf{x}_k + \mathbf{w}_k. \tag{5.44}$$

Where $\mathbf{u}_k$ is the input vector at time k, $\mathbf{v}_k$ is a random vector to express our uncertainty about the system model. The vector $\mathbf{v}_k$ is usually modeled as multivariate zero-mean Gaussian distribution, i. e., $\mathbf{v}_k \sim N(\mathbf{0}, \mathbf{R}_{vv})$. Since the equations are recursive, it is also possible to assume the initial vector $\mathbf{x}_0$ as a multivariate Gaussian distribution, i. e., $\mathbf{x}_0 \sim N(\bar{\mathbf{x}}_0, \mathbf{R}_{x0})$. You may notice that we emphasize that any added noise or uncertainty should be Gaussian distributed. I guess you should know why?

The question now is how to modify the optimal vector estimator discussed in the previous section to include the systematic dynamics of the state vector. With the formulation in (5.43) it is obvious that at time slot k we need to estimate not only $\mathbf{x}_k$, but also $\mathbf{x}_{k+1}$. For the derivation, since $\mathbf{u}_k$ is a deterministic input, we may assume it to be zero. However, it could be added later. Hence, we define $\tilde{\mathbf{x}}_{k+1} = E[\mathbf{x}_{k+1} \mid \mathbf{z}_k]$, and by substituting (5.43), we obtain

$$\tilde{\mathbf{x}}_{k+1} = E[\mathbf{A}_k\mathbf{x}_k + \mathbf{v}_k \mid \mathbf{z}_k]. \tag{5.45}$$

Since $\mathbf{v}_k$ has zero mean, then:

$$\tilde{\mathbf{x}}_{k+1} = \mathbf{A}_k E[\mathbf{x}_k \mid \mathbf{z}_k] = \mathbf{A}_k\hat{\mathbf{x}}. \tag{5.46}$$

Furthermore, we define the error of the two estimates as:

$$\tilde{\mathbf{e}}_k = \mathbf{x}_k - \tilde{\mathbf{x}}_k \tag{5.47}$$

$$\hat{\mathbf{e}}_k = \mathbf{x}_k - \hat{\mathbf{x}}_k. \tag{5.48}$$

We have already seen the covariance of the error vector $\hat{\mathbf{e}}_k$ in the previous section and was defined as:

$$\mathbf{P}_k = E[(\mathbf{x} - \hat{\mathbf{x}}_k)(\mathbf{x} - \hat{\mathbf{x}}_k)^T \mid \mathbf{z}_k].$$

However, here we define also:

$$\mathbf{M}_{k+1} = E[(\mathbf{x} - \tilde{\mathbf{x}}_{k+1})(\mathbf{x} - \tilde{\mathbf{x}}_{k+1})^T \mid \mathbf{z}_k]. \tag{5.49}$$

Observe that $\mathbf{M}_k$ can be considered as an estimation of the covariance matrix of the hidden state vector $\mathbf{x}$, $\mathbf{R}_{xx}$, at the time slot k.

We can find the relation between $\mathbf{M}_k$ and $\mathbf{P}_k$ as follows

$$\tilde{\mathbf{e}}_{k+1} = \mathbf{x}_{k+1} - \tilde{\mathbf{x}}_{k+1} = \mathbf{A}_k\mathbf{x}_k + \mathbf{v}_k - \mathbf{A}_k\hat{\mathbf{x}}_k = \mathbf{A}_k\hat{\mathbf{e}}_k + \mathbf{v}_k \tag{5.50}$$

$$\mathbf{M}_{k+1} = E[\tilde{\mathbf{e}}_{k+1}\tilde{\mathbf{e}}_{k+1}^T \mid \mathbf{z}_k] = E[(\mathbf{A}_k\hat{\mathbf{e}}_k + \mathbf{v}_k)(\hat{\mathbf{e}}_k^T\mathbf{A}_k^T + \mathbf{v}_k^T)]. \tag{5.51}$$

Therefore:

$$\mathbf{M}_{k+1} = \mathbf{A}_k\mathbf{P}_k\mathbf{A}_k^T + \mathbf{R}_{vv}. \tag{5.52}$$

By using the result in (5.30) the optimum estimate of the hidden state vector is given by:

$$\hat{\mathbf{x}}_k = \tilde{\mathbf{x}}_k + (\mathbf{M}_k^{-1} + \mathbf{H}_k^T\mathbf{R}_{ww}^{-1}\mathbf{H}_k)^{-1}\mathbf{H}_k^T\mathbf{R}_{ww}^{-1}(\mathbf{z}_k - \mathbf{H}_k\tilde{\mathbf{x}}_k). \tag{5.53}$$

Now we have all algorithms to state the recursive equations of the Kalman filter as follows

- Start with initial vector $\hat{\mathbf{x}}_0$ and initial matrix $\mathbf{P}_0$.
- For time slots $k = 1, 2, \ldots$
- $\tilde{\mathbf{x}}_{k+1} = \mathbf{A}_k\hat{\mathbf{x}}_k + \mathbf{B}_k\mathbf{u}_k$.
- $\mathbf{M}_{k+1} = \mathbf{A}_k\mathbf{P}_k\mathbf{A}_k^T + \mathbf{R}_{vv}$.
- $\hat{\mathbf{x}}_k = \tilde{\mathbf{x}}_k + \mathbf{P}_k\mathbf{H}_k^T\mathbf{R}_{ww}^{-1}(\mathbf{z}_k - \mathbf{H}_k\tilde{\mathbf{x}}_k) = \tilde{\mathbf{x}}_k + \mathbf{K}_k(\mathbf{z}_k - \mathbf{H}_k\tilde{\mathbf{x}}_k)$.
- $\mathbf{P}_k = (\mathbf{M}_k^{-1} + \mathbf{H}_k^T\mathbf{R}_{ww}^{-1}\mathbf{H}_k)^{-1}$.

The term $\mathbf{K}_k = \mathbf{P}_k\mathbf{H}_k^T\mathbf{R}_{ww}^{-1}$ is known as the Kalman gain.

Example 126. Consider the system given in Figure 5.1 with input $u_k = \cos(\frac{\pi}{5}k)$. We have sensor to measure only x_k^1 with uncorrelated additive noise with standard deviation $\sigma_m = 0.1$. Simulate the system and use the Kalman filter to estimate the states x_k^1 with sampling time of $T_s = 0.05$ s. Show both the idea state as well as the estimated state using Kalman. Assume that the modeling error is represented by additive Gaussian noise with a standard deviation of $\sigma_n = 0.05$.

```
% Octave Code
clear all
pkg load control
Ad = [1 0.1; -0.2 0.7];
Bd = [0;0.1];
Cd = [1 0];
Dd = [];
Ts = 0.05; %sampling time of the discrete
Tn= 20; % simulate from 0 up to Tn seconds
Sign=0.0025; % Standard deviasion of Noise Process
Sigm=0.01; % Standard deviasion of measurement noise
t=0:Ts:Tn; % Simulation time
SYSd = ss(Ad,Bd,Cd,Dd,Ts) % building discrete SS model
Len=length(t); %Simulation length
w=Sigm^.5*randn(Len,1); %measurement noise
Rww=cov(w); %Covariance of measurement noise
u=cos(pi*t/5); %input signal
n=Sign^.5*randn(Len,1); %process noise
Rnn=cov(n); %covariance of process noise
y=lsim(SYSd,u+n',t); %system simulation
z=y+w; %system output measurement

% Kalman Filter
xh=randn(2,1);
P=rand(2);
```

```
for k=1:Len
xt=Ad*xh(:,k)+Bd*u(k);
M=Ad*P*Ad'+Rnn;
yt(k)=z(k)-Cd*xt;
xh(:,k+1)=xt+P*Cd'*inv(Rww)*yt(k);
P=pinv(pinv(M)+Cd'*pinv(Rww)*Cd);
end
plot(t,xh(1,1:Len),'linewidth',1,t,y(1:Len),'linewidth',
      1,'linestyle','--')
axis([0 Tn -1 1])
xlabel('Time [s]','fontsize',18);
ylabel('ideal and estimated state','fontsize',18)
legend('estimated','ideal','fontsize',15)
```

The result of this example is shown in Figure 5.3.

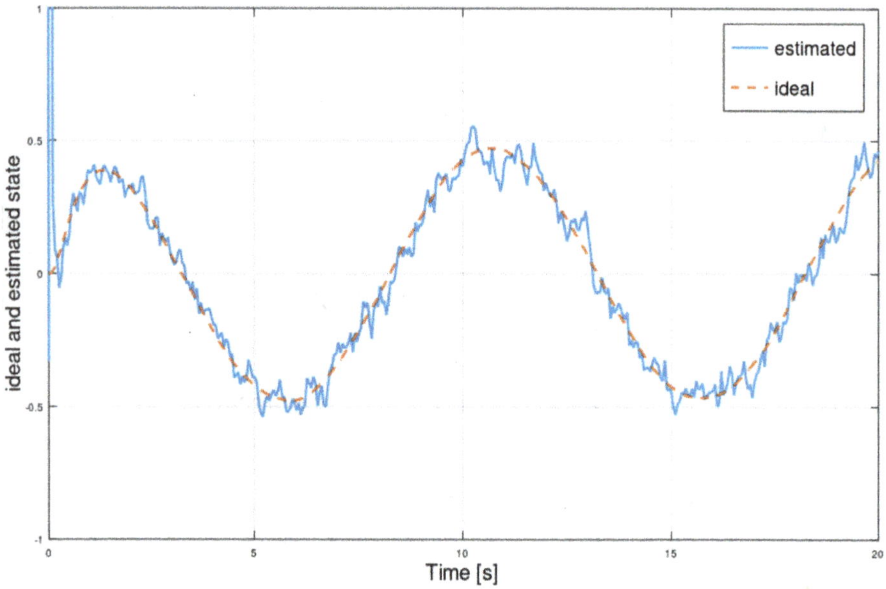

Figure 5.3: Estimated state.

5.4 Extended Kalman

The Kalman filter is widely regarded as the optimal estimator for linear systems with uncertainties that can be modeled as a Gaussian distribution. It has been demonstrated in the previous section that the Kalman filter is the best choice regardless of the evaluation criteria employed. However, it is important to note that many real-world systems exhibit nonlinear behavior, and ignoring these nonlinear characteristics can lead to significant errors in the models. In fact, relying solely on a linear model in such cases renders it ineffective in terms of achieving accurate modeling. There are several modified versions of Kalman filters in order to handle the nonlinearity problems. However, in the case of nonlinearity, we cannot claim the optimality of Kalman filters. Assume a nonlinear system represented as:

$$\mathbf{x}_{k+1} = \mathbf{f}(\mathbf{x}_k, \mathbf{u}_k) + \mathbf{v}_k \tag{5.54}$$

$$\mathbf{z}_k = \mathbf{h}(\mathbf{x}_k) + \mathbf{w}_k. \tag{5.55}$$

The concept of extended Kalman is based on approximating the nonlinear function as a linear one around and close to the operating point, $\mathbf{x}_{k+1}$. This approximation is carried out using the well known Taylor expansion. However, we assume here that both nonlinear functions are analytic functions, or at least first-order smooth differentiable functions. Assuming that $\mathbf{x}_{k+1}$ is close enough to $\mathbf{x}_k$, we can approximate the nonlinear functions as

$$\mathbf{f}(\mathbf{x}_{k+1}, \mathbf{u}_{k+1}) = \mathbf{f}(\mathbf{x}_k, \mathbf{u}_k) + \mathbf{Df}(\mathbf{x}_k, \mathbf{u}_k)(\mathbf{x}_{k+1} - \mathbf{x}_k) + \text{h. o. t.} \tag{5.56}$$

Assuming we have N states, the vector differentiation generates matrix as follows:

$$\mathbf{Df}(\mathbf{x}_k, \mathbf{u}_k) = \begin{bmatrix} \frac{\partial \mathbf{f}_1}{\partial x_1} & \frac{\partial \mathbf{f}_1}{\partial x_2} & \cdots & \frac{\partial \mathbf{f}_1}{\partial x_N} \\ \vdots & \vdots & \cdots & \vdots \\ \frac{\partial \mathbf{f}_N}{\partial x_1} & \frac{\partial \mathbf{f}_N}{\partial x_2} & \cdots & \frac{\partial \mathbf{f}_N}{\partial x_N} \end{bmatrix}_{\mathbf{x}_k, \mathbf{u}_k} . \tag{5.57}$$

The same expansion can be applied for the output nonlinear mapping $\mathbf{h}(\mathbf{x}_k)$. Substituting back in (5.54), the linear approximation becomes

$$\mathbf{x}_{k+1} = \mathbf{Df}_{\mathbf{x}_k, \mathbf{u}_k} \mathbf{x}_k + \mathbf{v}_k \tag{5.58}$$

$$\mathbf{z}_k = \mathbf{Dh}_{\mathbf{x}_k} \mathbf{x}_k + \mathbf{w}_k. \tag{5.59}$$

We now have the linearized version of the nonlinear system. However, we should remember again that the Kalman realization of the approximated linear system is generally not the optimal estimator as in the case of linear system. The approximation could be improved by considering more higher order terms, such as the second derivative (Hessian matrix). However, it has been observed that using a higher order approximation improves the estimation of the Kalman filter, but this improvement is only significant when dealing with very small noise. This is because the derivative process acts as a high-pass filter, which can amplify the effects of noise in the estimation. The extended Kalman filter is prone to divergence if the first-order linearization is not accurate enough.

To address this issue, several improvements have been proposed to the extended Kalman filter, such as the unscented Kalman filtering. The unscented Kalman filter employs an unscented transformation as an alternative to the first-order linearization used in the extended Kalman filter for nonlinear equations. This transformation has shown higher stability and improved accuracy in solving nonlinear problems compared to the extended Kalman filter.

However, discussing the unscented Kalman filter and its advantages goes beyond the scope of this chapter.

The extended Kalman filter algorithm in this case is given by the following steps (where $\mathbf{Df}_k = \mathbf{Df}_{\mathbf{x}_k,\mathbf{u}_k}$),
- Start with initial vector $\hat{\mathbf{x}}_0$ and initial matrix $\mathbf{P}_0$.
- For time slots $k = 1, 2, \ldots$
- $\tilde{\mathbf{x}}_{k+1} = \mathbf{f}(\hat{\mathbf{x}}_k, \mathbf{u}_k)$.
- $\mathbf{M}_{k+1} = \mathbf{Df}_k\mathbf{P}_k\mathbf{Df}_k^T + \mathbf{R}_{vv}$.
- $\hat{\mathbf{x}}_k = \tilde{\mathbf{x}}_k + \mathbf{P}_k\mathbf{Dh}_k^T\mathbf{R}_{ww}^{-1}(\mathbf{z}_k - \mathbf{Dh}_k\tilde{\mathbf{x}}_k) = \tilde{\mathbf{x}}_k + \mathbf{K}_k(\mathbf{z}_k - \mathbf{h}(\tilde{\mathbf{x}}_k))$.
- $\mathbf{P}_k = (\mathbf{M}_k^{-1} + \mathbf{Dh}_k^T\mathbf{R}_{ww}^{-1}\mathbf{Dh}_k)^{-1}$.

Example 127. Assume a nonlinear dynamic system given in Figure 5.4 with input u_k. The measurement noise is uncorrelated additive noise with standard deviation $\sigma_v = 0.01$. The process noise $\mathbf{w} = [w_1\ w_2]^T$ is uncorrelated Gaussian noise with covariance matrix $R_{ww} = 0.01\mathbf{I}$. Find the linear approximation of the system. Simulate the system and use the extended Kalman filter to estimate the states $x_{1,k}$ with sampling time of $T_s = 0.01$ s. Show the results for u_k as a single pulse. Show both the ideal state and the estimated state using the extended Kalman.

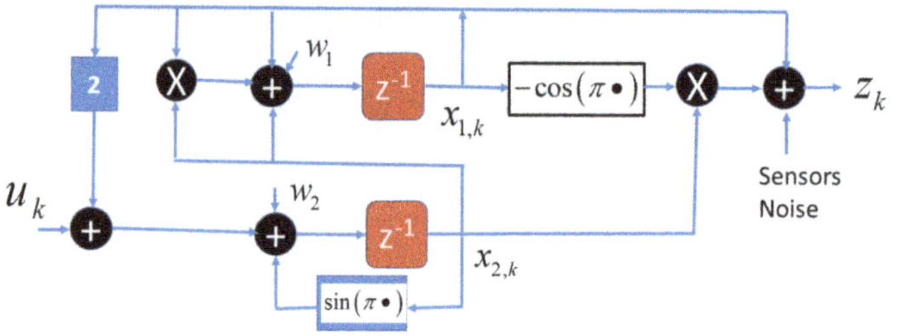

Figure 5.4: Example of a nonlinear dynamic system.

From figure we may express the system mathematical formula as

$$x_{1,k+1} = T_s(x_{1,k} + x_{2,k} + x_{1,k}x_{2,k}) + w_1 \tag{5.60}$$

$$x_{2,k+1} = T_s(2x_{1,k} + \sin(x_{2,k}\pi) + u_k) + w_2 \tag{5.61}$$

$$z_k = x_{1,k} - x_{2,k}\cos(x_{1,k}\pi) + v_k. \tag{5.62}$$

The matrix $\mathbf{Df}_{\mathbf{x}_k,\mathbf{u}_k}$ and vector $\mathbf{Dh}_k$ can be computed as

$$\mathbf{Df}_k = T_s \begin{bmatrix} 1 + x_{2,k} & 1 + x_{1,k} \\ 2 & \pi \cos(\pi x_{2,k}) \end{bmatrix} \tag{5.63}$$

$$\mathbf{Dh}_k = \begin{bmatrix} 1 + \pi x_{2,k} \sin(\pi x_{1,k}) \\ -\cos(\pi x_{1,k}) \end{bmatrix}. \tag{5.64}$$

We can see that the system matrices are functions in the states. This is not the case when we are dealing with a purely linear system. The reason is that, we linearize the system around the current states, so the resulting matrices depend on the slopes at those states. However, mathematically, we treat the system matrices just as time-varying matrices. If you are wondering about the multiplication in the sampling time, you may want to revisit Chapter 3.

The simulation result of this example is shown in Figure 5.5.

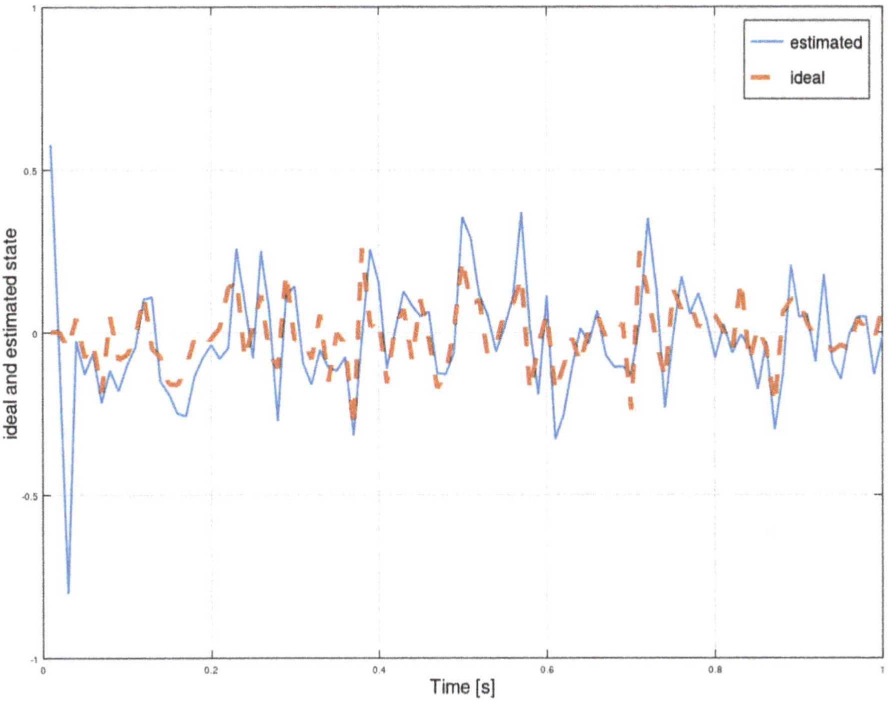

Figure 5.5: Estimated nonlinear state x_1.

5.5 Applications

Kalman filters have numerous applications in reducing uncertainties when estimating hidden or noisy states of dynamic systems. These systems can be modeled as either linear or nonlinear differential equations. In practice, most real-world problems can be represented as deterministic or stochastic differential equations that can be expressed or approximated in state space form. The dynamics of these systems are determined by the states, but our observations or measurements are usually based on only some of these states. Additionally, certain states may not be observable from the available observations, and there may be instances of missing measurements.

In such complex systems, the Kalman filter plays a critical role in optimizing the estimation of the system states. It is challenging to enumerate all the possible applications, but some of the most well-known applications include localization and positioning, navigation, real-time object tracking, optimal sensor fusion, and system identification. These applications highlight the versatility and importance of the Kalman filter in various domains.

6 Game theory

The field of game theory is vast, and it's impossible to cover it comprehensively in a single chapter. Therefore, this chapter offers a concise introduction to the topic, specifically within the context of modeling the uncertain responses of adaptive intelligent entities. For an excellent introductory resource on game theory, I recommend [7].

After completing this chapter and working through the provided examples, you will:

- Understand the fundamental concepts of game theory and its applications in modeling decision-making in competitive and cooperative scenarios.
- Differentiate between zero-sum and non-zero-sum games, and comprehend their implications for player strategies and outcomes.
- Analyze and solve bimatrix games, both cooperative and non-cooperative, to identify Nash equilibrium solutions and optimal strategies.
- Apply game theory concepts to continuous strategy games, where players have a range of possible actions, and understand the conditions for Nash equilibrium in such games.
- Explore the principles of evolutionary game theory.

In the previous chapters, we discussed the resolution of uncertainties related to physical problems. These problems can be characterized either by deterministic mathematical formulations or stochastic behavior that requires statistical analysis, such as finding the maximum likelihood. However, when it comes to the uncertainty surrounding the reaction or response of a rational systems within limited resources (material or moral), such as a human being, things can be entirely different. In this context, the response is influenced by a variety of factors, including the system's objectives, which are shaped by its current self-interest, vision, beliefs, environment, and its current mode. Game theory is a mathematical foundation to provide the best possible responses to rational, mature and intelligent players in case of conflicting objectives. To understand the difference between game theory and general optimization problems, let's start with this primitive example.

Example 128. Assume two companies, X and Y, with utilities of U_X and U_Y respectively. The utility here is the worth or value of a good or service that everyone wants to maximize. It can be objective, such as money, or subjective, such as reputation and trust, or hybrid. In this example, assume the utilities are functions in one parameter as $U_X = -x^2 + 10x + 1$, and $U_Y = -2y^2 + 40y + 20$. Companies X and Y can easily compute the optimal parameters, in order to maximize their utilities. This could be easily computed by solving for the order differentiation, and we get $x\star = 5$, and $y\star = 10$. Figure 6.1 shows the utility of each. The situation in this case is NOT related to game theory. We have no conflict of interest or competition between the two companies. The next example shows something completely different.

Example 129. Let's modify the previous example so that the selected parameter decision of each company will have direct impact on the other company. In this example,

https://doi.org/10.1515/9783111585055-006

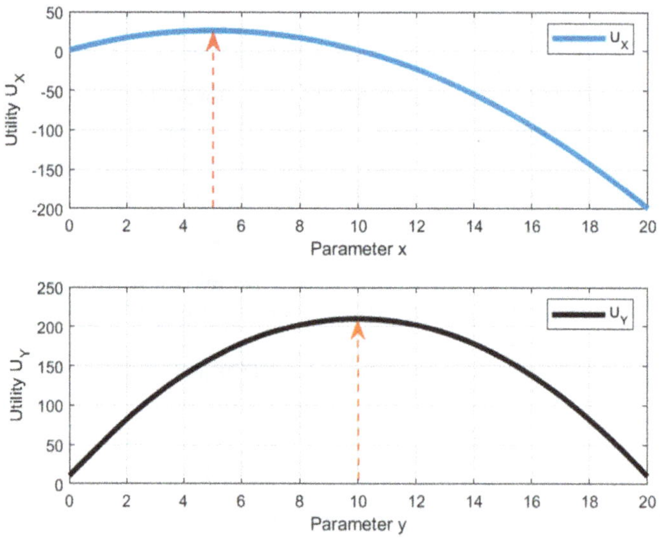

Figure 6.1: Achieved Utilities.

assume that the utilities of each company is given by $U_X = -x^2 + 10xy + 1 - y$ and $U_Y = -2y^2 + 40yx + 20 - x$. Let's also assume that the selected parameters are limited as $x \in [-20, 20]$, and the same for $y \in [-20, 20]$. Assume that you are the CEO of company X and I am the CEO of company Y. You may decide to use $x^* = 5$, which is the value that maximizes the utility value as in the previous example. In this case, the utility of your company will be $U_X = -26 + 49y$. You can see that your absolute utility depends on my decision as the CEO of company Y. If my objective is to minimize your utility regardless of my utility, I could chose $y = -20$, in which case $U_X = -1006$. However, the utility of company Y also depends on the parameter selection of company X. In this case, $U_Y = -4785$. Therefore, company Y will suffer even more than X because of its decision.

In the above simple example, we may call the decision maker of each company as the game *player*. The utility parameter is called the game *strategy*. In realistic games, each player's objective is to maximize their *utility*. The response of each player is generally uncertain. However, if the players are mature and intelligent, we can expect how they will respond.

Example 130. In the previous example, assume that you are an observer of the game, and you notice that player X is playing with strategy $x^* = 5$. You believe that player Y, should play with a strategy that maximizes their own utility. How will player Y respond? What will be the utilities for both players?

Solution. It is clear that with $x^* = 5$, we get $U_Y = -2y^2 + 200y + 15$. This utility is maximized at $\frac{dU_Y}{dy} = -4y^* + 200 = 0 \rightarrow y^* = 50$. But, unfortunately, this strategy is outside the maximum allowed parameters as shown in the previous example. Hence, they will stick with the maximum allowed value, which is $y^* = 20$. In this case, the utility of Y is $U_Y = 3215$. Moreover, the utility of X is $U_X = 956$. Indeed, it is better than the case in the previous example. However, are these strategies the best that both players can do? What if player X decides to reset their strategy based on the player Y's response (i. e., selecting $y = 20$)? What is the equilibrium solution, so that no player can do any better by deviating from it, if the other player plays it? What types of games do we have? What if there is uncertainty about the other players' utilities or strategies? These and many other questions will be addressed and discussed in the remainder of this chapter.

Exercise. In this example, if the game is non-cooperative, what is each player's best choice that no one could do better?

Generally speaking, game theory could be applied to at least the following four cases

- It can be used to infer the response of a rational player in certain situation and hence to resolve the response uncertainty. The players can be individuals, companies and institutes, or even countries and nations.
- It can be used to find the best possible response considering opportunities and risks.
- Looking at the responses of players, game theory could be used to reveal at least some of their objectives, strategies, and plans.
- It can provide the mathematical foundation for understanding the behavior of all living beings as individuals and societies. Living beings can be on the cell level up to human desires.

6.1 Rational uncertainty and game modeling

The word "game" in game theory should not be seen as just a word for "fun." For example, one country may start a war with another country, resulting in the loss of hundreds of thousands of lives and massive destruction, all to achieve certain objectives or change the rules of the game.

Some countries seek to obtain nuclear weapons, not with the intention of using them, but rather to create new strategies and achieve specific objectives by establishing new rules within their own game.

Consider the scenario of a cat trying to catch a mouse. The mouse, in turn, engages in a game of survival by evading the cat, taking the most challenging path. This can be seen as an evolutionary game.

In situations where our bodies are invaded by harmful bacteria or viruses, a game ensues between our immune system and the attacking enemies. This can also be viewed as an evolutionary game.

All of the previous examples involve players with conflicting objectives. For instance, the mouse wants to stay alive, while the cat aims to kill it. However, it is worth noting that the cat wants to stay alive by having food.

Now imagine you have a market in a particular area and another shop opens right in front of yours, selling similar goods. You would engage in a game with the other shop, using various strategies such as prices, product quality, marketing, special offers, and attracting customers with appealing decor. However, it might be more beneficial to initiate a cooperative game.

Let's consider a telecommunications example. Consider an ad hoc communication system that uses a channel with limited bandwidth. Multiple transmitters want to access it using their resources such as transmit power, usage time, and bandwidth (number of subcarriers). If one transmitter decides to continuously use its maximum power and bandwidth (which is generally an incorrect decision unless it has the best channel conditions), it will cause significant interference to the others. As a result, all other transmitters may also increase their transmit power. This situation can lead to communication breakdowns and congestion.

The above are just a few examples of the application of game theory. Generally, it has a wide range of applications, including politics, economics, business, psychology, evolutionary biology, warfare, social behavior, social engineering, project management, and risk assessment. In these days of developing systems with AI and machine learning, game theory plays a central role in AI-based decision making systems. It can be shown that game theory provides the mathematical framework that analyzes conflict and cooperation between intelligent and rational decision makers. In essence, game theory can be thought of as the study of conflict analysis.

Game theory consists of at least 4 elements such as:
- Players;
- Payoffs;
- Strategies;
- Game rules.

Players in game theory are rational, intelligent, and decision makers. However, rationality is not necessary in evolutionary games. Rationality of players implies that each player knows precisely their own payoff and is motivated to maximize it. This is generally true for human-based games. The intelligence of players implies that each player chooses the best possible strategy that can maximize their payoff, or at least minimize their losses. They should be intelligent enough to be able to predict or guess how their opponents will play. Finally, they should be decision makers, i. e., they will implement the selected strategy.

Each player in a game should have a clear goal or target to strive for. Without objectives, playing a game becomes meaningless. At its core, the aim of any player is to emerge victorious. In a broader sense, each player's objective should be to maximize their utility or minimize their losses. This objective can be referred to as maximizing the payoff.

Every player enveloped in a game should have a certain set of strategies. The strategy is not necessarily a single move, but it represents the procedure of how the player will proceed and respond during the game.

Finally, the game rules are the rules that each player should follow during the game. Game rules can be defined naturally (e. g., physical limits), by laws and regulations, or by agreements. It is expected that each player will strictly follow the game rules.

Mathematically, the game theory could be defined as $G = (\Omega, \{S_i\}, \{U_i\}, t)$, where Ω is the set of all players, S_i is the set of available strategies of each player, U_i is the set of utilities (payoffs) of each player, and the time t indicates that the game can be time dependent (dynamic games).

There are two main types of games: *cooperative* and *non-cooperative* games. In a non-cooperative game, players are not necessarily prohibited from cooperating, but any cooperation that occurs must be self-enforcing. This means that players cannot communicate with each other or coordinate their strategic choices.

Performing any game, the following steps should be taken:

- Gain a thorough understanding of the game, including its rules, rewards, potential losses, and the objectives of other players. It would be beneficial to anticipate the actions of other players.
- Define your targets or objectives with precision, considering that they may change over time as the game evolves.
- It is crucial to create a well-modeled representation of the game. A good model does not necessarily mean that it has to be extremely accurate. Highly accurate models can be overly complex, making it difficult to observe key aspects. Additionally, complex models require more processing time and effort to comprehend. Conversely, oversimplified models may overlook important game elements. When modeling the game, finding a balance between accuracy, complexity, and available computational resources is necessary.
- Use the game model to analyze potential outcomes, identify your best payoffs, anticipate worst-case losses, and select effective game strategies and plans. Evaluate and reassess your situation after each step.

6.1.1 Zero-sum and non-zero-sum games

Games can be classified into zero-sum and non-zero-sum games. Zero-sum games mean that any increase in the payoff of one player will be subtracted from the payoffs of other players. In the case of two players X and Y, the zero-sum game could be represented as

$$U_X(\mathbf{x}, \mathbf{y}) + U_Y(\mathbf{x}, \mathbf{y}) = C, \tag{6.1}$$

where C is a constant and can be $C = 0$. In the latter case, $U_X(\mathbf{x}, \mathbf{y}) = -U_Y(\mathbf{x}, \mathbf{y})$. Many situations can be described as zero-sum games. For instance, any game with a single winner falls into this category. On the other hand, non-zero-sum games allow both players to increase their benefits simultaneously, resulting in a win-win scenario.

Example 131. Consider a scenario where a particular state has two national mobile phone operators. According to local regulations, individuals can only choose one operator to subscribe to. Both operators strive to entice customers by offering attractive deals and high-quality services. Do you think this situation falls under a zero-sum or non-zero-sum game?

Solution. It is indeed a zero-sum game. If an individual decides to subscribe to one operator, he is essentially lost as a potential customer for the other operator.

We will study and analyze both types of zero-sum and non-zero-sum games later.

6.1.2 Cooperative and non-cooperative games

Players often engage in non-cooperative games for a variety of reasons, such as:
- Lack of communication.
- Lack of trust between players.
- Unfair distribution of coalition payoffs makes cooperation impractical.
- There are no benefits to be gained from cooperation, making non-cooperative play more advantageous.

However, in certain games, cooperation is both possible and beneficial. When players can establish contractual agreements with each other and there is a clear advantage to working together, it makes sense for them to form coalitions. By collaborating, players can achieve mutual benefits and maximize their outcomes.

6.1.3 Game modeling; strategic and extensive forms

As we mentioned earlier, games consist of players, payoffs, strategies, and rules. In order to analyze or to find the optimal response, we should have a model or a representation of the game. Here we briefly introduce two forms of game modeling. Strategic modeling is quite straightforward. If there are two players, the game is modeled as a matrix where player X is the row player and player Y is the column player. Each element in the matrix represents the payoffs of both players. Figure 6.2 shows an example of the strategic or normal form of games. In the figure, we have assumed two players, player X has n different strategies and player Y has m strategies. If player X plays their i^{th} strategy (i. e., the ith raw) and player Y plays their j^{th} strategy (i. e., the jth column), players X and Y will receive a_{ij} and b_{ij} as their payoffs respectively. In zero-sum games, it is sufficient to show only one player's payoff, because $b_{ij} = C - a_{ij}$.

Extensive form games illustrate all possible moves by each player throughout the game. When the number of potential moves is manageable, it becomes feasible to determine the best strategy to pursue. In this form, the game can be visualized as a tree with

Strategic Form	Player Y			
Player X	Strategy 1	Strategy 2	• • •	Strategy m
Strategy 1	(a_{11}, b_{11})	(a_{12}, b_{12})	• • •	(a_{1m}, b_{1m})
Strategy 2	(a_{21}, b_{21})	(a_{22}, b_{22})	• • •	(a_{2m}, b_{2m})
⋮	⋮	• • •	⋮	⋮
Strategy n	(a_{n1}, b_{n1})	(a_{n2}, b_{n2})	• • •	(a_{nm}, b_{nm})

Figure 6.2: Game Strategic Modeling.

numerous branches. However, it is worth noting that many real-world games present a challenge to represent in a comprehensive form due to their vast number of branches and potential moves. A prime example is the game of chess, where the sheer number of possible moves exceeds the number of atoms on Earth.

Example 132. Assume there are four cards placed in two places (A and B), each with two cards as shown in Figure 6.3. Player I chooses a location (A or B) and then decides to remove one or two cards from the selected location. Player II then chooses a location with at least one card and decides how many cards to remove. Then player I starts the second round with the same rules. When both locations have no cards, the game ends. The loser is the player who removed the last card(s). Model the game in the extensive form.

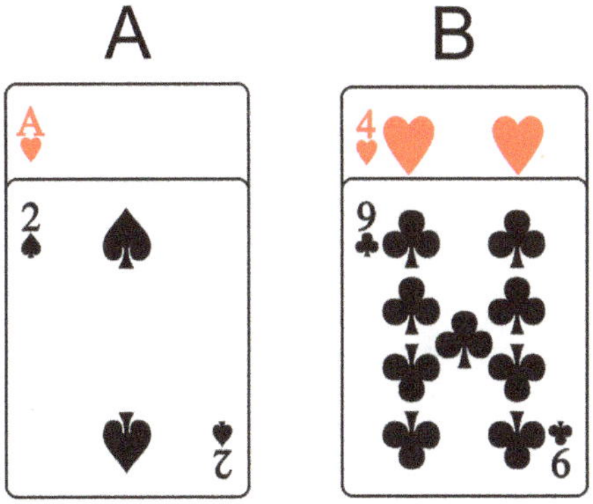

Figure 6.3: Four cards game.

Solution. The game could be constructed as the tree shown in Figure 6.4. It is possible to write down the strategies of each player. The extensive form of all the strategies of

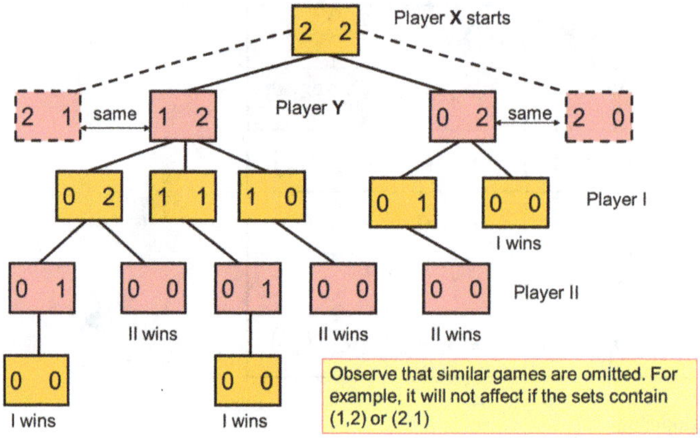

Figure 6.4: Game model tree.

players *I* and *II* is given in Figure 6.5. It is clear that the starting player has 3 strategies and the second player has 6 strategies to play the game. This game is a zero-sum game. Which is the best strategy for each player? We will analyze such games in the next section.

	Strategies for Player X
1	Play (1,2) then if at (0,2) play (0,1)
2	Play (1,2) then if at (0,2) play (0,0)
3	Play (0,2)

	Strategies for Player Y
1	If at (1,2) → (0,2); if at (0,2) → (0,1)
2	If at (1,2) → (1,1); if at (0,2) → (0,1)
3	If at (1,2) → (1,0); if at (0,2) → (0,1)
4	If at (1,2) → (0,2); if at (0,2) → (0,0)
5	If at (1,2) → (1,1); if at (0,2) → (0,0)
6	If at (1,2) → (1,0); if at (0,2) → (0,0)

Figure 6.5: Extensive model with all strategies.

6.2 Bimatrix games

Bimatrix games ARE one of the game models used to represent games played by two players. Each player has a finite number of strategies and a defined payoff, as shown in Figure 6.2. It is called bimatrix, because it can be represented by two matrices of each player. The payoff matrices of players *X* and *Y* are given respectively by

$$\mathbf{A} = \begin{bmatrix} a_{11} & a_{12} & \dots & a_{1m} \\ a_{21} & a_{22} & \dots & a_{2m} \\ \vdots & \dots & \dots & \vdots \\ a_{n1} & a_{n2} & \dots & a_{nm} \end{bmatrix} \tag{6.2}$$

$$\mathbf{B} = \begin{bmatrix} b_{11} & b_{12} & \dots & b_{1m} \\ b_{21} & b_{22} & \dots & b_{2m} \\ \vdots & \dots & \dots & \vdots \\ b_{n1} & b_{n2} & \dots & b_{nm} \end{bmatrix}. \tag{6.3}$$

A zero-sum game occurs when $\mathbf{B} = -\mathbf{A}$, making it a special case. In these games, each player aims to maximize their utility. However, maximizing one player's payoff implies minimizing the other player's payoff. This principle applies to all zero-sum games. In the next subsection, we will examine the behavior of zero-sum games.

6.2.1 Zero-sum games and MaxMin

When discussing zero-sum games in the context of a payoff matrix, we can approach the game theory problem from two perspectives. Firstly, we consider player X, who chooses strategies from the rows. The goal for player X is to maximize their own payoff. However, the specific payoff they receive depends on the column strategy chosen by player Y. Player Y, on the other hand, aims to minimize player X's payoff. It's important to remember that in a zero-sum game, minimizing your opponent's payoff results in maximizing your own payoff. The worst possible payoff that player X can receive is known as the lower value of the game v_l, and is given by

$$v_l = \max_{i=1,\dots n} \min_{j=1,\dots m} (a_{ij}). \tag{6.4}$$

Observe that the lower limit of the game is also the maximum payoff that player Y can achieve. On the other hand, the maximum payoff that player X can achieve is the same as the worst payoff that player Y may obtain and it is known as the upper limit of the game v_u and it is given by:

$$v_u = \min_{j=1,\dots m} \max_{i=1,\dots n} (a_{ij}). \tag{6.5}$$

The upper and lower values of the zero-sum payoff matrix is shown in Figure 6.6. Hence, it is always correct to say that

$$v_l \leq v_u. \tag{6.6}$$

In the case of deterministic zero-sum games, the game will have a saddle point in pure strategies if and only if

a_{11}	a_{12}	$\cdots$	a_{1m}	$\min_j a_{1j}$
a_{21}	a_{22}	$\cdots$	a_{2m}	$\min_j a_{2j}$
$\vdots$	$\vdots$	$\vdots$	$\vdots$	$\vdots$
a_{n1}	a_{n2}	$\cdots$	a_{nm}	$\min_j a_{nj}$
$\max_i a_{i1}$	$\max_i a_{i2}$	$\cdots$	$\max_i a_{im}$	

v_u = Smallest max v_l = Largest min

Figure 6.6: Upper and lower values of zero-sum games.

$$v_l = v_u. \tag{6.7}$$

When a game has a deterministic saddle point, it can be considered boring. This is because, in such games, at least one player (the winner) will always play the same strategy and will win. Deviating from the saddle point will not benefit the other player. In fact, it can even lead to a worse result, as will be shown by examples. However, the saddle point represents an equilibrium outcome where one player wins, and the other player has at least the minimum possible loss. However, each player can exploit any deviation from that point to improve their own payoffs.

Example 133. Consider a two-player zero-sum game with the payoff matrix given in Figure 6.7. Find the upper and lower bounds of the game. Does the game have at least one saddle point in pure strategy?

X/Y	A	B	C	
A	0.3	0.25	0.2	0.2
B	0.26	0.33	0.28	0.26
C	0.28	0.3	0.33	0.28 v_l
	v_u 0.3	0.33	0.33	

Figure 6.7: Game matrix of Example 133.

Solution. The matrix is 3 × 3, which means that both players X and Y have 3 strategies to play. If player X decides to play with strategy A, what should player Y play? It is a zero-sum game, hence, player Y should play the strategy which maximizes their payoff or at least minimizes their loss. Therefore, it is clear that if player Y is intelligent, he/she should play strategy C. The payoff if player X plays A and player Y plays C is $U_X = 0.2$, which implies that $U_Y = -0.2$. Now assume that you are player X. Which strategy (i. e., raw) will you choose to play? You know that player Y is intelligent and he/she will select the minimum of that row. The minimum payoff in the first row is 0.2, in the second row it's 0.26, and in the third row it's 0.28. Hence if you play the strategy C, you are guaranteed at least the payoff $U_X = 0.28$. This is the lower bound of the game. Now switch to player Y. Which column should you play? Obviously, you should choose the column that has the best payoff for you. If you select the first column, the maximum loss will be $U_Y = -0.3$, for the second column it will be $U_Y = -0.33$, and also for the third column $U_Y = -0.33$. Hence, it is logical for player Y to select strategy A. Therefore, the maximum that player X can obtain from this game is $U_X = 0.3$. Therefore, it depends on who starts the game. If player X starts the game, he/she should select strategy C to maximize their payoff. If player Y starts the game, he/she should select strategy A to minimize their loss. It is clear that in this example, there is no saddle point in pure strategy, because $v_u \neq v_l$.

Let us look at another example where there is a saddle point in the pure strategy.

Example 134. Construct the payoff matrix for the extensive game given in Example 132. Find the lower and upper bounds of the game.

Solution. Based on the strategies and the game rules, we can construct the payoff matrix as shown in Figure 6.8. It is zero-sum game where 1 means that player X wins and −1 that player Y wins. Looking at the payoff matrix, it is obvious that player Y will never play with strategy S_5, because player X will win with any strategy. In any case, player X is the starting player in this game. Regardless of the starting he/she will start with, if player Y plays strategy S_3, he/she will always win. This game has a saddle point at $v_u = v_l = -1$.

Player Y / Player X	S_1	S_2	S_3	S_4	S_5	S_6
S_1	1	1	-1	1	1	-1
S_2	-1	1	-1	-1	1	-1
S_3	-1	-1	-1	1	1	1

Figure 6.8: Game payoff matrix of Example 134.

From this example, it became evident that by listing all available strategies, one could identify the optimal strategy that maximizes the chances of winning, or even guarantees a win. In certain cases, it is feasible to determine a strategy that guarantees a 100 % chance of winning, given the option of either starting the game or playing next. An example of such a game is the well-known tic-tac-toe. However, in many games, it is not possible to represent the extensive form of the game. Chess serves as an example, where the number of possible strategies exceeds the number of atoms on Earth! Most real-life social games are even more complex than chess. Therefore, the extensive form is only used when the number of possibilities or strategies is manageable. If it were possible to model the game of chess entirely, it would lose its power and enjoyment!

Usually games in real scenarios are not deterministic. They have an element of uncertainty. For example, making a move may depend on some other event that might happen or not. To clarify this point, let's have the following simple example.

Example 135. Two players, X and Y, sit at a table where six cards are dealt face down. Of these cards, five are black and only one is red. The game follows a specific set of rules. Each player contributes 10 Euros into the pot, and has the option to pass the game to the other player. If they choose to pass, they must add another 10 Euros to the pot. Player X begins by selecting a card and revealing it. If the card is red, player X loses, and player Y claims the entire pot. If the card is black, it is player Y's turn. Before player Y plays, the cards are shuffled to conceal their colors. If the card player Y selects is red, player Y loses, and player X takes the pot. However, if the card is black, the game ends, and the pot is divided between the players. Construct the extensive form of the game. Moreover, show the payoff matrix of the game and determine the lower and upper values of the game.

Solution. Let's proceed by illustrating the extensive form of the game. A notable aspect of this game is the presence of winning and losing probabilities, which closely mirrors real-life situations involving uncertainty.

Before the game begins, each player will contribute 10 Euros to the pot. As shown in Figure 6.9, player X has two options at the start. They can choose to play or pass. If they decide to pass, they must add an additional 10 Euros to the pot. On the other hand, if they choose to play, they will draw one card. If the card is red, the game ends, and player X loses their initial 10 Euros. The probability of drawing a red card is 1/6. However, if player X draws a black card, it is player Y's turn. Player Y also has two options. They can pass, but they need to add 10 Euros to the pot first. If they pass, the game ends, and the pot is

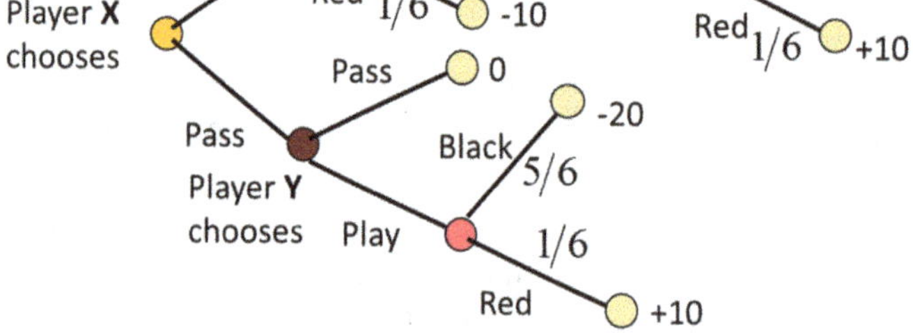

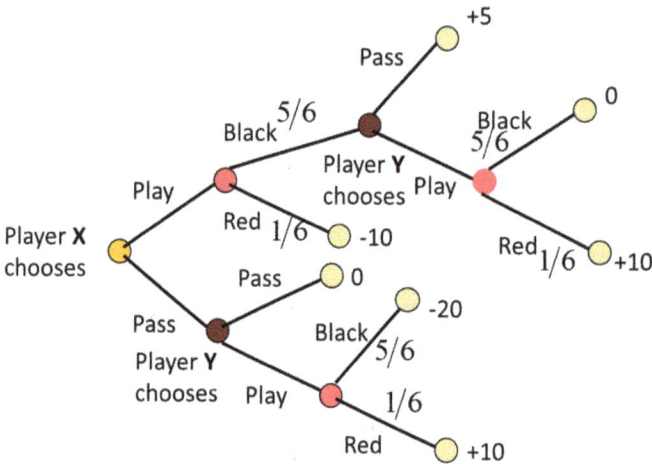

Figure 6.9: Game with uncertainties in Example 135.

divided, with player X receiving 5 Euros. Alternatively, player Y can choose to play and draw one card. If the card is red, the game ends, and player Y loses his/her 10 Euros, which goes to player X. If the card is black, the game ends with no winner. To further analyze the game, we need to construct the payoff game matrix. First, let's define all the strategies in the game. It is clear that player X has only two strategies which are X_1 = play or X_2 = pass. Furthermore, player Y has 4 (2 × 2) strategies as explained next:

- Y_1: If X_2 then Play; if X_1 then Pass.
- Y_2: If X_2 then Pass; if X_1 then Pass.
- Y_3: If X_1 then Play; if X_2 then Pass.
- Y_4: If X_1 then Play; if X_2 then Play.

We have to use the expectation in order to estimate the payoff matrix. The expectation is explained in Chapter 2 and corresponds to the average value obtain. Let's show two examples to compute the expected payoff of player X. Remember that this game is a zero-sum game, so it is enough to compute the payoff of one player. Before you look at the answer, try calculating player X's expected payoff if both players play their first strategies.

The first strategy of player X is to play X_1 =. The first strategy of player Y, as shown above, is to pass (if player X plays). Then we can compute the expected payoff as follows:

$$U_X(X_1, Y_1) = -10 \times P_r\,(\text{red}) + 5 \times P_r\,(\text{black}) \tag{6.8}$$

$$= -10 \times \frac{1}{6} + 5 \times \frac{5}{6} = \frac{15}{6} = 2.5\,\text{Euro}. \tag{6.9}$$

Next, we show how to calculate the expected payoff if player X plays their second strategy X_2 and the second player Y plays his/her first strategy Y_1.

$$U_X(X_2, Y_1) = 10 \times P_r\,(\text{red}) - 20 \times P_r\,(\text{black}) \tag{6.10}$$

$$= 10 \times \frac{1}{6} - 20 \times \frac{5}{6} = -\frac{90}{6} = -15\,\text{Euro}. \tag{6.11}$$

Remember that when $U_X(X_2, Y_1) = -15$ Euro implies that $U_Y(X_2, Y_1) = +15$ Euro. In a similar way, you can compute all elements of the payoff matrix as shown in Figure 6.10.

X/Y	Y_1	Y_2	Y_3	Y_4
X_1	2.5	2.5	−5/18	−5/18
X_2	−15	0	−15	0

Figure 6.10: Expected Payoff matrix of Example 135.

It is clear that the upper value of this game is the same as the lower value, i. e., $v_u = v_l = \frac{-5}{18}$. This means that there is a saddle point in pure strategies. On average player X (who starts) is a loser in the long run of the game.

We have previously mentioned the term *pure strategies* multiple times. It serves as a counterpoint to *mixed strategies*. Now, let's delve into the meaning of this distinction using the following example.

Example 136. Consider a two-player zero-sum game with 2 strategies for each player. The payoff matrix of the game is given by:

$$U_X = \begin{bmatrix} 3 & -1 \\ -1 & 9 \end{bmatrix}. \tag{6.12}$$

The lower value of this game is $v_l = -1$ and the upper value is $v_u = 3$, hence there is no saddle point in pure strategy. Consider yourself as player X, how you will you play this game? If it is sequential game, and your opponent knows which strategy you played, they can be always the winner. Because if you played the first strategy X_1, they will play Y_2 and you will lose -1. If you played the second strategy X_2, they will play the first strategy Y_1 and again you will lose -1. However, how you will play if neither of you knows which strategy the other player is playing until you receive the payoff? For example, if you played the first strategy and then you realized that you lost -1, then you know that your opponent played their second strategy. In this situation, both of you should both switch between the two strategies randomly. You should play the first strategy with probability x_1, where $0 \leq x_1 \leq 1$. Since we have only two strategies in this example, then the probability of the second strategy will be $x_2 = 1 - x_1$. Your opponent should also switch between their strategies with probabilities y_1, where $0 \leq y_1 \leq 1, y_2 = 1 - y_1$. The question here is how to find the best probability x_1 that you should select in order to maximize your expected payoff. Let's construct player X's expected payoff of as follows:

$$E[U_X(\mathbf{x}, \mathbf{y})] = \hat{U}_X(\mathbf{x}, \mathbf{y}) = 3x_1y_1 - x_1y_2 - x_2y_1 + 9x_2y_2, \tag{6.13}$$

where $\mathbf{x} = [x_1, x_2]^T$ and $\mathbf{y} = [y_1, y_2]^T$. The first term is 3 which is the payoff if both players play their first strategies. However, the will play it with probabilities x_1 and y_1 respectively. Since, both players are independent, then the probability that player X plays the first strategy AND that player Y also plays the first strategy is x_1y_1. How should each player play this game? There should be an equilibrium point or saddle point on the randomized strategy, so that no player could improve their expected payoff if they deviate from it, given that the other player is playing at this equilibrium point. More discussion about of randomized strategy games is given in the next subsection.

6.2.2 Randomized strategies

While not every game possesses a Nash equilibrium in pure strategies, it's worth noting that all finite games are guaranteed to possess at least one mixed (randomized) Nash equilibrium. The proof, although not particularly complex, falls beyond the purview of this book; however, it can readily be located within the pages of most theoretical game theory analysis texts. When there is no clearly dominant strategy, it is advisable to employ a randomized approach. In certain games, the most effective strategy to employ is one that introduces an element of randomness, making it difficult for your opponent to predict your next move. Let's consider a simple example to illustrate this point. Imagine that you are playing a game with another player in which both players simultaneously display one, two, or three fingers. The objective is to have the total number of fingers shown by both players result in an even number, ensuring a win for you; an odd number would result in your opponent winning, and vice versa. In this type of game, if your opponent senses a high probability that you will show only one finger, they may choose to show two fingers, thereby reducing your chances of winning. Hence, the optimal approach is to introduce randomness into your strategies when playing such a game. In randomized strategy games, we use expected utility or expected payoff as shown in Example 136.

The equilibrium point $[\mathbf{x}^*, \mathbf{y}^*]$ in any two-player game is a saddle point that satisfies the following characteristic:

$$U_X(\mathbf{x}^*,\mathbf{y}) \le U_X(\mathbf{x}^*,\mathbf{y}^*) \le U_X(\mathbf{x},\mathbf{y}^*) \tag{6.14}$$

$$U_Y(\mathbf{x},\mathbf{y}^*) \le U_Y(\mathbf{x}^*,\mathbf{y}^*) \le U_Y(\mathbf{x}^*,\mathbf{y}). \tag{6.15}$$

This point is called the *Nash Equilibrium*, and it is the major corner in the analysis of game theory. You can see that no player can improve their payoff by deviating from the equilibrium point if the other player has played the equilibrium point. In zero-sum games the equilibrium point is the point which maximizes the minimum of each player's payoff. Why is this? It is obvious that the second player will play with a strategy that minimizes the payoff of the first player (because it will be maximization for their payoff) and then the first player should maximize this minimum to obtain the highest possible payoff. Let's apply this result to the problem given in Example 136. Since there is no saddle point on the pure strategy, then it is always true to say that

$$U_X(\mathbf{x}^*,\mathbf{y}^*) \le U_X(\mathbf{x}^*,[1\,0]) \tag{6.16}$$

$$U_X(\mathbf{x}^*,\mathbf{y}^*) \le U_X(\mathbf{x}^*,[0\,1]). \tag{6.17}$$

It means that player X, with their equilibrium point, can achieve at least the same payoff as if the second player were to play any pure strategy, i. e., play their first strategy or play the second strategy. This result follows directly from the definition of the equilibrium point in equations (6.14). Consider that the payoff of player X at the equilibrium solution is $v_X = U_X(\mathbf{x}^*, \mathbf{y}^*)$. Applying this result to the expected payoff equation in (6.13), we obtain:

$$3x_1^* - x_2^* \geq v_X \tag{6.18}$$
$$-x_1^* + 9x_2^* \geq v_X \tag{6.19}$$
$$x_1^* + x_2^* = 1. \tag{6.20}$$

Since $x_2^* = 1 - x_1^*$, we obtain the following two inequalities in two unknowns:

$$4x_1^* - 1 \geq v_X \tag{6.21}$$
$$-10x_1^* + 9 \geq v_X. \tag{6.22}$$

We try to solve the above inequalities by replacing to quality. If the solution is achieved at $0 \leq x_1 \leq 1$, then the problem is solved. We now have

$$4x_1^* - v_X = 1 \tag{6.23}$$
$$10x_1^* + v_X = 9. \tag{6.24}$$

Solving the above two equations we obtain $x_1^* = \frac{5}{7}$, $x_2^* = \frac{2}{7}$ and $v_X = \frac{13}{7}$. Since v_X is the equilibrium point, it is possible to solve for $y_1^* = 1 - y_2^*$ using (6.13). Show that the optimal probability of player Y is given by $y_1^* =$ and $y_2^* =$. Figure 6.11 shows a graphical representation of the game from player X's point of view. The x-axis is dedicated to the probability of playing the first strategy and the y-axis is dedicated to the expected payoff of player X. Observe that player X wants to maximize their payoff and player Y aims to minimize this payoff (zero-sum game). Hence, if player Y plays the second strategy, player X will play their second strategy to obtain a payoff equal to 9. If player Y plays their first strategy, player X will play their first strategy to receive a payoff equal to 3. The best thing that player Y can do is to switch between the two strategies randomly.

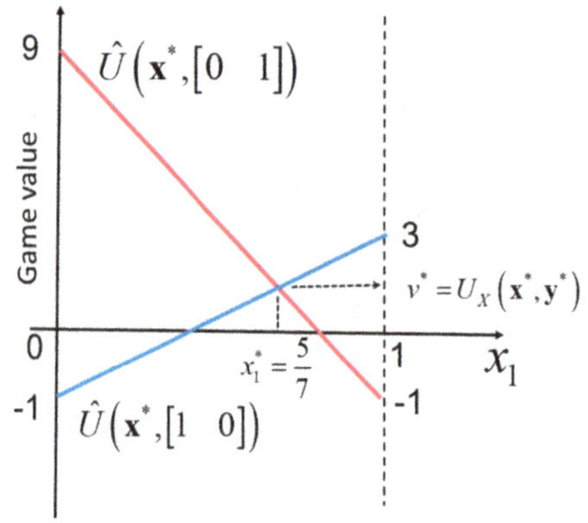

Figure 6.11: Graphical presentation of Example 136.

The best response of player X is to also switch between their two strategies randomly. The saddle point, where neither player can do better than the other, is the intersection of the two lines as shown in the figure. If player X moves in their switching probability such that $x_1 > x_1^* = \frac{5}{7}$, it is clear that player Y can exploit this deviation and switch to their second strategy and then player Y will improve their payoff and player X will receive $\hat{U}_X < v^*$. Same thing will happen if player X decided to play with $x_1 < x_1^*$. This is the game from the perspective of player X. We can repeat all this analysis from the representative of player Y. However, in zero-sum games, this is not necessary.

Theorem 1. *For any zero-sum two-player normal form game, there exists a Maximin strategy for each player. In other words, there must be at least one Nash equilibrium.*

The proof is straightforward and relies on the Von Neumann minimax theorem, although it won't be presented here.

Example 137. The following payoff matrix is for a two-player zero-sum game. Find the lower and upper bounds. Is there a saddle point for pure strategies? Find the equilibrium point for randomized strategies.

$$U_X = \begin{bmatrix} 2 & -2 & 2 \\ -4 & 10 & 2 \end{bmatrix}. \tag{6.25}$$

As mentioned earlier, we can determine the lower value of the game by identifying the minimum value in each row and then selecting the maximum of these two values. The first row has a minimum of −2, and the second row has a minimum of −4. Taking the maximum of these values gives us −2, denoted as $v_l = -2$.

To find the upper bound of the game, we calculate the maximum value in each column, which are 2, 10, and 2 respectively. Taking the minimum of these values yields 2, denoted as $v_u = 2$.

Since there is no pure strategy saddle point in this game, it is advisable for both players to adopt randomized strategies. Player X should implement a strategy that involves randomly selecting between two strategies, while player Y should do the same, choosing between three strategies. To identify the equilibrium point for the randomized strategies, we can apply the same concept as before. Player X will always achieve a game value of v^* or higher when playing their equilibrium point $[x_1^*, 1 - x_1^*]$, regardless of the pure strategy chosen by player Y. For example, the expected payoff of player X when player Y plays their first strategy is $U_X(\mathbf{x}^*, [1,0,0]) = 2x_1^* - 4(1-x_1^*) = 6x_1^* - 4 \geq v^*$, applying the same concept with other 2 column strategies we get, $U_X(\mathbf{x}^*, [0,1,0]) = -12x_1^* + 10 \geq v^*$, and $U_X(\mathbf{x}^*, [0,0,1]) = 2 \geq v^*$. Figure 6.12 shows a graphical representation of these inequalities. The equilibrium point is the maximum of the minimum intersection within the allowed range. The equilibrium point will be at $x_1^* = \frac{7}{9}, x_2^* = \frac{2}{9}$, and the saddle point at $v^* = \frac{2}{3}$. Solving the game from player Y side, we obtain $y_1^* = \frac{2}{3}, y_2^* = \frac{1}{3}$, and $y_3^* = 0$. This means that player Y must not play the third strategy at all. This is obvious since they have no chance of getting a positive payoff.

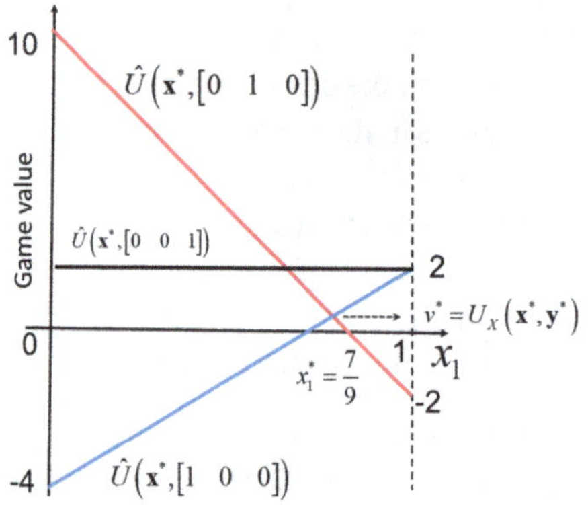

Figure 6.12: Graphical presentation of Example 137.

6.2.3 Linear programming formulation for zero-sum games

The above sections introduce the concepts of two-player zero-sum games, encompassing both pure and randomized strategies. However, in order to effectively solve problems of various sizes within this domain, it is necessary to establish a mathematical framework. Consider a zero-sum game with a payoff matrix of player X as $\mathbf{A}$ with size $n \times m$. Player X has n strategies and player Y has m strategies. It is zero-sum game, so that the payoff of player Y is $\mathbf{B} = -\mathbf{A}$. We employ the same method as in the previous section. The expected payoff of player X when they play at their equilibrium point $\mathbf{x}^*$, will ensure that the attained value is at least v^*, the game value. Furthermore, the payoff achieved by player X can be higher if player Y deviates from their equilibrium point $\mathbf{y}^*$. Let's define $\mathbf{S}_j^y$ as player Y plays their j^{th} column (i. e. strategy). In other words, $\mathbf{y} = [0\ 0\ \ldots 1\ 0 \ldots 0]^T$, where the 1 is at the jth position of the vector. Therefore, it is always true to say

$$\hat{U}_X(\mathbf{x}^*, \mathbf{S}_j^y) = \sum_{i=1}^{n} x_i^* a_{ij} \geq v^*, \tag{6.26}$$

where a_{ij} are values of the payoff matrix as shown in Figure 6.6. Remember that in pure or randomized strategies, we must always achieve $\sum_{i=1}^{n} x_i^* = \sum_{j=1}^{m} y_j^* = 1$. Let's define a new parameter $z_i^X = \frac{x_i^*}{v^*}$. The main objective of player X is to maximize their payoff v^*. However, $\sum_{i=1}^{n} z_i^X = \sum_{i=1}^{n} \frac{x_i^*}{v^*} = \frac{1}{v^*}$. Hence, to maximize v^*, we need to minimize $\sum_{i=1}^{n} z_i^X$. The equation (6.26) can be reformulated by dividing both sides by v^* as

$$\hat{U}_X(\mathbf{x}^*, \mathbf{S}_j^y) = \sum_{i=1}^{n} z_i^X a_{ij} \geq 1. \tag{6.27}$$

Based on the above settings, we can formulate the problem of solving the two-player zero-sum game as the following optimization formulation (linear programming):

$$\min \sum_{i=1}^{n} z_i^X \tag{6.28}$$

$$\text{Subject to:} \tag{6.29}$$

$$\sum_{i=1}^{n} z_i^X a_{ij} \geq 1 \quad \forall j = 1, \ldots m \tag{6.30}$$

$$z_i^X \geq 0 \quad \forall i = 1, \ldots n. \tag{6.31}$$

Solving the above linear programming problem will solve the zero-sum games when the game value $v^* \neq 0$. If the game value is zero, we simply have to add any number of all elements of the payoff matrix. This does not affect the equilibrium point. In matrix and vector form, with $\mathbf{z}^X = [z_1^X\, z_2^X \ldots z_n^X]^T$ and $\mathbf{1}_n$ as column vector of n ones, the above LP problem can be represented as:

$$\min \mathbf{1}_n^T \mathbf{z}^X \tag{6.32}$$

$$\text{Subject to:} \tag{6.33}$$

$$\mathbf{A}^T \mathbf{z}^X \geq \mathbf{1}_m \tag{6.34}$$

$$z_i^X \geq 0 \quad \forall i = 1, \ldots n. \tag{6.35}$$

Perhaps all programming languages have ready-made packages for solving linear programming problems. Octave has a useful function called *glpk* that we will use it later to solve the zero-sum games. Once we solve for z_i^X, we can find the probability of selecting strategy i^{th} as $x_i = z_i^X \times v^*$.

We have shown how to compute the equilibrium point of player X. What about player Y? Player Y will play to minimize v^*. It is a zero-sum game, so minimizing the game value of player X leads to maximize player Y's payoff. If player Y plays with their equilibrium point and player X plays with any pure strategy then:

$$\hat{U}_Y(\mathbf{S}_i^X, \mathbf{y}^*) = \sum_{j=1}^{m} y_i^* a_{ij} \leq v^*. \tag{6.36}$$

Similarly $\mathbf{S}_i^X$ is defined as player X is playing their i^{th} raw. Following similar steps as before, we will have the following linear optimization

$$\max \mathbf{z}^y \mathbf{1}_m^T \tag{6.37}$$

$$\text{Subject to:} \tag{6.38}$$

$$\mathbf{A}(\mathbf{z}^y)^T \leq \mathbf{1}_n \tag{6.39}$$

$$z_j^y \geq 0 \quad \forall j = 1, \ldots m. \tag{6.40}$$

Example 138. Use any linear programming function to find the equilibrium point and the game value of the following zero-sum game

$$U_X = \begin{bmatrix} 2 & -2 & 2 & -10 \\ -4 & 10 & 2 & 10 \\ 4 & 0 & -6 & -1 \end{bmatrix}. \tag{6.41}$$

"The following code demonstrates how to compute the equilibrium point in randomized strategies for any given size of the payoff matrix. The obtained results are as follows:

- For player X: $\mathbf{x}^* = [0.364, 0.397, 0.238]$.
- For player Y: $\mathbf{y}^* = [0.529, 0, 0.331, 0.150]$.
- The game value is $v^* = 0.093$.

Based on the results, it is evident that player Y will never use their second strategy. The reason for this is that the fourth column outperforms the second column. Comparing the individual elements, it becomes apparent that all values in the fourth strategy are superior to those in the second strategy of player Y. Therefore, there is no incentive for player Y to consider employing the second strategy."

```
% Octave code
function [X,Y,v]=zerosum(A)
siz=size(A);
b1=ones(siz(2),1);
c1=ones(siz(1),1);
x=glpk(c1,A',b1,[],[],repmat('L', 1, siz(2)));
y=glpk(-b1,A,c1,[],[],repmat('U', 1, siz(1)));
v=1/sum(x); % Game value
X=x.*v; % Mixed Probability of Row Player
Y=y.*v; % Mixed Probability of Column Player
```

6.2.4 Non-zero-sum games

Not all games are zero-sum. In fact, there are several games where both players can gain payoffs that are not necessarily at the expense of the other player. In these games, there can be correlations between the payoffs of the two players, but they do not conform to the traditional zero-sum structure. To illustrate this, consider the scenario where customers are allowed to possess multiple SIM cards. In such a case, the payoff functions of the service providers are not required to be zero-sum. Similarly, in a game where different shops sell the same gadgets, it is not necessary for the game to be zero-sum. Customers may choose to purchase from different shops based on various factors such as product quality or service. In non-zero-sum games, it is possible to have outcomes where both players achieve a win-win situation or even a lose-lose situation. The game's structure allows for more diverse possibilities beyond the strict zero-sum paradigm. In the case of two players, each player will have their payoff matrix as shown in Figure 6.2.

The payoff matrix of player X can be defined as $\mathbf{A}$, and that of player Y as $\mathbf{B}$. The expected payoffs for each player are given by

$$\hat{U_X}(\mathbf{x},\mathbf{y}) = E[U_X(\mathbf{x},\mathbf{y})] = \mathbf{x}^T\mathbf{A}\mathbf{y}, \tag{6.42}$$

$$\hat{U}_Y(\mathbf{x},\mathbf{y}) = E[U_Y(\mathbf{x},\mathbf{y})] = \mathbf{x}^T\mathbf{B}\mathbf{y}, \tag{6.43}$$

where $\mathbf{x} = [x_1\, x_2\, \ldots\, x_n]^T$, and $\mathbf{y} = [y_1\, y_2\, \ldots\, y_m]^T$.

Example 139. Consider a two-player non-zero-sum game defined by the following two payoff matrices for players X and Y respectively:

$$\mathbf{A} = \begin{bmatrix} 2 & -2 & 2 \\ -4 & 10 & 2 \end{bmatrix} \tag{6.44}$$

$$\mathbf{B} = \begin{bmatrix} 3 & 8 & 9 \\ 1 & -4 & -5 \end{bmatrix}. \tag{6.45}$$

Find the expected payoff of each player when
- $\mathbf{x}_1 = [1\,0]^T$, and $\mathbf{y}_1 = [0\,1\,0]^T$;
- $\mathbf{x}_2 = [0.1\,0.9]^T$ and $\mathbf{y}_2 = [0.2\,0.8\,0]^T$.

In the first case, player X and Y both play with pure strategy, player X plays their first strategy (first raw) and player Y plays the second strategy (second column). Hence $U_X(\mathbf{x}_1,\mathbf{y}_1) = -2$, and $U_Y(\mathbf{x}_1,\mathbf{y}_1) = 8$. In the second scenario, the players are playing with a randomized strategy, and in this case we may compute the expected payoff as:

$$\hat{U}_X = 0.1 \times (2 \times 0.2 - 2 \times 0.8) + 0.9 \times (-4 \times 0.2 + 10 \times 0.8) = 6.36 \tag{6.46}$$

$$\hat{U}_Y = 0.1 \times (3 \times 0.2 + 8 \times 0.8) + 0.9 \times (1 \times 0.2 - 4 \times 0.8) = -2. \tag{6.47}$$

The Nash equilibrium solution, denoted as $[\mathbf{x}^*, \mathbf{y}^*]$, in non-zero-sum games is defined similarly to before. It represents the solution where neither player can gain any advantage by changing their strategy if the other player remains unchanged. In other words, it is the state where both players have reached a stable outcome, and any deviation by one player would not yield any immediate benefits in terms of their individual payoffs unless the other player also deviates.

In certain non-zero-sum games, the Nash equilibrium can be found in pure strategies. The assignment of the Nash equilibrium can be determined using the following approach: player X examines each column and identifies their highest payoff, while player Y assesses each row and identifies their highest payoff. If there is a point that is marked as the highest payoff by both players, it represents the Nash equilibrium point. To illustrate this concept further, let's consider an iconic example in game theory known as the Prisoner's Dilemma.

Prisoner's dilemma

Two students missed the last exam in a particular course because they were at a party, and they informed the teacher that they encountered car trouble while on the way to the exam. Surprisingly, the teacher agreed to arrange a make-up exam for them. On the day of the makeup exam, the teacher assigned each student to a separate classroom. The students were presented with a single question: 'Are you certain that your absence from the original exam was genuinely caused by car trouble?' The consequences were made clear: if they lied, they would be dismissed from the school! How should the two students play the game? The students were provided with payoff matrices for the game, as shown in Figure 6.13. The letter *D* means dismissed from the school. It could be represented as a very large negative number. Moreover, it's important to note that they were unable to communicate with each other. Each student (player) has to make his/her decision without communicating or knowing the other's decision.

Student X/Y	Confess	Don' t confess
Confess	(0,0)	(3,D)
Don' t confess	(D,3)	(1,1)

Figure 6.13: Bimatrix of prisoner's dilemma.

In the scenario where both students confess to attending the party, they will both receive a grade of 0 in the course. If one student confesses and tells the truth, while the other lies, the truth-teller will be rewarded with a grade of 3, while the one who conceals the truth will face dismissal from the school, represented by a significant negative value. Alternatively, if neither student confesses, the teacher will give them both a grade of 1, assuming they were honest.

Does this example have a Nash equilibrium point in pure strategy?

It's evident that each student aims to maximize their expected payoff. However, it's important to note that there is no communication or cooperation between them as they answer the questions. To determine if there is a Nash equilibrium in the pure strategy, we can follow a systematic approach: Identify the highest number in each column and mark it with an asterisk. Identify the second highest number in each row and mark it with an asterisk. Any pair of numbers that both have an asterisk represents a Nash equilibrium in pure strategies.

The result for this example is provided in Figure 6.14. Analyzing the previous slide, it becomes evident that we have a single pure Nash equilibrium at (confess, confess). If either student deviates from confess while the other student continues to play confess,

Student X/Y	Confess	Don' t confess
Confess	(0*,0*)	(3*,D)
Don' t confess	(D,3*)	(1,1)

Figure 6.14: Nash equilibrium of prisoner's dilemma.

the deviating student's payoff transitions from failing one exam to being expelled from school entirely! However, according to the previous table, both students can achieve better outcomes if they both choose not to confess, resulting in a passing grade of 1 for both. Nevertheless, there is an incentive for each student to play the confess strategy. If the other student chooses not to confess, they will receive a grade of 3 in the course. This situation implies that players are rewarded for betraying the other player. The payoff pair $(1, 1)$ is considered unstable because one player can improve their outcome by deviating, assuming the other player does not. Conversely, the payoff pair $(0, 0)$ is stable, since neither player can increase their own payoff if they both stick to it. This simplified non-zero-sum game provides some insights on how we think in similar situations. This gives some mathematical modeling that could be used to predict the behaviors of an intelligent decision maker.

Example 140. Find the Nash equilibrium in pure strategies of a two-player non-zero-sum game with payoff matrices given by:

$$\mathbf{A} = \begin{bmatrix} 2 & 3 \\ 1 & 4 \end{bmatrix} \tag{6.48}$$

$$\mathbf{B} = \begin{bmatrix} 2 & 1 \\ 3 & 4 \end{bmatrix}. \tag{6.49}$$

To apply the same rule as before, we examine the first number in each column and identify the largest number. Similarly, we assess the second number in each row and identify the largest numbers. This example yields two Nash equilibrium points as shown in Figure 6.15. Naturally, the point $(4, 4)$ stands out as the better option for both players.

Not all non-zero-sum games possess pure Nash equilibrium points. In numerous instances, players must adopt mixed strategies. Determining the optimal probability vector for each player resembles the approach used in zero-sum games. However, in non-zero-sum games, each player has their own payoff matrix. For example, you can check that players with the following payoff matrices, they do not have a Nash equilibrium in pure strategy:

Players X/Y	Strategy 1	Strategy 2
Strategy 1	**(2*,2*)**	**(3,1)**
Strategy 2	**(1,3)**	**(4*,4*)**

Figure 6.15: Bimatrix of Example 140.

$$\mathbf{A} = \begin{bmatrix} 2 & 1 \\ 0 & 3 \end{bmatrix} \tag{6.50}$$

$$\mathbf{B} = \begin{bmatrix} 0 & 2 \\ 1 & 0 \end{bmatrix}. \tag{6.51}$$

The expected payoffs of each player are given in equations (6.42). In playing this non-zero-sum game, players have at least two options. The first approach involves each player solely considering their own payoff matrix and assuming the game is a zero-sum game. Consequently, each player engages in their own independent zero-sum game, receiving a payoff equal to their game value. However, this approach is not the most optimal strategy. Although the obtained payoff by assuming a zero-sum game is called the safety value, it does not represent the best way to play the game. This value is guaranteed for each player, as the payoff matrix remains fixed. The second and superior method involves a joint optimization approach, where both players strive to find the Nash equilibrium point by optimizing both payoff matrices.

To elucidate these methods, let's begin with the following example

Example 141. Consider a non-zero game given by the following matrices

$$\mathbf{A} = \begin{bmatrix} 2 & 1 \\ -3 & 3 \end{bmatrix} \tag{6.52}$$

$$\mathbf{B} = \begin{bmatrix} 0 & 7 \\ 3 & 0 \end{bmatrix}. \tag{6.53}$$

Check if there are any pure Nash equilibrium points. If each player assumed it was a zero-sum game, find the optimal mixed strategy and the expected payoff of each player.

Solution. You may put the two payoff matrices together as shown before, and you will not find a Nash equilibrium in the pure strategy as shown in Figure 6.16.

If we apply each matrix to our *zerosum* code in octave shown earlier in the previous section, we receive the following results for each player.
- $x_1 = 0.85, x_2 = 0.15, v_X(\mathbf{A}) = 1.29.$
- $y_1 = 0.7, y_2 = 0.3, v_Y(\mathbf{B}^T) = 2.10.$

Players X/Y	Strategy 1	Strategy 2
Strategy 1	**(2*,0)**	**(1,7*)**
Strategy 2	**(-3,3*)**	**(3*,0)**

Figure 6.16: Bimatrix of Example 141.

The results above provide safety values for each player, indicating what they can achieve. However, it's important to note that these values were obtained assuming a zero-sum game. In non-zero-sum games, players should generally adopt better strategies to maximize their outcomes and exceed the safety values (if possible).

Nash equilibrium for non-zero sum games

At the Nash equilibrium solution $(\mathbf{x}^*, \mathbf{y}^*)$ of non-zero-sum games, we have

$$\hat{U}_X(\mathbf{x}^*, \mathbf{y}^*) \geq \hat{U}_X(\mathbf{x}, \mathbf{y}^*) \tag{6.54}$$

$$\hat{U}_Y(\mathbf{x}^*, \mathbf{y}^*) \geq \hat{U}_Y(\mathbf{x}^*, \mathbf{y}). \tag{6.55}$$

In the case of a bimatrix, the above equations can be represented as follows:

$$\hat{U}_X(\mathbf{x}^*, \mathbf{y}^*) = \mathbf{x}^{*T}\mathbf{A}\mathbf{y}^* \geq \mathbf{x}^T\mathbf{A}\mathbf{y}^* \tag{6.56}$$

$$\hat{U}_Y(\mathbf{x}^*, \mathbf{y}^*) = \mathbf{x}^{*T}\mathbf{B}\mathbf{y}^* \geq \mathbf{x}^{*T}\mathbf{B}\mathbf{y}. \tag{6.57}$$

To find the equilibrium vectors $(\mathbf{x}^*, \mathbf{y}^*)$ that meet the aforementioned conditions, there are various straightforward methods, especially for small payoff matrices. However, for the sake of brevity, we will skip these methods and dive directly into the general mathematical framework. It is possible to solve a general two-player non-zero-sum game by solving the following nonlinear optimization problem

$$\begin{aligned}
&\max_{\mathbf{x},\mathbf{y},\alpha,\beta} \mathbf{x}^T(\mathbf{A}+\mathbf{B})\mathbf{y} - \alpha - \beta \\
&\text{Subject to:} \\
&\quad \mathbf{A}\mathbf{y} \leq \alpha \mathbf{1}_n \\
&\quad \mathbf{B}^T\mathbf{x} \leq \beta \mathbf{1}_m \\
&\quad x_i \geq 0 \quad \forall i = 1, \ldots n \\
&\quad y_j \geq 0 \quad \forall j = 1, \ldots m \\
&\quad \sum_{i=1}^{n} x_i = \sum_{j=1}^{m} y_j = 1,
\end{aligned} \tag{6.58}$$

where $\alpha = \hat{U}_X(\mathbf{x}^*, \mathbf{y}^*)$, and $\beta = \hat{U}_Y(\mathbf{x}^*, \mathbf{y}^*)$. There are several software packages to solve such nonlinear programming problems. A possible code using octave is shown next.

```
% Octave code
function XX=nonzerosum(A,B)
pkg load optim
if size(A) != size(B)
printf('Matrices sizes must be identical')
Return
end
[n,m] = size(A);
x0=randn(n+m+2,1); % initial values of x, y, alpha, and
    beta
function e = g (x)
E=[ones(1,n) zeros(1,m);zeros(1,n) ones(1,m)];
e=E*x(1:n+m)-ones(2,1);
endfunction
function r = h (x)
r=-([B' zeros(m,m);zeros(n,n) A]*x(1:n+m)-[x(n+m+2)*
    ones(m,1);x(n+m+1)*ones(n,1)]);
endfunction
function obj = phi (x)
obj = -x(1:n)'*(A+B)*x(n+1:n+m)+x(n+m+1)+x(n+m+2);
endfunction
lp=[zeros(n+m,1);-inf;-inf];
up=[ones(n+m,1);inf;inf];
[XX, obj, info, iter, nf, lambda] = sqp (x0, @phi, @g,
    @h,lp,up);
endfunction
```

Example 142. Using the above *nonzerosum* code, find the Nash equilibrium solutions for the following payoff matrices:

$$\mathbf{A} = \begin{bmatrix} 2 & 1 & -6 & 2 \\ -3 & 3 & -6 & 2 \\ 0 & 4 & 2 & 8 \\ 7 & -1 & 0 & 9 \end{bmatrix} \tag{6.59}$$

$$\mathbf{B} = \begin{bmatrix} 0 & 1 & -4 & -2 \\ -3 & 7 & 7 & 10 \\ 6 & -3 & 0 & 1 \\ -1 & 2 & 6 & -2 \end{bmatrix}. \tag{6.60}$$

You can repeat the run with different initial values to find more Nash equilibria points (if any!).

Solution. After running the code several times, we obtained the following results:
- $\mathbf{x}^* = [0, 0, 0.54, 0.46]$, $\mathbf{y}^* = [0.22, 0, 0.78, 0]$, $U_X^* = 1.55$, $U_Y^* = 2.77$.
- $\mathbf{x}^* = [0.42, 0, 0.18, 0.4]$, $\mathbf{y}^* = [0.42, 0.58, 0, 0]$, $U_X^* = 2.33$, $U_Y^* = 0.68$.

You may easily check that this game does not have a Nash equilibrium in pure strategies. The safety value of both players are 1.55 and 0.6 respectively.

Randomized strategies in incomplete information games

In the previous section, we assumed that both players possessed complete knowledge of each other's precise payoff matrices. However, in more realistic scenarios, players often need to make estimates about the elements within their opponent's payoff matrix. Consequently, the estimated probability associated with each strategy chosen by the opposing player may not align with the actual probability of its use. To illustrate, consider player Y, who may opt to play strategy S_{y1} with a probability denoted as y_1, but player X believes that player Y should employ the same strategy with a probability denoted as $\pi_X(S_{y1})$. When player X's estimate of player Y's payoff matrix is inaccurate, discrepancies may arise, leading to instances where $y_j \neq \pi_X(S_{yj})$ for certain or all strategies $j \in \{1, 2, \dots, N_y\}$. Incomplete information in games can manifest itself not only in the uncertainty about the opponent's payoff matrix but also in variables such as the number of strategies or hidden objectives. When dealing with incomplete games, probabilistic methods become indispensable tools for refining our understanding and improving our estimates of the unknown parameters in our opponent's strategies. This necessity leads to the use of tools such as estimation, elaborated upon in the chapter on estimation theory. In this context, the equilibrium solution that emerges is known as the *Bayesian Nash Equilibrium*.

N-player games

In real-life games, more than two players are typically involved. For instance, numerous companies may bid on a particular project. In another example, utility companies compete to acquire more customers. Additionally, game coordination between multiple parties often occurs in politics. As an example, let's consider a scenario where three players are involved in a non-zero-sum game.

Example 143. Suppose three players X, Y, and Z are involved in a non-zero-sum game with 3, 4, and 2 possible strategies respectively. We have to define two bimatrices for each strategy of player Z as

$$\mathrm{A} = \begin{bmatrix} (2,1,3) & (-2,2,8) & (2,2,2) & (3,0,-3) \\ (-9,10,-7) & (20,10,30) & (-8,7,2) & (-8,0,0) \\ (11,12,-16) & (-1,-2,-7) & (16,4,-4) & (22,11,33) \end{bmatrix} \tag{6.61}$$

$$\mathbf{B} = \begin{bmatrix} (16,5,-22) & (2,3,27) & (-2,11,-3) & (0,0,0) \\ (-4,-4,-1) & (-9,-7,2) & (8,2,4) & (11,11,31) \\ (7,10,12) & (21,-1,-3) & (-8,-5,0) & (0,11,-3) \end{bmatrix}. \tag{6.62}$$

Bimatrix **A** represents the payoffs for each player if player Z chooses their first strategy, while bimatrix **B** represents the payoffs for each player if player Z chooses their second strategy. If player Z has k different strategies, we will have k bimatrices. If players X, Y, and Z play the second, third, and second strategies respectively, each player's payoff will be $(8,2,4)$.

Example 144. Use the same payoff matrices of the previous example, show the bimatrices based on player X.

Solution. In this case we will have three bimatrices, one corresponding to each strategy of player X. Let's call them matrices **A**, **B**, **C**.

$$\mathbf{A} = \begin{bmatrix} (2,1,3) & (-2,2,8) & (2,2,2) & (3,0,-3) \\ (16,5,-22) & (2,3,27) & (-2,11,-3) & (0,0,0) \end{bmatrix} \tag{6.63}$$

$$\mathbf{B} = \begin{bmatrix} (-9,10,-7) & (20,10,30) & (-8,7,2) & (-8,0,0) \\ (-4,-4,-1) & (-9,-7,2) & (8,2,4) & (11,11,31) \end{bmatrix} \tag{6.64}$$

$$\mathbf{C} = \begin{bmatrix} (11,12,-16) & (-1,-2,-7) & (16,4,-4) & (22,11,33) \\ (7,10,12) & (21,-1,-3) & (-8,-5,0) & (0,11,-3) \end{bmatrix}. \tag{6.65}$$

6.2.5 Continuous strategy games

In ad hoc wireless networks, multiple transceivers often share the same channel. Each device has the ability to adjust its transmit power. When there's only one transmitter-receiver pair in the channel, increasing the transmit power enhances the quality of the received signal. However, the scenario changes when there are multiple independent transmitter-receiver pairs. In such cases, increasing the transmit power of one transmitter raises the interference level for other receivers. Consequently, this prompts other transmitters to increase their transmit power as well. This situation can potentially disrupt communication for most pairs. To address this issue, game theory provides a valuable framework for analysis. Here, each player (transmitter) adopts a strategy represented by the transmit power, which is a continuous real number ranging from 0 to the maximum allowable transmit power, denoted as $p_{\max}$. The utility of each pair can be defined as the signal quality received at their respective receivers. For instance, for the ith pair, we denote the utility as $u_i(x_1, \ldots, x_i, \ldots, x_n)$, where $0 \le x_i \le p_{\max}$ is the transmit power of transmitter i. In such problems we can explore the Nash equilibrium solution in both cooperative and non-cooperative games. It is possible to analyze these games in pure strategies as well as in randomized strategies. Not all continuous strategy games have a Nash equilibrium in pure strategy.

Assuming K players where the strategy of each player k is represented by a continuous real number x_k, the Nash equilibrium solution (if there is at least one in pure strategy) should be reached:

$$u_k(x_1^*,\dots,x_{k-1}^*,x_k,\dots,x_K^*) \le u_k(x_1^*,\dots,x_k^*,\dots,x_K^*) \quad \forall k=1,\dots K. \tag{6.66}$$

It is the same concept as before, no player can achieve a better utility by deviating from the equilibrium solution if all other players stick to it. In other words, a Nash equilibrium is a mutual best response. It is not necessary for every continuous game to have a Nash equilibrium in the pure strategy. The following are *sufficient but not necessary* conditions to have a pure equilibrium solution:

$$\frac{\partial u_k(x_1,\dots,x_K)}{\partial x_k} = 0, \quad \forall k=1,\dots K \tag{6.67}$$

$$x_k \rightarrowtail u_k \quad (x_1,\dots,x_K) \tag{6.68}$$

$$\frac{\partial^2 u_k(x_1,\dots,x_K)}{\partial x_k^2} < 0, \quad \forall k=1,\dots K. \tag{6.69}$$

The first condition guarantees that each utility is smooth and has an extreme point with its strategy. The second condition is for the uniqueness, and the third condition is to guarantee that the extreme point is maximum.

Example 145. Assume that there are two players with continuous positive strategies x_1, x_2 and their payoffs are given respectively by

$$u_1(x_1,x_2) = -\frac{1}{2}x_1^2 - x_2^2 + 10x_1 - \frac{1}{2}x_1x_2 + 10 \tag{6.70}$$

$$u_2(x_1,x_2) = -x_1^2 - \frac{1}{2}x_2^2 + 5x_2 + x_1x_2 + 2. \tag{6.71}$$

Is there a Nash equilibrium solution? Show what would happen if any player deviated from the equilibrium point.

Solution. The first derivatives for both utilities are:

$$\frac{\partial u_1}{\partial x_1} = -x_1 + 10 - \frac{1}{2}x_2 = 0 \tag{6.72}$$

$$\frac{\partial u_2}{\partial x_2} = -x_2 + 5 + x_1 = 0. \tag{6.73}$$

Solving the above two linear equations we obtain $x_1 = 5$ and $x_2 = 10$. To check that both solutions are maximum, we take the second derivatives as

$$\frac{\partial^2 u_1}{\partial x_1^2} = -1 \tag{6.74}$$

$$\frac{\partial^2 u_2}{\partial x_2^2} = -1. \tag{6.75}$$

Therefore, the Nash equilibrium point is $(x_1^*, x_2^*) = (5, 10)$. If it is a non-cooperative game, both players should stick to this solution regardless of their satisfaction. At this point, the utilities are $u_1 = -77.5$ and $u_2 = 27$. We may guess that player 1 should be very sad with such low a utility in this game. Assume that player 1 decided to increase their strategy to $x_1 = 10$. Suppose that player 2 decided to keep their strategy at the equilibrium value. In this case, the utility of player 1 will be even worse and becomes $u_1 = -90$. The utility of player 2 will be dropped as well to $u_1 = 2$. However, player 2 can even improve their utility due to the deviation of player 1 by setting their strategy to $x_2 = 15$. Now the utility of player 2 will be elevated to $u_2 = 14.5$ and player 1 will be further dropped to $u_1 = -240$. If you draw the utilities of both players, you can see that the best strategy both can adopt if they play non-cooperatively is the Nash equilibrium. Remember that the Nash equilibrium is generally not the Pareto optimal solution. For example, if both players have decided to play with a cooperative procedure, the optimal strategy will be different.

In continuous games, the Nash equilibrium can be an infinite number of points. The next example demonstrates this case.

Example 146. Assume there are two players with continuous strategies x_1, x_2, where both are $\in [0, 10]$, and their utilities are respectively given by

$$u_1(x_1, x_2) = x_1 x_2 - \frac{1}{2}x_1^2 \tag{6.76}$$

$$u_2(x_1, x_2) = x_1 x_2 - \frac{1}{2}x_2^2. \tag{6.77}$$

Using the same approach as previously demonstrated, we can ascertain that the Nash equilibrium solution to this dilemma is reached when $x_1^* = x_2^*$. This implies that any value within the between 0 and 10 has the potential to be a Nash equilibrium, given that $x_1 = x_2$. Both will receive identical payoffs at the Nash equilibrium point as $u_1^* = u_2^* = \frac{x^2}{2}$. The maximum that both can achieve is $u = 50$ at $x_1^* = x_2^* = 10$. Figure 6.17 shows the payoff of player 2 when player 1 plays the strategy $x_1 = 5$. It is clear that player 2 should play with an identical strategy to maximize their payoff.

Previous sections on discrete strategy games have demonstrated that players can adopt a randomized approach to strategically enhance the expected value of their utility function. Building on this insight, a natural extension arises: how can we effectively apply randomization to the realm of continuous strategy games? In this context, the approach shifts from seeking the best instance probability to a more comprehensive objective. Here, the focus for each player is on discovering the optimal or most advantageous probability density function that aligns with the maximization of their utility functions. The resulting strategic value is then derived by sampling from this optimal probability density function. Assume that we have N players, and each player i has a real number

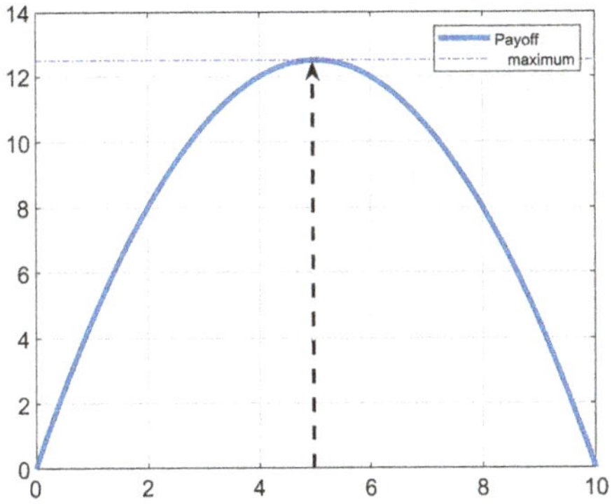

Figure 6.17: Payoff of player 2.

strategy x_i, which is limited within closed interval as $x_i \in [x_i^l, x_i^u]$, and expected utility $u_i(x_1, \ldots, x_i, \ldots, x_N)$. The expected payoff of the first player, if he/she plays with a certain strategy x_1^s will be

$$E_1[u_1] = \int_{x_2^l}^{x_2^u} \cdots \int_{x_N^l}^{x_N^u} u_1(x_1^s, x_2, \ldots, x_N) f_{X_2,\ldots,X_N}(x_2, \ldots, x_N) dx_2 \ldots dx_N. \tag{6.78}$$

Other players may choose to play a correlated strategy (coalitions) in addition to the first player. However, in the absence of dependencies between players, the joint probability density function transforms into a product of individual probability density functions: $f_{X_2,\ldots,X_N}(x_2, \ldots, x_N) = f_{X_2}(x_2) \times \cdots \times f_{X_N}(x_N)$.

It is crucial to note that in real-world scenarios, such as business or strategy games, players rarely make random decisions devoid of specific objectives. Their choices are typically informed by strategic considerations that may depend on uncertain variables, such as market demand. In such cases, we can interpret the probability density function as a tool for quantifying the uncertainty that the first player has about the strategies of others. The expected utility of each player is the best that the player can predict from the outcome of the game, and it is also the average over many rounds of the game. If players benefit from engaging in games with strategies that incorporate an element of randomness, particularly randomness in the strategies employed by others, then it logically follows that player 1 should also introduce an element of randomness into their own strategy, particularly when there is no established equilibrium in pure strategies. This adaptability allows player 1 to respond effectively to the unpredictability of their counterparts' actions. The current challenge lies in determining the optimal probability density function that each player should adopt.

Example 147. In the last example, if player 1 decided to play with fixed strategy $x_1 = 10$, but player 2 plays with random strategies such as x_2 is a uniform distribution from 0 to 10. Compute the expected payoffs for each player. Is the result good for both players?

Solution. The payoffs of both players are now random variables given by $u_1 = 10x_2 - 50$ and $u_2 = 10x_2 - \frac{x_2^2}{2}$ respectively. It is straightforward to compute the expected payoffs as

$$E[u_1] = 10E[x_2] - 50 = 10 \times 5 - 50 = 0 \quad (6.79)$$

$$E[u_2] = 10E[x_2] - \frac{E[x_2^2]}{2} = 50 - \frac{100}{6} = 33.3. \quad (6.80)$$

It becomes evident that player 1 should not adhere to a fixed strategy when facing player 2, who employs a randomized approach.

Example 148. In the previous example, player 1 decided to switch to the randomized strategy x_1 that is also uniformly distributed from 0 to 10, but the samples are selected independently of player 2's choice. Compute the expected payoffs for each player.

Solution. The expected payoffs for both players are

$$E_{1,2}[u_1] = E_{1,2}[x_1x_2] - \frac{E_1[x_1^2]}{2} \quad (6.81)$$

$$E_{1,2}[u_2] = E_{1,2}[x_1x_2] - \frac{E_2[x_2^2]}{2}. \quad (6.82)$$

Since x_1 and x_2 are independent, then $E[x_1x_2] = E_1[x_1]E_2[x_2] = 25$. Therefore, each player will achieve an expected payoff of $E_{1,2}[u_1] = E_{1,2}[u_2] = 8.34$

In the previous example, we can see that the cross-correlation (and the multiplication of the means if both players play independently) and the mean squares are important in determining the Nash equilibrium. Nevertheless, since we have a symmetric game in the previous example, which leads to $E_{1,2}[u1^*] = E_{1,2}[u2^*]$, the main condition required for both players is to have $E_1[x_1^2] = E_2[x_2^2]$.

To achieve success in the game, it is imperative to carefully assess the uncertainties surrounding players' strategies and payoffs. While in certain games, accurately estimating other player's payoffs may be feasible, other games pose significant challenges in predicting the opponents' reactions. A prime illustration of this complexity can be found in the battle between physicians and cancer.

Physicians employ various strategies in their quest to defeat cancer, including chemotherapy, radiotherapy, and radical surgery. In response, cancer cells employ their own set of strategies to develop resistance against medications and to proliferate within the patient's body, among other tactics. Understanding and navigating these intricate dynamics is essential to optimizing the treatment procedure and maximizing our chance of success. The problem should be formulated as a zero-sum game. When the opponent is not human, we encounter what is referred to as an evolutionary game, which we will delve discuss later.

6.3 Cooperative games

Players often choose to engage in non-cooperative games for various reasons, including:

- Impossibility of cooperation: Cooperation may be impossible due to factors such as inadequate communication channels, a lack of trust between players, or an unfair distribution of coalition payoffs.
- Lack of incentives: In some cases, there may be no tangible benefits to cooperating. Consequently, opting for a non-cooperative strategy may be more advantageous.

However, in certain game scenarios, the possibility of cooperation can be both viable and beneficial. This occurs when players can make binding agreements with each other, and there is a clear advantage to collaboration. In such cases, it becomes rational for players to form coalitions that work together to secure mutual advantages.

In the context of cooperative games, assume there are more than one ($n > 1$) players, each of whom is assigned a unique number from the set $N = \{1, 2, \dots, n\}$. The game involves the players' ability to collaborate by forming coalitions, i. e., there is a possibility of communication and negotiation. A coalition is essentially a group consisting of some or all of the players, represented as a subset $S \subseteq N$. The purpose of forming these coalitions is to mutually benefit each member involved, allowing them to collectively get more than they could individually. The total number of possible coalitions is 2^N. For instance, if we have 8 players, there can be a total of $2^8 = 256$ possible coalitions, including $\{\Phi\}$ (the empty set), $\{1\}$, $\{1, 2\}$, $\{1, 3\}$, $\{1, 4\} \dots$ up to $\{1, 2, 3, 4, 5, 6, 7, 8\}$. The aim is to determine an equitable distribution of benefits among the members of a coalition. To achieve this, we begin by quantifying the benefits of a coalition using a real-valued function known as the *characteristic function*. This function is used to measure the overall reward that a coalition can attain. More specifically, the characteristic function of a coalition $S \subseteq N$ represents the maximum guaranteed payoff that can be achieved by that coalition. In other words, the characteristic function of cooperative game is the projection of each element of the set S onto real numbers. To express on possible form of characteristic functions, let's assume a game consists of 3 players. Therefore there are $2^3 = 8$ possible subsets as $\{\Phi\}$, $\{1\}$, $\{2\}$, $\{3\}$, $\{1, 2\}$, $\{1, 3\}$, $\{2, 3\}$, $\{1, 2, 3\}$. The characteristic function $b(\{1\})$ returns a real number representing the benefits obtained by player 1 if when playing non-cooperatively. However, $b(\{1, 2, 3\})$ returns the maximum total benefits if all players play cooperatively. Characteristic functions should have two properties, $b(\{\Phi\}) = 0$, and $b(\{1, 2, \dots, K\}) \geq \sum_{k=1}^{K} b(\{k\})$. The only requirement imposed on a characteristic function is that it assigns a benefit of zero to the empty coalition, and the benefit of the grand coalition (which includes all players) must be at least equal to the sum of the benefits of the individual players if no coalitions are formed.

The following example shows how to determine the characteristic function of a non-zero bimatrix game.

Example 149. Assume a two-player non-zero-sum game given by the following bimatrix

$$A = \begin{bmatrix} (2,8) & (3,-3) \\ (-2,15) & (10,8) \end{bmatrix}. \tag{6.83}$$

The minimum payoff attainable by player 1, irrespective of the strategies chosen by player 2, is determined by the safety value within their payoff matrix. Remarkably, this value aligns with player 1's characteristic function, $b(\{1\}) = 2$. It is clear that the minimum payoff attainable by player 2 is given by $b(\{2\}) = 8$. Finally, if both players play cooperatively, $b(\{1,2\}) = 18$. Player 1 has a compelling incentive to cooperate, since their payoff would significantly increase from 2 to 10. On the other hand, player 2 may initially see no benefit in cooperating with player 1, as their own payoff remains unaffected. Nevertheless, the prospect of sharing the profits stemming from collaboration could potentially entice player 2 to consider cooperation. Do not forget that the characteristic value of no player is zero, i. e., $b(\{\Phi\}) = 0$.

Example 150. Find the characteristic function of the three-player game given in Example 143.

Solution. In the example, players X, Y, and Z have three, four, and two strategies, respectively. In this case, we have the concept of coalition. It means that $b(\{X\})$ returns the benefits of player X when plays non-cooperatively against coalition of players Y and Z. However, $b(\{X,Y\})$ returns the benefits of the coalition of players X and Y against player Z. Let's call the ith strategy of each player as S_i. The next tables show the collation of XY vs. Z, XZ vs. Y and YZ vs. X, respectively.

Using the data provided in the tables, we can determine the characteristic function values which represent the game value of the matrix. To perform this calculation, as an example, consider the bimatrix corresponding to the coalition between players X and Y versus Z, as presented in Table 6.1 as:

$$\mathbf{M_{XY}} = \begin{bmatrix} 3 & 21 \\ 0 & 5 \\ 4 & 9 \\ 3 & 0 \\ 1 & -8 \\ 30 & -16 \\ -1 & 10 \\ -8 & 22 \\ 23 & 17 \\ -3 & 20 \\ 20 & -13 \\ 33 & 11 \end{bmatrix}. \tag{6.84}$$

Table 6.1: Players (X, Y) vs Player (Z).

Players (X, Y)	Player Z	Player Z
s^X, s^Y	s_1^Z	s_2^Z
s_1^X, s_1^Y	(3, 3)	(21, −22)
s_1^X, s_2^Y	(0, 8)	(5, 27)
s_1^X, s_3^Y	(4, 2)	(9, −3)
s_1^X, s_4^Y	(3, −3)	(0, 0)
s_2^X, s_1^Y	(1, −7)	(−8, −1)
s_2^X, s_2^Y	(30, 30)	(−16, 2)
s_2^X, s_3^Y	(−1, 2)	(10, 4)
s_2^X, s_4^Y	(−8, 0)	(22, 31)
s_3^X, s_1^Y	(23, −16)	(17, 12)
s_3^X, s_2^Y	(−3, −7)	(20, −3)
s_3^X, s_3^Y	(20, −4)	(−13, 0)
s_3^X, s_4^Y	(33, 33)	(11, −3)

The game value is $b(\{XY\}) = 18$. The characteristic function of player Z, i. e., the game value of the transpose of the matrix from Table 6.1 as

$$\mathbf{M}_Z = \begin{bmatrix} 3 & -22 \\ 8 & 27 \\ 2 & -3 \\ -3 & 0 \\ -7 & -1 \\ 30 & 2 \\ 2 & 4 \\ 0 & 31 \\ -16 & 12 \\ -7 & -3 \\ -4 & 0 \\ 33 & -3 \end{bmatrix} \tag{6.85}$$

can be computed as $b(\{Z\}) = -5.96$. Similarly, we can compute the characteristic function values of all the possibilities in this three-player game. The results are $b(\{XZ\}) = 4.43$, $b(\{Y\}) = 1.45$, $b(\{YZ\}) = 5.93$, $b(\{X\}) = 1.08$, and $b(\{\Phi\}) = 0$. Finally, the value $b(\{XYZ\}) = 66$ is derived from maximizing the collective payoff achieved when all three players cooperate, forming what is known as a *Grand Coalition*. In this scenario, the largest sum of payoffs is attained.

From the previous example, it is clear that any coalition leads to higher payoffs than each player alone.

All characteristic functions are required to adhere to a set of specific properties outlined below

- $b(\{\Phi\}) = 0$.
- Superadditivity $b(\{A \cup B\}) \geq b(\{A\}) + b(\{B\})$, for all $A, B \subset N$. This property asserts that the benefits generated by the larger coalition $\{A \cup B\}$ must be no less than the combined benefits of the individual coalitions $\{A\}$ and $\{B\}$.

Tables 6.2 and 6.3 can be used to compute the other characterstic functions.

Table 6.2: Players (X, Z) vs Player (Y).

Players (X, Z)	Player Y	Player Y	Player Y	Player Y
s^X, s^Z	s_1^Y	s_2^Y	s_3^Y	s_4^Y
s_1^X, s_1^Z	(5, 1)	(6, 2)	(4, 2)	(0, 0)
s_1^X, s_2^Z	(−6, 5)	(29, 3)	(−5, 11)	(0, 0)
s_2^X, s_1^Z	(−16, 10)	(50, 10)	(−6, 7)	(−8, 0)
s_2^X, s_2^Z	(−5, −4)	(−7, −7)	(12, 2)	(42, 11)
s_3^X, s_1^Z	(−5, 12)	(−8, −2)	(12, 4)	(55, 11)
s_3^X, s_2^Z	(19, 10)	(18, −1)	(−8, −5)	(−3, 11)

Table 6.3: Players (Y, Z) vs Player (X).

Players (Y, Z)	Player X	Player X	Player X
s^Y, s^Z	s_1^X	s_2^X	s_3^X
s_1^Y, s_1^Z	(2, 4)	(−9, 3)	(11, −4)
s_1^Y, s_2^Z	(16, −17)	(−4, −5)	(7, 22)
s_2^Y, s_1^Z	(−2, 10)	(20, 40)	(−1, −9)
s_2^Y, s_2^Z	(2, 30)	(−9, −5)	(21, −4)
s_3^Y, s_1^Z	(2, 4)	(−8, 9)	(16, 0)
s_3^Y, s_2^Z	(−2, 8)	(8, 6)	(−8, −5)
s_4^Y, s_1^Z	(3, −3)	(−8, 0)	(22, 44)
s_4^Y, s_2^Z	(0, 0)	(11, 42)	(0, 8)

If the benefits of a coalition are greater than the sum of the individual benefits, we call it as an *essential game*. In contrast, if this condition is not met, it is termed an *inessential game*. The previous example indeed falls under the category of an essential game.

A crucial aspect of cooperative games lies in determining the suitable distribution (or allocation) of coalition outcomes. To illustrate, consider the grand coalition presented in Example 150, where the total benefit amounts to 66. The challenge is how to fairly allocate this sum among the participating players. Nevertheless, the concept of fairness can be somewhat ambiguous, as it may vary depending on the specific criteria or standards used to evaluate it.

Before we delve into this topic, it's imperative that any allocation meet two fundamental criteria. Assuming there are N players, and we denote the allocation to the ith player as h_i, the first criterion demands that h_i must be greater than or equal to the player's individual payoff, represented as $b(\{i\})$, i. e., $h_i \geq b(\{i\})$. Naturally, the second criterion follows suit, requiring that the sum of all allocations across players must be equal to the payoff of the grand coalition, denoted as $b(\{N\})$, i. e., $\sum_{i=1}^{N} h_i = b(\{N\})$. Generally speaking, no rational players will play cooperatively if their allocated shares are less than if they play individually, i. e., non-cooperatively. Observe that the allocation h_i can be negative. In Example 150 player Z may accept the share of $h_Z = -1$, because it is still better than playing individually as $b(\{Z\}) = -5.96$. It is important to have a systematic procedure for allocating shares among players (it is also called imputation). Systematic allocation procedures are typically based on specific criteria. However, it's important to note that these criteria can sometimes lead to multiple allocation vectors that fulfill the same criterion, or in certain cases, there may be no feasible allocation vector that satisfies the criterion.

Nevertheless, a crucial aspect of cooperative games involves reaching a consensus among all participating players on the allocation-sharing criteria. Additionally, there should be a clear strategy in place for optimizing the selection when the allocation process generates more than one possible allocation vector that meets the established criteria.

Shapley criteria

We will present only one method for allocating coalition benefits. The method is called the Shapley value, and it is based on the following four axioms:

- Efficiency: The sum of the allocated values should be equal to the grand coalition, i. e., $\sum_{i=1}^{N} h_i = b(\{N\})$.
- Symmetry: It means that players with the same contribution to the coalition should have the same the share. For two different players i, j, if $b(\{U + i\})\forall U \subset \{N - i\} = b(\{U + j\})\forall U \subset \{N - j\}$, then $h_i = h_j$.
- Dummy: Players whose coalition with other players will not have any benefits, should receive nothing. Mathematically this means that if $b(\{U+i\}) = b(\{U\})\forall U \subseteq N$, then $h_i = 0$.
- Additivity: The value should exhibit additivity over the entire spectrum of possible coalitions.

The Shapley value of an individual player i is defined by the following formula

$$h_i = \sum_{U \in \Xi_i} (b(\{U\}) - b(\{U - i\})) \frac{(\mid U \mid -1)!(\mid N \mid - \mid U \mid)!}{\mid N \mid!}. \tag{6.86}$$

Where Ξ_i is the set of all coalitions $U \subseteq N$ containing player i, i. e., $i \in U$, $\mid U \mid$ is the number of elements in $\{U\}$, and $\mid N \mid$ is the total number of players.

Example 151. Find the optimal share in the three-player game in Example 150 using the Shapley formula.

Solution. In this game we have three players X, Y, and Z. Assume their numbers are 1, 2, and 3 respectively. For player 1 (i. e., X) we have the following set of possibilities $\Xi_1 = \{1, 12, 13, 123\}$. Moreover, $| N |= 3$. The share of player X is

$$h_1 = (1.08 - 0)\frac{0!2!}{3!} + (18 - 1.45)\frac{1!1!}{3!} + \frac{10.39}{6} + \frac{60.07}{3} = 24.87. \tag{6.87}$$

In the same manner, we can compute the other shares of player 2 (i. e., Y), $\Xi_2 = \{2, 12, 23, 123\}$

$$h_2 = (1.45 - 0)\frac{1}{3} + (18 - 1.08)\frac{1}{6} + \frac{(11.89)}{6} + \frac{61.57}{3} = 25.81. \tag{6.88}$$

Finally, the share of player 3 (i. e., Z) with $\Xi_3 = \{3, 13, 23, 123\}$ will be

$$h_3 = (-5.96 - 0)\frac{1}{3} + (4.43 + 5.96)\frac{1}{6} + \frac{4.48}{6} + \frac{48}{3} = 16.49. \tag{6.89}$$

All players should be happy with this share.

6.4 Game models

Game theory offers a robust framework for analyzing the decision-making dynamics of rational individuals or entities facing conflict and uncertainty. The study of game models allows us to delve deeply into the intricate web of players' decisions and their responses to different strategies. While game theory boasts a multitude of standard models, this section introduces three renowned game models that hold applicability across a spectrum of multidisciplinary domains. These models serve as versatile tools, that can be applied to a wide range of real-world scenarios and contexts.

6.4.1 Cournot duopoly game

Consider two companies that produce a similar gadget. It is possible to sell the gadget at price p_r per unit. However, this price is governed by the law of supply and demand. We may model this relation as

$$p_r = D(K), \tag{6.90}$$

where $K \geq 0$ is the number of gadgets produced and $D(\cdot) \geq 0$ is a positive decreasing function with $D(\hat{K}) = 0$, where $\hat{K}$ is the number of gadgets where the gadget loses its value. Every factory wants to maximize their profits, however just producing a large number of gadgets will reduce the selling price and this leads to a drop in their profit. Assume that the cost of producing each gadget for each company is $C_1(k_1)$ and $C_2(k_2)$ respectively, where k_i is the number of gadgets produced by the ith firm. In general, the cost of manufacturing gadgets is directly influenced by the quantity produced. The key question here is: what is the optimal production quantity for each firm to reach a Nash equilibrium? This equilibrium point is reached when neither firm can improve their situation by deviating from this production quantity, assuming the other firm maintains the same level. Assume the quantities produced by firms 1 and 2 are k_1 and k_2 respectively, where $K = k_1 + k_2$. Given that the quantities are integers, it would be natural to model and solve this problem as a combinatorial problem. However, for the sake of simplicity, let's approach the problem with the assumption that the quantities can be treated as real numbers. In some scenarios, such as wireless telecommunications, the variables represented by terms like k_1 and k_2 are inherently real numbers. An example of this is transmit power in wireless communications, where exceeding a certain threshold should incur penalties due to increased interference and power consumption. The payoff functions of each firm is represented as their profit given by:

$$u_1(k_1, k_2) = k_1 \times (D(k_1 + k_2) - C_1(k_1)) \tag{6.91}$$

$$u_2(k_1, k_2) = k_2 \times (D(k_1 + k_2) - C_2(k_2)). \tag{6.92}$$

The Nash equilibrium depends on the demand and cost functions and could be found by taking the partial derivatives as shown before:

$$\frac{\partial u_1(k_1, k_2)}{\partial k_1} = D(K) - C_1(k_1) + k_1 \frac{\partial D(K)}{\partial k_1} - k_1 \frac{\partial C_1(k_1)}{\partial k_1} = 0 \tag{6.93}$$

$$\frac{\partial u_2(k_1, k_2)}{\partial k_2} = D(K) - C_2(k_2) + k_2 \frac{\partial D(K)}{\partial k_2} - k_2 \frac{\partial C_2(k_2)}{\partial k_2} = 0. \tag{6.94}$$

In order to have a Nash equilibrium, the second derivatives of the above equations should be less than zero.

Example 152. In the Cournot duopoly game, the demand function is assumed to be $D(K) = 100 \times e^{-0.01K}$, for $0 \le K \le 400$ and $D(K) = 0$ for $K > 400$. The production costs of the firms are fixed and given as $c_1 = 2$ and $c_2 = 5$, respectively. Find the Nash equilibrium production quantity.

Solution. By a direct substitution of the demand function and the costs, we have

$$100e^{-0.01K} - k_1 e^{-0.01K} - 2 = 0 \tag{6.95}$$

$$100e^{-0.01K} - k_2 e^{-0.01K} - 5 = 0. \tag{6.96}$$

We have two nonlinear equations in two unknowns. You may solve them using the *fsolve* command in Octave to obtain $k_1 \approx 90$, and $k_2 \approx 74$. The total production is $K = 164$ gadgets. The unite selling price is $100e^{-1.64} = 19.4$. The total profits of firms one and two are $u_1 = 1566$, and $u_2 = 1065$. No one can do better than these values by deviating from the equilibrium production quantities. For example, assume that firm 1 decides to increase its production to 120 units and the second firm stick with the equilibrium value of 74. In this case, the profit of the first firm will decrease to 1484 and for the second firm to 694. Both will lose by deviating from the Nash equilibrium.

Example 153. In the previous example, the demand function could be expanded as $D(K) = 100 - K + \frac{0.01K^2}{2} - \frac{0.01^2K^3}{3!} + \cdots$. The first order approximation could be given as $D(K) = \max(0, 100 - K)$. Repeat Example 152 and find the Nash equilibrium solution in the case of the first order linear approximation.

Solution. It is straightforward to apply the sufficient conditions given earlier to obtain the following two linear equations:

$$2k_1 + k_2 = 100 - c_1 \tag{6.97}$$

$$k_1 + 2k_2 = 100 - c_2. \tag{6.98}$$

By solving the aforementioned pair of linear equations (without the need to substitute the specific gadget cost values to assess their influence), we arrive at the following Nash equilibrium quantities:

$$k_1^* = \frac{100 - 2c_1 + c_2}{3} \tag{6.99}$$

$$k_2^* = \frac{100 - 2c_2 + c_1}{3}. \tag{6.100}$$

It's fascinating and intuitive to note that when a firm faces higher production costs, it tends to decrease its optimal production quantity while simultaneously increasing the optimal production quantity of the opponent firm. For the given production costs, the optimal quantities are $k_1^* = 34$ and $k_2^* = 30$. The achieved utilities (profits) of each firm are $u_1 \approx 1156$ and $u_1 \approx 930$.

6.4.2 Stackelberg games

The Stackelberg model addresses a scenario in which two factories do not determine their production quantities simultaneously, but rather make these decisions sequentially. This model falls into the category of sequential games. Sequential games offer a compelling framework for modeling radio resource scheduling in wireless communications. In this context, when one transmitter initiates the allocation of radio resources such as power and subcarriers, it's important to consider that another transmitter will soon begin its own resource allocation process. If the second transmitter were to transmit with excessive power due to a lack of available slots, it could potentially disrupt the communication link. To address this challenge, a key objective is to guide both transmitters toward reaching a Nash equilibrium solution. This equilibrium would ensure that both transmitters make strategic decisions that optimize their resource allocation without adversely affecting the overall performance of the communication link. Achieving this equilibrium requires careful coordination and intelligent strategies to ensure efficient radio resource utilization and minimize interference, ultimately enhancing the wireless communication system's overall effectiveness.

Let's continue the discussion considering the example of gadget production quantities. In the framework of this model, one firm takes the initiative by publicly disclosing its intended production quantity. Subsequently, the second firm (referred to as Firm *II*) formulates its production strategy based on this initial disclosure. The primary distinction between the Stackelberg and previous Cournot duopoly games lies in their strategic dynamics. In the Stackelberg model, one firm assumes the role of the leader, initiating actions, while the other firm assumes a follower position. For instance, let's assume that Firm *I* openly declares its intention to produce a quantity k_1, incurring a cost of c_1 per unit. It's important to note that c_1 can either be a fixed, deterministic value in terms of number of units produced or uncertain and subject to probability distributions. Uncertainty may arise from either a lack of knowledge about other firms or fluctuations in raw material prices. The payoffs of both firms follow the same equations in (6.91). However, in the Stackelberg game, it is assumed that Firm *II* knows the production quantity of Firm *I*. How do you think that Firm *II* should respond to that? Of course, it will try to maximize its payoff based on the given k_1^*, which leads to

$$\frac{\partial u_2(k_1,k_2)}{\partial k_2} = D(K) - C_2(k_2) + k_2\frac{\partial D(K)}{\partial k_2} - k_2\frac{\partial C_2(k_2)}{\partial k_2}\bigg|_{k_1=k_1^*} = 0. \tag{6.101}$$

Hence, Firm *I* knows that Firm *II* will produce k_2^* that achieves the maximum of its payoff given in (6.101). Therefore, Firm *I* should determine k_1^* that maximizes their payoff based on k_2^*, i. e.,

$$\frac{\partial u_1(k_1,k_2)}{\partial k_1} = D(K) - C_1(k_1) + k_1\frac{\partial D(K)}{\partial k_1} - k_1\frac{\partial C_1(k_1)}{\partial k_1}\bigg|_{k_2=k_2^*} = 0. \tag{6.102}$$

Example 154. In the following example, we will reexamine the scenario presented in Example 153, with the premise that Firm *I* initiated the production of a particular gadget before Firm *II* entered the same business. We aim to determine the Nash equilibrium production levels for both firms and to explore the advantages enjoyed by the firm that initiated the business.

Solution. First, Firm *I* has declared its production quantity k_1^*, hence, the best production quantity of Firm *II* is the one that maximizes its payoff as

$$\frac{\partial u_2(k_1,k_2)}{\partial k_2} = D(K) - C_2(k_2) + k_2\frac{\partial D(K)}{\partial k_2} - k_2\frac{\partial C_2(k_2)}{\partial k_2}\bigg|_{k_1=k_1^*} = 0. \tag{6.103}$$

This leads to

$$k_2^* = \frac{100 - k_1^* - c_2}{2}. \tag{6.104}$$

However, what should be the optimal quantity value, denoted as k_1^*, that Firm I should produce? Firm I is aware that the decision-maker at Firm II, who will make their choice later, is astute and will maximize their own payoff based on their quantity selection. Consequently, the payoff equation of Firm I becomes:

$$u_1 = 100k_1 - k_1^2 - k_1 k_2^* - c_1 k_1 \tag{6.105}$$

$$\rightarrow u_1 = 50k_1 - \frac{k_1^2}{2} + \left(\frac{c_2}{2} - c_1\right)k_1. \tag{6.106}$$

The maximum of the above payoff utility is achieved at

$$k_1^* = \frac{100 + c_2 - 2c_1}{2}. \tag{6.107}$$

By substituting (6.107) back in (6.104), we get

$$k_2^* = \frac{100 - 3c_2 + 2c_1}{4}. \tag{6.108}$$

The utility achieved by each firm at the Nash equilibrium quantities are given

$$u_1 = \frac{1}{2}\left(\frac{100 + c_2 - 2c_1}{2}\right)^2 \tag{6.109}$$

$$u_2 = \left(\frac{100 - 3c_2 + 2c_1}{4}\right)^2. \tag{6.110}$$

The optimum total generated quantity is

$$K^* = k_1^* + k_2^* = \frac{300 - c_2 - 2c_1}{4}. \tag{6.111}$$

Example 155. In the previous example compute the achieved utility as well as the optimum quantities of each firm if the costs are $c_1 = 2$ and $c_2 = 5$.

Solution. Just applying the equations in the previous example, we obtain $K = \frac{300-5-4}{4} \approx 73$, $k_1 = \frac{100+5-4}{2} \approx 51$, $k_2 = \frac{100-15+4}{4} \approx 22$, $u_1 \approx 1275$, and $u_2 \approx 495$. When comparing the findings presented here with those in Example 153, a clear demonstration emerges of the advantages associated with being an industry leader. By establishing itself as the market leader, Firm 1 reaps the rewards and achieves a level of that exceeds that of the Cournot model case.

6.5 Evolution game theory and semi-rational responses

One fundamental condition in classical game theory is the assumption that all players are rational and consistently select the optimal strategy while predicting (with varying accuracy depending on the situation) how other players will behave. However, there are situations where this assumption of complete rationality does not hold. In such cases, players may resort to trial-and-error approaches or other unconventional strategies to navigate the game.

In nature, we see a fascinating balance between small and large creatures, herbivores and carnivores, weak and beautiful insects like butterflies, and powerful and fierce animals like lions, and other remarkable biodiversity with a more astonishing and wondrous ecological balance. In the light of this reality, we find that the classical game theory does not work as it should, and it may fail in many cases to model the mathematical foundations of this ecological and biological balance. Therefore, an important development has been made to the classical game theory to incorporate these biological and environmental observations, which can be used in many applications beyond biology, including economics and more. Examples of such scenarios:

- In the realm of biology, organisms operate without full awareness of formal games, yet they engage in the game of survival using pre-programmed rules and strategies. Therefore, it is of paramount importance in the field of biology to explore how organisms respond to challenging environments, adapting and evolving in the process.Cancer cells or bacteria can be studied as they dynamically adapt their survival strategies in response to medical interventions, striving to outmaneuver treatment efforts.
- In the domain of autonomous technologies, such as self-driving cars, it is not realistic to assume that the entities involved always act with complete rationality. This leads us to recognize that in various emerging applications across economics, social systems, and technical domains, evolutionary games are gaining prominence as a more suitable framework compared to classical games.

These evolutionary game models account for the complex and often non-rational behavior exhibited by players, offering a more realistic representation of many real-world situations. Evolutionary games operate on the principle of *Evolutionary Stable Strategy* (ESS) as opposed to the classical game theory concept of Nash equilibrium. An Evolutionary Stable Strategy can be defined as a robust strategy that cannot be outperformed by mutant strategies, thus preventing unfavorable mutations from dominating the population. Every Evolutionary Stable Strategy (ESS) is a subset of the Nash equilibrium, but it's important to note that not every Nash equilibrium qualifies as an Evolutionary Stable Strategy. The next example demonstrates the concept of ESS.

Example 156. Imagine a large network of ad hoc transceivers deployed over a specific geographical area. These devices simply transmit their messages without engaging in any form of sensing or optimizing the scheduling of their transmissions. These devices fall into two categories of possible alternatives (or strategies) based on their transmit power: 'Large' utilize their maximum transmit power, which can potentially lead to higher interference with other devices, while 'Small' use very low transmit power to minimize interference. The performance of each communication link is quantified on a scale from 0 to 100, where 100 is the highest quality. Each transceiver may play the first strategy of Small or the second strategy of Large power. The bimatrix of this game is represented as follows:

$$\mathbf{A} = \begin{bmatrix} (\mathbf{50,50}) & (10,50) \\ (50,10) & (\mathbf{50,50}) \end{bmatrix}. \tag{6.112}$$

It is evident that this game exhibits two Nash equilibrium solutions, both yielding identical payoffs when both players choose 'Large' or when both players choose 'Small.' Now, let's assess the stability of these equilibrium points. Let's assume that a certain percentage, denoted as p, of devices opt for the 'Small' power strategy, while the remaining fraction, $1 - p$, choose the 'Large' power strategy. Notably, we can interpret this p as the probability that a randomly selected device will employ the Small strategy. The expected payoff of a device decided to play Small strategy will be

$$E[U_S] = 50p + 10(1 - p) = 40p + 10. \tag{6.113}$$

However, the expected payoff when playing the Large strategy will be

$$E[U_L] = 50p + 50(1 - p) = 50. \tag{6.114}$$

Therefore, the device should play the Small strategy if $E[U_S] > E[U_L]$. However, this will never happen because the maximum is achieved when $p = 1$, and even at this value $E[U_S] = E[U_L]$. Therefore, the Small-Small solution cannot be an Evolutionary Stable Strategy (ESS).

In the previous example, we observed that there is only one Evolutionary Stable Strategy (ESS) in the pure strategy space, which corresponds to the Large-Large strategy. However, while the Small-Small strategy is a Nash equilibrium, it lacks stability and can be vulnerable to being overtaken by the Large-Large strategy.

To illustrate this, consider a transceiver device that chooses to employ the Small power strategy. In this case, this device receives a payoff of $U = 50$ only if all other transceiver devices also opt for the Small strategy. However, if any of the other transceiver devices deviates from Small and switches to the Large strategy, our first device's payoff drops significantly to $U = 10$ without affecting the payoffs of the other Large strategy players, i. e., they will remain to receive $U = 50$. Since this is a non-cooperative game, no external enforcement can compel transceiver devices to remain committed to the Small strategy. Therefore, while Small-Small is indeed a Nash equilibrium, its lack of stability makes it susceptible to being easily overtaken by the alternative Nash equilibrium of Large-Large. On the other hand, if a transceiver device chooses to adopt the 'Large' strategy, this strategy proves unbeatable regardless of the strategies chosen by other players, ensuring its stability and dominance.

For K-player games, x_k^* is the a Nash equilibrium strategy of the kth player and x_k is any other strategy that does not achieve Nash equilibrium conditions. Hence, we can define Evolutionary Stable Strategies as the ones achieves the following two conditions:

$$E[u_k(x_1^*,\dots,x_k^*,\dots,x_K^*)] \geq E[u_k(x_1^*,\dots,x_k,\dots,x_K^*)] \tag{6.115}$$

$$E[u_k(x_1,\dots,x_k^*,\dots,x_K)] > E[u_k(x_1,\dots,x_k,\dots,x_K)]. \tag{6.116}$$

The first condition is indeed the classical definition of a Nash equilibrium, however, the second condition adds stronger constraints. Both conditions are required for an ESS solution. It follows that every Evolutionary Stable Strategy (ESS) must also be a Nash equilibrium, but the reverse is not necessarily true. When dealing with a bimatrix symmetric game that possesses at least one Nash equilibrium in pure strategies, one can determine whether any of these Nash equilibria qualify as Evolutionary Stable Strategies (ESS) by evaluating the following conditions: assume that the kth strategy (i. e., row of the first player) is the Nash equilibrium strategy of the first player then

- the kth strategy (i. e., column of the second player) is also a Nash equilibrium strategy. In other words, the Nash equilibrium is at (k, k).
- The payoff $a_{kj} > a_{jj}\ \forall\ k \neq j$.

Example 157. Is there any ESS in the following symmetric bimatrix game?

$$\mathbf{A} = \begin{bmatrix} (10,50) & \mathbf{(50,50)} \\ \mathbf{(50,50)} & (10,50) \end{bmatrix}. \tag{6.117}$$

Solution. Even this bimatrix game has two Nash equilibrium solutions in pure strategies, however, non of them is ESS. The first condition for a Nash equilibrium to be an ESS is that it must be symmetric. But why? Assume player I plays Small, player II can deviate from Small to Large without loss, however, player I's payoff will be dropped. Therefore Small-Large strategy is not ESS, even it is NE. The same goes for the Large-Small strategy.

The same concept of ESS can be applied in mixed strategy games as well.

Example 158. Is there any ESS in pure or mixed strategies?

$$\mathbf{A} = \begin{bmatrix} (0,0) & \mathbf{(-1,1)} \\ \mathbf{(1,-1)} & (-10,-1y10) \end{bmatrix}. \tag{6.118}$$

Solution. It is clear that both Nash equilibrium solutions cannot be ESS because they lack the symmetry condition. However, this game matrix has another Nash equilibrium solution in mixed strategies at $X = [0.9,\ 0.1]^T$ and since the game is symmetric, therefore, the Nash equilibrium strategy of the opponent will be identical, i. e., $Y = [0.9,\ 0.1]^T$. Is this NE strategy also an ESS? To answer this question, we should test the extra condition required for ESS given in (6.116). Since the bimatrix is symmetric, then it is always true to say that $x_1 = y_1$, and $x_2 = y_2$. The expected payoff of the player X when they play with $X^* = [0.9,\ 0.1]^T$ and player Y with $Y = [y_1,\ (1-y_1)]^T$ is

$$\begin{aligned} E_X[X^*, Y] &= -0.9(1-y_1) + 0.1y_1 - (1-y_1) \\ &= 2y_1 - 1.9. \end{aligned} \tag{6.119}$$

If $E_X[X^*, Y] > E_X[X, Y]$, for any $X \neq X^*$ and $Y \neq Y^*$ solutions, then $[X^*, Y^*]$ is an Evolutionary Stable Strategy. The expected utility of player X for any symmetric strategy is given by:

$$\begin{aligned} E_X[X, Y] &= -x_1(1-y_1) + x_2y_1 - 10x_2(1-y_1) \\ &= -y_1(1-y_1) + (1-y_1)y_1 - 10(1-y_1)^2 \\ &= -10(1-y_1)^2. \end{aligned} \tag{6.120}$$

For $Y \neq Y^*$, equation (6.119) is always greater than (6.120), as shown in Figure 6.18. Hence, the mixed Nash equilibrium strategy is an Evolutionary Stable Strategy.

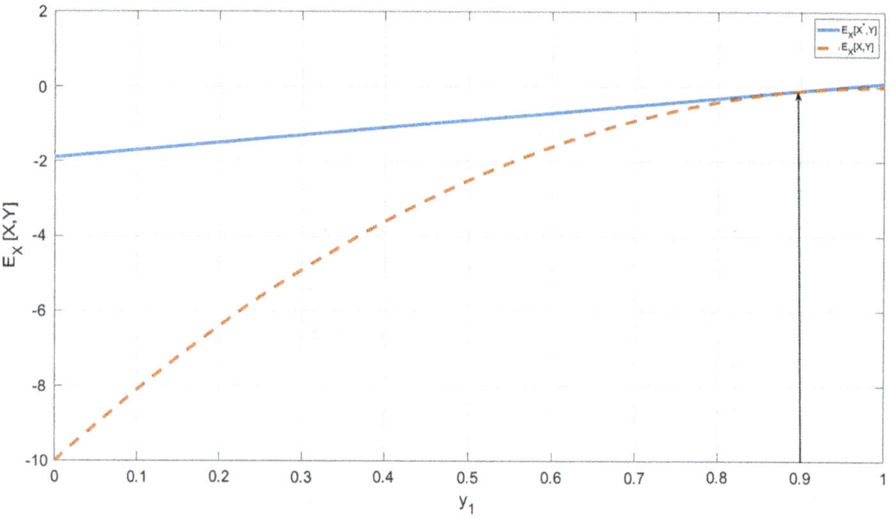

Figure 6.18: Evolutionary stable strategy.

6.6 Stochastic games

Uncertainty is a pervasive element in virtually all games, encompassing various factors such as uncertain opponent preferences, fluctuating payoffs, concealed strategies, undisclosed plans, and underlying objectives, among others. Stochastic games emerge when these inherent uncertainties are integrated into both the game's modeling and its resolution. However, the term "stochastic games" is specifically applicable when the games are played repeatedly, or when there is a history of previous similar games to refer to.

For instance, consider a game model in which a player must determine the Nash equilibrium for the quantity of gadgets to supply to the market. In this scenario, there is a lack of precise information concerning the opponent's factory production costs or gadget quality. Consequently, the players must make an initial estimate of these unknown parameters. This initial estimate can be based on market conditions and any available historical data.

However, if the game is played sequentially over time, it becomes possible to employ estimation theory to enhance the accuracy of these initial estimates. Methods such as maximum likelihood estimators can be utilized, leveraging the outcomes of prior rounds to refine the estimates. Some game models are well-suited for representation as hidden Markov models (HMM), where game outcomes are used to estimate the concealed states of the opponent's strategies and preferences.

Bibliography

[1] A. Papoulis (2002). *Probability, Random Variables and Stochastic Processes*, 4th Edition, McGraw-Hill.
[2] B. P. Lathi (2017). *Linear Systems and Signals*, Oxford University Press.
[3] M. Krystek (2024). *Measurement Uncertainties: Error Propagation, Probabilistic Modelling, Statistical Methods*, De Gruyter.
[4] I. Elishakoff (2024). *Multifaceted Uncertainty Quantification*, De Gruyter.
[5] S. Kay (2017). *Fundamentals of Statistical Signal Processing*, Pearson.
[6] J. Speyer and W. Chung (2008). *Stochastic Processes, Estimation and Control*, SIAM.
[7] E. N. Barron (2024). *Game Theory: An Introduction*, 3ed Edition, Wiley.

https://doi.org/10.1515/9783111585055-007

Index

https://doi.org/10.1515/9783111585055-008

www.ingramcontent.com/pod-product-compliance
Lightning Source LLC
Chambersburg PA
CBHW082107220326
41598CB00066BA/5663